Intelligent Control and Learning Systems

Volume 12

Series Editor

Dong Shen, School of Mathematics, Renmin University of China, Beijing, Beijing, China

The Springer book series Intelligent Control and Learning Systems addresses the emerging advances in intelligent control and learning systems from both mathematical theory and engineering application perspectives. It is a series of monographs and contributed volumes focusing on the in-depth exploration of learning theory in control such as iterative learning, machine learning, deep learning, and others sharing the learning concept, and their corresponding intelligent system frameworks in engineering applications. This series is featured by the comprehensive understanding and practical application of learning mechanisms. This book series involves applications in industrial engineering, control engineering, and material engineering, etc.

The Intelligent Control and Learning System book series promotes the exchange of emerging theory and technology of intelligent control and learning systems between academia and industry. It aims to provide a timely reflection of the advances in intelligent control and learning systems. This book series is distinguished by the combination of the system theory and emerging topics such as machine learning, artificial intelligence, and big data. As a collection, this book series provides valuable resources to a wide audience in academia, the engineering research community, industry and anyone else looking to expand their knowledge in intelligent control and learning systems.

Wenjun Xiong · Zijian Luo · Daniel W. C. Ho

Iterative Learning Control for Network Systems Under Constrained Information Communication

Springer

Wenjun Xiong
School of Computing and Artificial Intelligence
Southwestern University of Finance and Economics
Chengdu, Sichuan, China

Zijian Luo
School of Science
Civil Aviation Flight University of China
Guanghan, Sichuan, China

Daniel W. C. Ho
Department of Mathematics
City University of Hong Kong
Hong Kong, China

ISSN 2662-5458 ISSN 2662-5466 (electronic)
Intelligent Control and Learning Systems
ISBN 978-981-97-0925-0 ISBN 978-981-97-0926-7 (eBook)
https://doi.org/10.1007/978-981-97-0926-7

This Springer imprint is published by the registered company Springer Nature Singapore Pte Ltd.
The registered company address is: 152 Beach Road, #21-01/04 Gateway East, Singapore 189721, Singapore

Preface

Traditional control methods often fail to meet the needs of real-world applications as networked systems continue to grow in size and complexity. The dynamic control problems of networked systems become very difficult when information and communication are limited. To handle these challenges, iterative learning control (ILC) may provide some effective controlling strategies. The ILC also attracts the attention of researchers as an advanced control strategy. For this reason, it makes sense to discuss control problems in constrained network systems with the help of iterative learning control approaches.

As an important branch of intelligent control, ILC is a type of high-precision control technology. This method repeatedly uses information from previous trials to obtain the desired control inputs, improving the quality of system operation. After more than three decades of development, ILC has achieved a wealth of theoretical results and has been applied in some real systems. However, the traditional iterative learning method is mainly applicable to control systems with a point-to-point structure. Furthermore, the existing theories assume that system data can be measured in its entirety and the data does not need to be transmitted. In fact, a number of systems have complex network structures. These network structures inevitably lead to data dropout, data quantization, and other incomplete information phenomena. Hence, it motivates us to investigate the ILC problems under some network communication constraints.

This book is dedicated to applying iterative learning strategies to networked systems under various communication constraints. The book focuses on the design and analysis of ILC in the presence of information quantization, switching topology, unmeasurable state, data dropout, network attacks, etc. There are 12 chapters in this book. Chapter 1 introduces the foundations and current research in iterative learning control and networked systems. Chapters 2–4 mainly explore the consensus problem under quantization, limited information communication, and switching topology. Chapters 5–7 investigate the tacking problem under different state information, saturated finite interval, and discontinuous control strategies. Chapters 8 and 9 further consider the finite-iteration tracking problem under data packet losses and the FlexRay communication protocol, respectively. Chapters 10 and 11 study the problem of sampled data tracking and the stability of multilayer network systems with constraints. Finally, Chap. 12 discusses the security problem of network systems under cyber-attacks. This book is written for students, academics, and engineers working in the areas of networked systems, multi-agent systems, and iterative learning algorithms.

The authors would like to express their sincere gratitude to all their collaborators, as some of the chapters in this monograph are based on joint contributions with them. The authors would like to thank all ILC experts around the world as they benefited greatly from the previous foundational works when preparing this book.

Chengdu, China Wenjun Xiong
Guanghan, China Zijian Luo
Hong Kong, China Daniel W. C. Ho
December 2023

Acknowledgments This work was supported by the National Natural Science Foundation of China (No. 62373308), the Central Leading Local Science and Technology Development Fund (No. 2023ZYD0009), and the Financial Intelligence and Financial Engineering Key Laboratory of Sichuan Province.

Contents

Symbols

$\mathbb{Z}_+$	The set of nonnegative integers
$\mathbb{R}$	The set of real numbers
$\mathbb{R}_+$	The set of nonnegative real numbers
$\mathbb{R}^n$	The n-dimensional Euclidean space
$\mathbf{0}$	Zero vector with appropriate dimension
$\mathbf{1}$	Vector $[1, \cdots, 1]^{\mathrm{T}}$ with an appropriate dimension
O	Zero matrix with appropriate dimension
I_N	$N \times N$ identity matrix
$x \in \mathbb{R}^n$	n-dimensional column vector $(x_1, \cdots, x_n)^{\mathrm{T}}$
y^{T}	The transpose of vector or matrix y
$\|x\|$	Euclidean norm of the vector x
$\|A\|$	Spectral norm of the matrix A
$\|\cdot\|_1$	1-norm for both vector and matrix
$\lvert z \rvert$	The absolute value of number z
$\rho(A)$	The spectral radius of matrix A
$\lambda_m(A)$	The minimal eigenvalues of matrix A
$\lambda_M(A)$	The maximal eigenvalues of matrix A
$\lambda_i(A)$	The ith eigenvalue of A
$\otimes$	The Kronecker product
$\lfloor z \rfloor$	The largest integer part of number z
$X > Y$	The matrix $X - Y$ is positive definite
$X \geq Y$	The matrix $X - Y$ is positive semi-define

Chapter 1
Introduction

1.1 Iterative Learning Control

Iterative learning control (ILC), as a highly efficient control technology, was formally proposed in 1984 (Arimoto et al. 1984). This control technology is a powerful methodology that aims to improve the performance of dynamic systems by learning from repetition and refining control actions. This technique has applications in various fields, ranging from industrial automation to robotics (Bristow et al. 2006; Li et al. 2013; Riaz et al. 2021), allowing systems to achieve higher accuracy, precision, and stability. Through the process of repeated execution and learning from past experiences, ILC enables systems to adapt and enhance their performance over time.

ILC uses past data to adjust and gradually improve control actions. The process involves three essential components: learning, repetition, and adaptation. A control action is applied repeatedly to a dynamic system, and the resulting outputs are compared with the desired trajectory. Deviations between the actual and desired outcomes adjust the control inputs in subsequent iterations. The system refines its control actions in each iteration, reducing errors and gradually converging toward the desired performance. The principle of operation is shown in Fig.1.1, where t denotes the time, and k represents the kth iteration process. $u(t, k)$ is the system input, $y(t, k)$ is the system output, and $y_d(t)$ is the desired output or trajectory. The learning error (or tracking error) is defined as $e(t, k) \triangleq y_d(t) - y(t, k)$. In particular, the system's control task is to seek the appropriate control strategy $u(t, k)$ through learning, which in turn makes the system output $y(t, k)$ approximates the desired trajectory $y_d(t)$. After years of development and improvement, ILC has grown into an important branch in the field of intelligent control.

W. Xiong et al., *Iterative Learning Control for Network Systems Under Constrained Information Communication*, Intelligent Control and Learning Systems 12,
https://doi.org/10.1007/978-981-97-0926-7_1

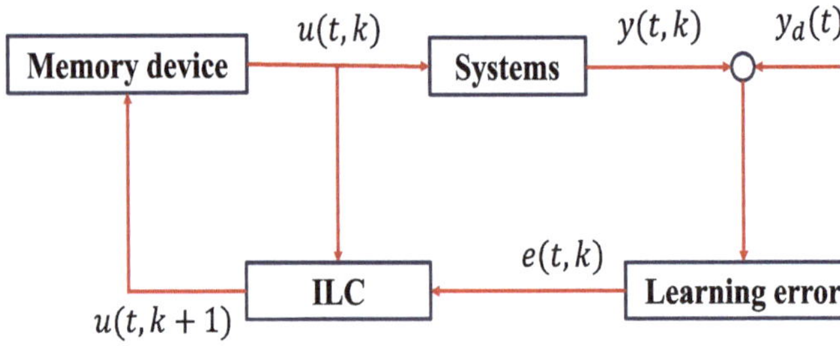

Fig. 1.1 The principle of ILC

1.1.1 Current Status of ILC Research

In recent years, researchers have made significant strides in advancing the theoretical foundations of ILC (Riaz et al. 2021; Shen 2018; Shen and Li 2019; Yu and Chai 2019). The development of sophisticated update laws and convergence analysis techniques has led to improved performance and robustness, such as the event-triggered ILC strategies (Lin et al. 2021; Xiong et al. 2016; Zhang and Li 2018), data-driven-based compensation ILC (He et al. 2022), neural-network-based ILC approaches (Patan and Patan 2020; Zhang et al. 2021), adaptive ILC schemes (Bu and Hou 2016; Huang et al. 2021a, b), etc. These research results enrich the theoretical system of ILC and broaden the scope of practical applications. As a result, a wide variety of ILC issues have attracted widespread attention. Researchers also conduct studies on the following topics:

(1) ILC studies in non-strictly repeating systems. This type of research primarily focuses on the non-strict repetition of the dynamic properties of system objects (Meng and Moore 2017a, b; Meng and Zhang 2021a, b), strictly non-repetitive desired trajectory (Boudjedir et al. 2019; Jin 2018), randomly varying iteration intervals (Jin 2021; Li and Shen 2017; Shen and Li 2019), and so on.

(2) Norm optimal ILC and predictive ILC schemes. This type of problem considers the derivation of the corresponding norm-optimal ILC algorithms by defining suitable performance metrics and combining them with optimization theory, when the information about the system is completely known (Barton and Alleyne 2011; Yu et al. 2018; Zhuang et al. 2023). In addition, researchers have proposed predictive ILC strategies to deal with some specific problems that are caused by the use of non-causal system information in the solution process (Chu et al. 2016; Lin et al. 2021).

(3) ILC studies in constrained systems. The system state and control are generally subject to various constraints due to actual physical factors. As a result, it is necessary to investigate and analyze the problem of ILC for constrained systems. For state-constrained control problems, researchers propose a range of effective ILC schemes (Jin and Xu 2013; Qian et al. 2023; Xu and Jin 2013; Yovchev et al. 2018).

Furthermore, for system input limitations, the novel ILC algorithms are designed to handle the tracking problem (He et al. 2018a, b; Sebastian et al. 2019; Zhuang et al. 2023), etc.

(4) ILC algorithms for incomplete information systems. The limitation of network transmission in the current system frequently results in data loss, transmission delays, and data quantization, which leads to problems such as information insecurity. In response to these problems, researchers design reasonable ILC algorithms to regulate the system and realize the control objectives. For example, ILC has been considered under data dropouts (Bu et al. 2013; Huo et al. 2023; Shen 2018; Shen and Xu 2017; Shen et al. 2016), with delays (Chen et al. 1998; Liu and Ruan 2017; Meng et al. 2016), with quantization (Bu et al. 2018; Shen and Xu 2016), and so on.

(5) Distributed ILC schemes in networked systems. Network systems are currently a hot topic of research in the control field. For example, consensus tracking is investigated by applying the ILC approach and delay exponential matrix in Luo et al. (2018). As for the bipartite consensus problem, ILC methods are designed to achieve the consensus tracking in Liang et al. (2021). In addition, ILC algorithms are also used in cooperative trajectory tracking (Meindi et al. 2022), formation control (Bu et al. 2018), distributed optimization (Luo et al. 2023; Song et al. 2022), and so on. Compared with the traditional control algorithms, ILC approaches can realize high-accuracy tracking in a finite-time interval.

(6) Applications of ILC technique. Due to the advantages of ILC, this control approach has also been applied in many practical systems. For example, ILC has been discussed in robotic manipulators (Yovchev et al. 2018), robotic fish control (Li et al. 2016), stroke rehabilitation (Freeman 2017), etc. Here, we only list some representative results, and one can view the rest of the application results in some related literature.

1.2 Network Systems

With the increasing development of computer technology and network communication technology, the structure of the control system is becoming increasingly complex, and the spatial distribution is becoming increasingly prominent. With the increase of equipment, the hardware arrangement of traditional control methods has become increasingly complex, and a series of problems, such as high overall maintenance difficulty and high cost, have become increasingly prominent. As a result, the control problem of networked systems has developed in the past years (Ge et al. 2017; Xu and Ma 2019; Zhang et al. 2020). Networked systems have the advantages of sharing information resources, easy expansion and maintenance, increased system flexibility and reliability, etc. Hence, networked systems have become one of the hot spots in the control and system research field, and they are increasingly applied in industry (Mu et al. 2021; Zhang et al. 2019). It is worth noting that networked systems present new challenges for system analysis and design, particularly in the context of constrained information communication.

1.2.1 Constrained Information Communication

Thus, networked systems need to transmit control signals and feedback information in real time to achieve precise control. However, in real-world applications, information transmission is frequently subject to technical limitations, communication equipment failures, congestion, and other factors. The constrained information communication may bring the following challenges:

(1) Data quantification and transmission. Network signals usually need to be quantized before transmission due to bandwidth limitations. The quantized data contains less information than the system's original signal, which may create an obstacle to the design of control strategies (Jiang and Liu 2013; Liang et al. 2020; Liu et al. 2016). Furthermore, the quantized information has a nonlinear relationship with the system's actual signal, which may make the theoretical analysis difficult.

(2) Network data dropout. Data loss is a major problem in networked systems. It is caused by communication network blocks or terminal-induced communication limitations (Liu et al. 2017). Retransmission mechanisms are commonly incorporated into networks, but their transmission time is limited. Exceeding this time limit may result in loss of data. When data is lost, there is a pressing need to find a solution to achieve effective control of the networked system.

(3) Network-induced delay. Delays are common in the information transmission of network systems. They are caused by network bandwidth limitations and network congestion. The time delay usually affects the operation of the control system. For example, it can reduce tracking performance and affect the system's stability under discussion (Ge et al. 2017; Milano and Anghel 2011; Qian et al. 2014). In the field of control and engineering, the study of time-delayed systems has become a popular research problem. The key to this research is designing reasonable and simple control schemes to suppress the negative effects of the time delay. The theoretical analysis is also one of the difficulties.

(4) System state is not measurable. In reality, the state of the network system may not be directly measurable due to packet loss, network delays, sensor failures, and so on. The immeasurability of the system state leads to poor control performance and system instability (Dehghanpour et al. 2019; Montanari and Aguirre 2020). Thus, to improve the network system's stability and performance, it is desirable to take special measures to address the problem of the unobservable state. Therefore, it is important to work out how to design appropriate observers and control strategies for different situations to achieve the system target.

(5) Network security. This problem has received widespread attention because devices in a networked environment are vulnerable to interception and attack (Li et al. 2015). Furthermore, cyber-attacks can result in loss of control, destabilization, or other security threats to the entire system (Volkova et al. 2019). Hence, effective control strategies are employed to suppress malicious attacks to achieve the security and stability of the discussed system (He et al. 2020; Xiong et al. 2022; Yu et al. 2022; Zhao et al. 2021). As a result, there is an urgent requirement to design appropriate security control strategies according to the characteristics of network attacks.

Ultimately, this monograph focuses on the control problem study of network systems under network communication constraints. Iterative learning methods are applied to investigate the relevant problems in networked systems. The theoretical results are presented in the following chapters.

1.2.2 Basic Knowledge

I. Graphs

Assume that a multi-agent system is described by a directed graph $\mathcal{G} = (\mathcal{V}, \mathcal{E}, \mathcal{A})$. $\mathcal{V} = \{1, \ldots, N\}$ represents the set of vertices, $\mathcal{E} \subseteq \mathcal{V} \times \mathcal{V}$ represents the set of edges, and $\mathcal{A} = (\alpha_{ij})_{N\times N}$ represents the adjacency matrix. An order pair $(j, i) \in \mathcal{E}$ denotes directed edge from j to i, which means that node i can receive the information from node j. The elements of the adjacency matrix $\mathcal{A}$ are nonnegative, i.e., $\alpha_{ii} = 0$, $\alpha_{ij} > 0$ if $(j, i) \in \mathcal{E}$ and $\alpha_{ij} = 0$ otherwise. Define $\Delta_{ii} = \sum_{j=1}^{N} \alpha_{ij}$, which denotes the in-degree of vertex i, and $\mathcal{D} = \mathrm{diag}\{\Delta_{11}, \ldots, \Delta_{NN}\}$. Hence, Laplacian matrix of $\mathcal{G}$ is described as $\mathcal{L} = \mathcal{D} - \mathcal{A}$. In reality, the desired state information can be obtained by part of nodes. Then, the desired state is treated as a (virtual) leader, denoted by vertex 0 in the graph representation. Specifically, $q^i = 1$ if node i can get the information from the desired state, and $q^i = 0$ otherwise. $\tilde{\mathcal{G}} = (\mathcal{V} \bigcup \{0\}, \tilde{\mathcal{E}}, \tilde{\mathcal{A}})$ describes the information flow among leader and followers. The edge set is denoted as $\tilde{\mathcal{E}}$, and the new weighted adjacency matrix is denoted as $\tilde{\mathcal{A}}$.

II. Time-Varying Graphs

Analogously, the communications between agents at time t are represented by the undirected graph $\mathcal{G}(t) = \{\mathcal{V}, \mathcal{E}(t), \mathcal{A}(t)\}$, where $\mathcal{V} = \{1, 2, \ldots, N\}$ is the set of nodes, $\mathcal{E}(t) = \{(i, j) : i, j \in \mathcal{V}\}$ denotes the edge set at time t. And $\mathcal{A}(t) = (\alpha_{ij}(t))_{N\times N}$ denotes the coupling configuration of the network at the time t. If there is an active communication channel from node j to node i $(i \neq j)$ at time t, then $\alpha_{ij}(t) = 1$; otherwise, $\alpha_{ij}(t) = 0$ $(i \neq j)$. Moreover, $\alpha_{ii}(t) = 0,\ i = 1, 2, \ldots, N$. The Laplacian of $\mathcal{A}(t)$ is defined as $L(t) = \mathcal{D}(t) - \mathcal{A}(t)$, and $\mathcal{D}(t) = (\Delta_{ij}(t))_{N\times N}$ is a diagonal matrix with $\Delta_{ii}(t) = \sum_{j=1}^{N} \alpha_{ij}(t)$. The neighborhood of agent i at the time t is defined as $N_i(t) = \{j \in \mathcal{V} | \alpha_{ij}(t) = 1\}$. The cardinal number of $N_i(t)$ is called the degree of i at time t and denoted by $d_i(t)$. Denote $\Pi_i = \bigcup_{t=1}^{T} N_i(t)$.

III. Signed Graphs

A signed digraph $\tilde{\mathcal{G}} = (\mathcal{V}, \mathcal{E}, \tilde{\mathcal{A}})$ is a graph with the vertex set $\mathcal{V} = \{v_1, v_2, \ldots, v_n\}$, the edge set $\mathcal{E} \subseteq \mathcal{V} \times \mathcal{V}$, and the directed weighted matrix $\tilde{\mathcal{A}} = (\tilde{\alpha}_{ij})_{n\times n} \in \mathbb{R}^{n\times n}$. Edge $(v_i, v_j) \in \mathcal{E}$ if and only if node j sends information to node i. The set of neighbors of node v_i is defined as $\mathcal{N}_i = \{v_j \in \mathcal{V} : (v_i, v_j) \in \mathcal{E}\}$. The signed digraph is specified by the weighted matrix $\tilde{\mathcal{A}}$, which is defined as

(1) $\tilde{\alpha}_{ii} = 0,\ \forall\, i \in \{1, \ldots, n\}$;
(2) $\tilde{\alpha}_{ij} > 0\, (i \neq j,\ i, j \in \{1, \ldots, n\})$ means the interaction between the ith and the jth nodes is cooperative;

(3) $\tilde{\alpha}_{ij} < 0\,(i \neq j,\ i, j \in \{1, \ldots, n\})$ means the interaction between the ith and the jth nodes is competitive.

In the graph, a digon means a pair of edges having the same nodes (v_i, v_j), $(v_j, v_i) \in \mathcal{E}$. A signed digraph $\tilde{\mathcal{G}}$ is called digon sign-symmetric if $\tilde{\alpha}_{ij}\tilde{\alpha}_{ji} \geq 0$, $\forall\, i, j \in \{1, 2, \ldots, n\}$; otherwise, the digraph $\tilde{\mathcal{G}}$ is called digon sign-unsymmetric. A signed digraph $\tilde{\mathcal{G}}$ is called structurally balanced if it has a bipartition $\mathcal{V}_1,\ \mathcal{V}_2,\ \mathcal{V}_1 \cup \mathcal{V}_2 = \mathcal{V},\ \mathcal{V}_1 \cap \mathcal{V}_2 = \varnothing$, to satisfy that $\tilde{\alpha}_{ij} \geq 0,\ \forall\ v_i, v_j \in \mathcal{V}_a,\ a \in \{1, 2\}$; $\tilde{\alpha}_{ij} \leq 0,\ \forall\ v_i \in \mathcal{V}_a,\ v_j \in \mathcal{V}_b,\ b \neq a,\ a, b \in \{1, 2\}$, otherwise, the signed digraph $\tilde{\mathcal{G}}$ is called structurally unbalanced. The Laplacian L of the signed matrix $\tilde{\mathcal{A}}$ is defined as $\mathcal{L} = (l_{ij})_{n\times n}$ with

$$l_{ij} = \begin{cases} \sum\limits_{v_j \in N_i} |\tilde{\alpha}_{ij}|, & j = i, \\ -\tilde{\alpha}_{ij}, & j \neq i. \end{cases} \tag{1.1}$$

1.2.3 *Notations*

In this monograph, $x \in \mathbb{R}^n$ denotes the n-dimensional column vector, $\|x\|$ denotes the Euclidean norm of the vector x, and $\|x\|_1$ denotes the 1-norm of the vector x. For a symmetric matrix A, $\lambda_m(A)$ and $\lambda_M(A)$ represent the minimal and maximal eigenvalues of matrix A, respectively. The spectral norm of A is defined by $\|A\| \triangleq (\lambda_M(A^{\mathrm{T}}A))^{\frac{1}{2}}$, and the spectral radius of A is $\rho(A) \triangleq \max\limits_i |\lambda_i(A)|$. For real symmetric matrices X and Y, $X \geq Y$ (or $X > Y$) means that matrix $X - Y$ is positive semi-definite (or positive definite).

1.3 Structure of This Monograph

This book mainly discusses the ILC problem under constrained information communication. The book is organized as follows:

Chapter 1: This chapter begins with the background of ILC and network systems. By reviewing the current status of ILC research, researchers have a better understanding of research progress and potential issues. Furthermore, it also provides some basic knowledge of network systems.

Chapter 2: This chapter studies the iterative learning consensus problem for discrete-time systems with an event-triggered scheme and quantization. The event-triggered scheme is considered in our iterative learning controllers to reduce the number of iteration steps to be updated. Our trigger mechanism is designed to depend on time instant and learning steps. Quantization is then introduced in the event-triggered controllers, and some relaxed conditions are presented to guarantee the tracking problem by using some interval matrix properties.

Chapter 3: This chapter mainly discusses the problem of quantized iterative learning for digital networks with time-varying topologies. The information is first encoded as symbolic data and then transmitted. After receiving the data, the receiver uses a decoder to estimate the state of the sender. Iterative learning quantized communication is considered during the encoding and decoding process. Then, a sufficient condition for achieving the consensus tracking problem in an interval using quantized iterative learning controllers is given.

Chapter 4: This chapter investigates the consensus tracking problem of discrete linear systems in the presence of two kinds of switching topologies. The two independent switching topologies are considered in the control and observation networks, respectively. Observer-based learning control strategies are proposed to solve the consensus problem when the state information of each node is unavailable. Some sufficient conditions are derived to guarantee the asymptotic convergence of the consensus tracking error.

Chapter 5: This chapter studies the tracking problem of T-S fuzzy systems with measurable and unmeasurable state information by building synchronous and asynchronous iterative learning controllers. In the beginning, the T-S fuzzy model was created to explain dynamic systems with changeable structures. To solve the tracking issue of T-S fuzzy systems, iterative learning controllers are then developed. Here, synchronous and asynchronous iterative learning control schemes are planned. Additionally, due to network transmission bandwidth capabilities and/or internal and/or external disturbances, system states are thought to be both measurably and immeasurably variable.

Chapter 6: This chapter considers the interval iterative learning problem for dynamic systems with HNN (hierarchical neural network)-structural output. The initial goal is to design the output of a dynamic system with an HNN structure. A necessary and sufficient condition is stated to implement ILC and achieve interval tracking in a finite interval. The saturated ILC is then taken into account for the system under discussion, and a less stringent criterion is proposed to enable tracking in a finite interval utilizing a technique for network structure decomposition.

Chapter 7: This chapter investigates the tracking problem of coupled continuous networks by discontinuous ILC method. Considering the difficulties of instantaneous disturbance and continuous information transmission, two discontinuous ILC schemes with impulse and zero-order hold (ZOH) sampling data are introduced. Then, using the Bellman–Gronwall lemma, some criteria were proposed to solve the tracking problem of continuous networks within finite intervals.

Chapter 8: This chapter discusses finite-iteration tracking for singular coupled systems with packet-dropping learning controllers. Iterative learning controllers are then developed to achieve the finite-iteration tracking of singular systems. Singular coupled systems are produced to characterize systems with some algebraic limitations. First, the definition of finite-iteration tracking is put out, and the explicit calculation of the settling iteration follows. The iterative learning controllers are also made to take packet losses into account.

Chapter 9: This chapter mainly proposes a novel periodic iterative learning control scheme to examine the finite-iteration tracking of discrete networks. The FlexRay communication protocol with both static and dynamic segments is introduced in the design of the proposed ILC scheme. The design has the benefit of lowering the demand on the communication channel's bandwidth, which enhances the efficiency of finite-iteration tracking. Based on the dynamic FlexRay communication channels, this chapter presents the first integrative approach to iterative learning design that is applicable to different learning processes in network control design.

Chapter 10: This chapter addresses the tracking for discrete systems by designing two kinds of multi-layered iterative learning schemes with cooperative–antagonistic interactions. By using the signed graph theory, we design iterative learning schemes that are multi-layered and have cooperative–antagonistic interactions. Furthermore, considering the limited bandwidth of information storage, the state information of these controllers is updated in light of previous learning iterations but does not depend only on the last iteration. Then, two simple criteria are proposed for discussing the tracking of discrete systems with multi-layered and cooperative–antagonistic iterative schemes.

Chapter 11: This chapter discusses the stability and increase of the convergence rate of a multi-layer supply chain network (SCN) by creating a new learning control strategy. First, a three-layer multi-layer SCN is constructed, and several state limitations are resolved. The control inputs are the quantities of items sent between distinct nodes. Then, to explore the stability of the states, a new type of learning control technique is developed. In addition, the scheme's convergence rate of all states is demonstrated to be faster than the standard schemes. Our method can conserve a scarce control resource, and it also increases the convergence rates of every node.

Chapter 12: This chapter mainly investigates the security problem of discrete-time systems under random deception and DoS attacks. These two types of network attacks (deception and DoS attacks) may occur alternately, resulting in the infeasibility of obtaining actual output measurements. As a result, an estimation output formula is proposed to evaluate the actual output. According to the obtained estimation formula, iterative learning strategies are designed and applied to achieve system security. Also, with appropriate iterative learning strategies, the convergence conditions we achieved are simpler than the existing literature.

1.4 Summary

In this chapter, the concepts and definitions of ILC and graphs have been first introduced. Some dynamic research on network systems has been further summarized. Furthermore, it provides the main issues of the monograph study. The structure of this monograph is presented at the end of this chapter.

References

Arimoto S, Kawamura S, Miyazaki F (1984) Bettering operation of robots by learning. J Robotic Syst 1(2):123–140

Barton KL, Alleyne AG (2011) A norm optimal approach to time-varying ILC with application to a multi-axis robotic testbed. IEEE Trans Control Syst Tech 19(1):166–180

Boudjedir CE, Boukhetala D, Bouri M (2019) Iterative learning control of multivariable uncertain nonlinear systems with nonrepetitive trajectory. Nonlinear Dyn 95:2197–2208

Bristow DA, Tharayil M, Alleyne AG (2006) A survey of iterative learning control. IEEE Control Syst Mag 26(3):96–114

Bu XH, Hou ZS (2016) Adaptive iterative learning control for linear systems with binary-valued observations. IEEE Trans Neur Net Lear Syst 29(1):232–237

Bu XH, Yu FS, Hou ZS, Wang FZ (2013) Iterative learning control for a class of nonlinear systems with random packet losses. Nonlinear Anal 14(1):567–580

Bu XH, Hou ZS, Yu QX, Yang Y (2018) Quantized data driven iterative learning control for a class of nonlinear systems with sensor saturation. IEEE Trans Syst Man Cybern Syst 50(12):5119–5129

Bu XH, Cui LZ, Hou ZS, Qian W (2018) Formation control for a class of nonlinear multiagent systems using model-free adaptive iterative learning. Inter J Robust Nonlinear Control 28(4):1402–1412

Chen YQ, Gong ZM, Wen CY (1998) Analysis of a high-order iterative learning control algorithm for uncertain nonlinear systems with state delays. Automatica 34(3):345–353

Chu B, Owens DH, Freeman CT (2016) Iterative learning control with predictive trial information: convergence, robustness, and experimental verification. IEEE Trans Control Syst Tech 24(3):1101–1108

Dehghanpour K, Wang ZY, Wang JH, Yuan YX, Bu FK (2019) A survey on state estimation techniques and challenges in smart distribution systems. IEEE Trans Smart Grid 10(2):2312–2322

Freeman CT (2017) Robust ILC design with application to stroke rehabilitation. Automatica 81:270–278

Ge X, Yang F, Han QL (2017) Distributed networked control systems: a brief overview. Inform Sci 380:117–131

He W, Meng TT, He XY, Ge Sam SZ (2018a) Unified iterative learning control for flexible structures with input constraints. Automatica 96:326–336

He W, Meng TT, Huang DQ, Li XF (2018b) Adaptive boundary iterative learning control for an Euler-Bernoulli beam system with input constraint. IEEE Trans Neur Netw Learn Syst 29(5):1539–1549

He WL, Luo TH, Tang Y, Du WL, Tian YC, Qian F (2020) Secure communication based on quantized synchronization of chaotic neural networks under an event-triggered strategy. IEEE Trans Neur Net Lear Syst 31(9):3334–3345

He SY, Chen WB, Li DW, Xi YG, Xu YW, Zheng PY (2022) Iterative learning control with data-driven-based compensation. IEEE Trans Cybern 52(8):7492–7503

Huang DQ, Chen Y, Meng DY, Sun PF (2021) Adaptive iterative learning control for high-speed train: a multi-agent approach. IEEE Trans Syst Man Cybern Syst 51(7):4067–4077

Huang JS, Wang W, Su XJ (2021) Adaptive iterative learning control of multiple autonomous vehicles with a time-varying reference under actuator faults. IEEE Trans Neur Netw Learn Syst 32(12):5512–5525

Huo N, Jiang H, Shen D, Wang JR (2023) Finite-level uniformly quantized learning control with random data dropouts. Inter J Robust Nonlinear Control 33(7):4056–4075

Jiang ZP, Liu TF (2013) Quantized nonlinear control-a survey. Acta Auto Sinica 39(11):1820–1830

Jin X (2018) Fault-tolerant iterative learning control for mobile robots non-repetitive trajectory tracking with output constraints. Automatica 94:63–71

Jin X (2021) Iterative learning control for MIMO nonlinear systems with iteration-varying trial lengths using modified composite energy function analysis. IEEE Trans Cybern 51(12):6080–6090
Jin X, Xu JX (2013) Iterative learning control for output-constrained systems with both parametric and nonparametric uncertainties. Automatica 49(8):2508–2516
Li XF, Shen D (2017) Two novel iterative learning control schemes for systems with randomly varying trial lengths. Syst Contr Lett 107:9–16
Li Y, Chen YQ, Ahn HS, Tian GH (2013) A survey on fractional-order iterative learning control. J Optim Theory Appl 156:127–140
Li AS, Li XC, Pan YC, Zhang W (2015) Strategies for network security. Sci China Inf Sci 58(1):012107
Li XF, Ren QY, Xu JX (2016) Precise speed tracking control of a robotic fish via iterative learning control. IEEE Trans Ind Electron 63(4):2221–2228
Liang H, Zhang Y, Huang T, Ma H (2020) Prescribed performance cooperative control for multiagent systems with input quantization. IEEE Trans Cyber 50(5):1810–1819
Liang JQ, Bu XH, Cui LZ, Hou ZS (2021) Finite time asymmetric bipartite consensus for multi-agent systems based on iterative learning control. Int J Robust Nonlinear Control 31(12):5708–5724
Lin N, Chi RH, Huang B, Hou ZS (2021) Event-triggered nonlinear iterative learning control. IEEE Trans Neur Netw Learn Syst 32(11):5118–5128
Lin N, Chi RH, Huang B (2021) Auxiliary predictive compensation-based ILC for variable pass lengths. IEEE Trans Syst Man Cybern Syst 51(7):4048–4056
Liu J, Ruan XE (2017) Networked iterative learning control design for discrete-time systems with stochastic communication delay in input and output channels. Inter J Syst Sci 48(9):1844–1855
Liu K, Fridman E, Johansson KH, Xia YQ (2016) Quantized control under round-robin communication protocol. IEEE Trans Ind Electron 63(7):4461–4471
Liu Y, Candell R, Moayeri N (2017) Effects of wireless packet loss in industrial process control systems. ISA Trans 68:412–424
Luo ZJ, Xiong WJ, Yu XH (2018) Consensus tracking in multi-node systems using iterative learning control based on delay exponential matrix. Unmanned Syst 6(3):209–219
Luo ZJ, Xiong WJ, Huang TW, Duan J (2023) Distributed quadratic optimization with terminal consensus iterative learning strategy. Neurocomputing 528:12–19
Meindl M, Molinari F, Lehmann D, Seel T (2022) Collective iterative learning control: exploiting diversity in multi-agent systems for reference tracking tasks. IEEE Trans Control Syst Tech 30(4):1390–1402
Meng DY, Moore KL (2017) Robust iterative learning control for nonrepetitive uncertain systems. IEEE Trans Auto Control 62(2):907–913
Meng DY, Moore KL (2017) Convergence of iterative learning control for SISO nonrepetitive systems subject to iteration-dependent uncertainties. Automatica 79:167–177
Meng DY, Zhang JY (2021) Robust optimization-based iterative learning control for nonlinear systems with nonrepetitive uncertainties. IEEE/CAA J Auto Sin 8(5):1001–1014
Meng DY, Zhang JY (2021) Convergence analysis of robust iterative learning control against nonrepetitive uncertainties: System equivalence transformation. IEEE Trans Neur Net Lear Syst 32(9):3867–3879
Meng DY, Jia YM, Du JP (2016) Consensus seeking via iterative learning for multi-agent systems with switching topologies and communication time-delays. Inter J Robust Nonlinear Control 26(17):3772–3790
Milano F, Anghel M (2011) Impact of time delays on power system stability. IEEE Trans Circ Syst I 59(4):889–900
Montanari AN, Aguirre LA (2020) Observability of network systems: a critical review of recent results. J Control Auto Elec Syst 31:1348–1374
Mu CX, Wang K, Sun CY (2021) Learning control supported by dynamic event communication applying to industrial systems. IEEE Trans Ind Inform 17(4):2325–2335

Patan K, Patan M (2020) Neural-network-based iterative learning control of nonlinear systems. ISA Trans 98:445–453
Qian YF, Wu XQ, Lü JH, Lu JA (2014) Consensus of second-order multi-agent systems with nonlinear dynamics and time delay. Nonlinear Dyn 78(1):495–503
Qian K, Li Z, Chakrabarty S, Zhang ZQ, Xie SQ (2023) Robust iterative learning control for pneumatic muscle with uncertainties and state constraints. IEEE Trans Ind Electron 70(2):1802–1810
Riaz S, Hui L, Aldemir MS, Afzal F (2021) A future concern of iterative learning control: a survey. J Stat Manag Syst 24(6):1301–1322
Sebastian G, Tan Y, Oetomo D (2019) Convergence analysis of feedback-based iterative learning control with input saturation. Automatica 101:44–52
Shen D (2018) Iterative learning control with incomplete information: a survey. IEEE/CAA J Autom Sin 5(5):885–901
Shen D, Li XF (2019) A survey on iterative learning control with randomly varying lengths: model, synthesis, and convergence analysis. Annu Rev Control 48:89–102
Shen D, Xu Y (2016) Iterative learning control for discrete-time stochastic systems with quantized information. IEEE/CAA J Auto Sinica 3(1):59–67
Shen D, Xu JX (2017) A framework of iterative learning control under random data dropouts: mean square and almost sure convergence. Inter J Adapt Control Signal Process 31(12):1825–1852
Shen D, Zhang W, Wang YQ, Chien CJ (2016) On almost sure and mean square convergence of P-type ILC under randomly varying iteration lengths. Automatica 63:359–365
Song Q, Meng DY, Liu F (2022) Consensus-based iterative learning of heterogeneous agents with application to distributed optimization. Automatica 137:110096
Volkova A, Niedermeier M, Basmadjian R, Meer DH (2019) Security challenges in control network protocols: a survey. IEEE Commun Sur Tutor 21(1):619–639
Xiong WJ, Yu XH, Patel R, Yu WW (2016) Iterative learning control for discrete-time systems with event-triggered transmission strategy and quantization. Automatica 72:84–91
Xiong WJ, Gong K, Wen GH, Lü JH (2022) Security analysis of discrete nonlinear systems with injection attacks under iterative learning schemes. IEEE Trans Syst Man Cybern Syst 52(2):927–935
Xu JX, Jin X (2013) State-constrained iterative learning control for a class of MIMO systems. IEEE Trans Auto Control 58(5):1322–1327
Xu L, Ma H (2019) Distributed output-feedback cooperative tracking control of non-linear multi-agent systems with unknown actuator failures. IET Control Theory Appl 13(5):2465–2475
Yovchev K, Delchev K, Krastev E (2018) State space constrained iterative learning control for robotic manipulators. Asian J Control 20(3):1145–1150
Yu M, Chai S (2019) A survey on high-order internal model based iterative learning control. IEEE Access 7:127024–127031
Yu QX, Hou ZS, Xu JX (2018) D-type ILC based dynamic modeling and norm optimal ILC for high-speed trains. IEEE Trans Control Syst Tech 26(2):652–663
Yu W, Bu XH, Hou ZS (2022) Security data-driven control for nonlinear systems subject to deception and false data injection attacks. IEEE Trans Net Sci Eng 9(4):2910–2921
Zhang T, Li JM (2018) Event-triggered iterative learning control for multi-agent systems with quantization. Asian J Control 20(3):1088–1101
Zhang F, Kodituwakku HADE, Hines JW, Coble J (2019) Multilayer data-driven cyber-attack detection system for industrial control systems based on network, system, and process data. IEEE Trans Ind Inform 15(7):4362–4369
Zhang XM, Han QL, Ge X, Ding DR, Ding L, Yue D, Peng C (2020) Networked control systems: a survey of trends and techniques. IEEE/CAA J Auto Sinica 7(1):1–17
Zhang DL, Wang ZN, Masayoshi T (2021) Neural-network-based iterative learning control for multiple tasks[J]. IEEE Trans Neur Netw learn Syst 32(9):4178–4190
Zhao D, Wang ZD, Ho DWC, Wei GL (2021) Observer-based PID security control for discrete time-delay systems under cyber-attacks. IEEE Trans Syst Man Cybern Syst 51(6):3926–3938

Zhuang ZH, Tao HF, Chen YY, Stojanovic V, Paszke W (2023) An optimal iterative learning control approach for linear systems with nonuniform trial lengths under input constraints. IEEE Trans Syst Man Cybern Syst 53(6):3461–3473

Chapter 2
Consensus Under Event-Triggered Transmission and Quantization

2.1 Introduction

It is well known that ILC is an effective technique that focuses on improving the performance of systems over a fixed time interval by learning from previous executions (trials, iterations, passes). This control strategy can tackle dynamic systems with a high level of uncertainty in a simple way. As a result, ILC has been widely applied to industrial robots (Arimoto et al. 1984), chemical reactors (Mezghani et al. 2002), input saturation (Tan et al. 2011; Xiong et al. 2016; Xu et al. 2004; Zhang et al. 2015), heat equations (Huang et al. 2013), sampled-data systems (Abidi and Xu 2011), and multi-agent systems (Li and Li 2014; Meng et al. 2015a, b, 2013, 2014; Meng and Moore 2016). For example, Li and Li (2014) showed that all followers could track the leader uniformly on the finite interval for the consensus problem and keep the desired distance from the leader to achieve the velocity consensus uniformly on $[1, T]$. In Meng et al. (2014), tackled the formation control problems for multi-agent systems with nonlinear dynamics and switching network topologies. It was shown in Meng et al. (2015a) that these uncertainties of multi-agent systems are dynamically changing not only along the time axis but also along the iteration axis.

In the above literature, the proposed ILC algorithms are updated in each iteration step. In our opinion, it is costly and unnecessary to update the control algorithms in each iteration when the control strategy varies less in some successive iteration steps. Therefore, a natural idea is to use the event-triggering mechanism to reduce the number of iterative steps that need to be updated. The design of an appropriate measurement error threshold is the key to the event-triggered mechanism. The control algorithm can be updated when the measurement error reaches or exceeds the threshold. As a good control approach that can be used to reduce the communication load, event-triggered strategies have been receiving great attention in wireless sensor/actuator systems (Mazo and Tabuada 2011), networked systems (Yue et al. 2013), fuzzy systems (Peng et al. 2013), sampled-data control systems (Peng and Han

W. Xiong et al., *Iterative Learning Control for Network Systems Under Constrained Information Communication*, Intelligent Control and Learning Systems 12,
https://doi.org/10.1007/978-981-97-0926-7_2

2013; Zou et al. 2015), multi-agent systems (Fan et al. 2013; Hu et al. 2016; Zhang et al. 2014), etc. Due to the characteristics and advantages of the event-triggered strategy, this chapter will develop novel ILC algorithms which are different from the existing literature. Specifically, our event-triggered strategies are related to time t and iterative learning step k.

In some practical applications, such as industrial control systems, it is necessary to control multiple dynamic systems by exchanging information on multiple communication networks. However, due to storage and inter-node communication bandwidth constraints, the original information needs to be quantized. As a result, it is necessary to analyze the quantizer and then determine the degree of quantization's influence on the dynamic system. In the meantime, the problem of quantized control for dynamic systems has been discussed and studied in the literature (Delchamps 1990; Fu and Xie 2005; Liu et al. 2012; Wang et al. 2012; Zhang et al. 2015). However, to the best of our knowledge, the quantized ILC problem for dynamic systems has not been fully investigated despite its potential in practical applications. Furthermore, in order to reduce the requirements for storage and communication bandwidth, it is necessary to design ILC strategies with event-triggered and quantized mechanisms. Research in this area has not been fully explored. As a result, this chapter will analyze and address the above research problems.

2.2 Problem Formulation

In this chapter, we mainly consider a discrete-time network system which consists of N nodes. In this system, each node has two independent dynamic processes: (1) the process shows the system dynamics about time t and (2) the process describes the system dynamics of node about the learning step k. Thus, the system dynamics can be described by

$$x_i(t+1,k) = c_i x_i(t,k) + b_i u_i(t,k), \tag{2.1}$$

where $x_i(t,k)$ is the state of node i, $i \in \{1, \ldots, N\}$, and $x(t,k) = (x_1(t,k), \ldots, x_N(t,k))^{\mathrm{T}} \in \mathbb{R}^N$; $t \in \{0, 1, \ldots, T\}$, $T > 0$ denotes the terminal time; and $k \in \mathbb{Z}_+$ denotes the iterative learning step. $C = \mathrm{diag}(c_1, \ldots, c_N) \in \mathbb{R}^{N\times N}$ and $B = \mathrm{diag}(b_1, \ldots, b_N) \in \mathbb{R}^{N\times N}$ are constant matrices. $u_i(t,k)$ denotes the iterative learning controller of node i, then $u(t,k) = (u_1(t,k), \ldots, u_N(t,k))^{\mathrm{T}} \in \mathbb{R}^N$. For every node i in system (2.1), it is said to achieve the consensus or tracking of a desired trajectory if

$$\lim_{k\to+\infty} x_i(t,k) = x^*(t),\ t \in \{0, 1, v, T\}, \tag{2.2}$$

where $x^*(t) \in \mathbb{R}$ denotes the desired trajectory. Let $e_i(t,k) = x_i(t,k) - x^*(t)$ be the tracking error of node i, then $e(t,k) = (e_1(t,k), \ldots, e_N(t,k))^{\mathrm{T}} \in \mathbb{R}^N$. Note that the objective (2.2) holds if and only if $\lim\limits_{k\to+\infty} e(t,k) = \mathbf{0}, \forall\, t \in \{0, 1, \ldots, T\}$.

2.3 ILC Scheme

2.3.1 Event-Triggered Transmission and Quantization

In the existing research (Meng et al. 2014, 2015a), the learning controller $u(t,k)$ is always updated in each iteration k. In fact, to save computational resources, ILCs will not be updated temporarily when the tracking error changes are minor in some iteration steps. Thus, the ILCs with event-triggered strategy will be considered. This chapter defines the state measurement error of node i by $\delta_i(x_i(t,k)) \triangleq \widehat{x}_i(t,k) - x_i(t,k),\ i \in \{1, \ldots, N\},\ t \in \{0, \ldots, T\},\ k \in \mathbb{Z}_+$. $\widehat{x}_i(t,k)$ represents the latest sampled state of node i, which will be given later. In addition, the transmission information needs to be coded or quantized due to the limitation of storage and communication bandwidth among nodes. Hence, according to the measurement error of node i, the event-triggered strategies without and with quantization are given as follows:

$$|\delta_i(x_i(t,k))| = \gamma_i \cdot \left|\sum_{j=1}^{n} l_{ij}\widehat{x}_j(t,k)\right|, \tag{2.3}$$

$$|\delta_i(x_i(t,k))| = \gamma_i \cdot \left|\sum_{j=1}^{n} l_{ij} q(\widehat{x}_i(t,k))\right|, \tag{2.4}$$

where parameter $\gamma_i > 0$, matrix $L = (l_{ij})_{N\times N} \in \mathbb{R}^{N\times N}$, and $\|L\| \neq 0$. $q(\cdot) : \mathbb{R} \to \Lambda_{\varpi}$ is a logarithmic quantizer, for a given accuracy parameter $\varpi \in (0,1)$, then we define the logarithmic set of quantization levels

$$\begin{aligned}\Lambda_{\varpi} = &\Big\{\pm\,\omega_{(i)} : \omega_{(i)} = \varpi^{i}\omega_{(0)},\ i = \pm1, \pm2, \cdots\Big\} \\ &\bigcup\Big\{\pm\,\omega_{(0)}\Big\}\bigcup\Big\{0\Big\},\ \omega_{(0)} > 0.\end{aligned} \tag{2.5}$$

Thus, the associated quantizer $q(\cdot)$ is defined as follows:

$$q(x) = \begin{cases} \omega_{(i)}, & \frac{1}{1+\sigma}\omega_{(i)} < x \le \frac{1}{1-\sigma}\omega_{(i)}; \\ 0, & x = 0; \\ -q(-x), & x < 0, \end{cases} \tag{2.6}$$

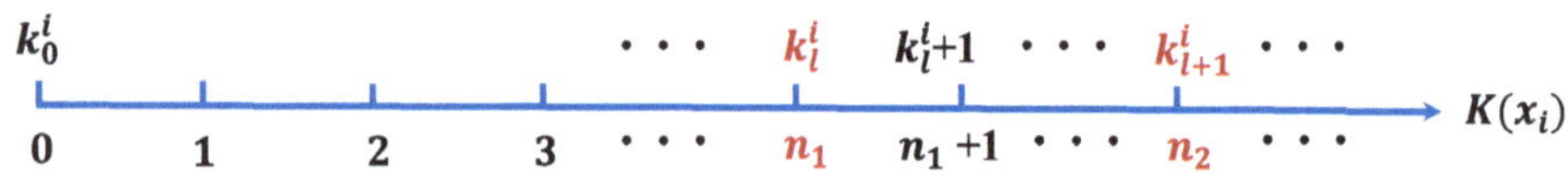

Fig. 2.1 The information broadcasting iteration sequence $\{k_l^i\}$ of node i

where $\sigma = \frac{1-\varpi}{1+\varpi}$ is named sector bound in Fu and Xie (2005). When event-triggered condition (2.3) or (2.4) is satisfied, node i will broadcast its state and update its control protocol. The information broadcasting iteration sequence of node i is $\{k_l^i\}$ ($i \in \{1, \ldots, N\}$, $l \in \mathbb{Z}_+$) is shown in Fig. 2.1. Also, $\widehat{x}_i(t, k) = x_i(t, k_l^i)$, $k_l^i \le k < k_{l+1}^i$, $l \in \mathbb{Z}_+$.

According to the definition of logarithmic quantizer, the quantization error satisfies the following equality:

$$q(\widehat{x}_i(t, k)) = \widehat{x}_i(t, k) + \widetilde{\Lambda}(t, k)\widehat{x}_i(t, k), \tag{2.7}$$

where $\widetilde{\Lambda}(t, k) \in [-\sigma, \sigma]$, $t \in \{0, \ldots, T\}$, $i \in \{1, \ldots, N\}$ and $k \in \mathbb{Z}_+$.

2.3.2 ILC Scheme Design

According to (2.3) and (2.4), we further consider the ILCs with and without quantization, which can be expressed as

$$\widehat{u}_i(t, k_{l+1}^i) = \widehat{u}_i(t, k_l^i) + \Gamma_1^i \sum_{j=1}^{n} l_{ij}\widehat{x}_i(t + 1, k), \tag{2.8}$$

and

$$\widehat{u}_i(t, k_{l+1}^i) = \widehat{u}_i(t, k_l^i) + \Gamma_2^i \sum_{j=1}^{n} l_{ij} q(\widehat{x}_i(t + 1, k)), \tag{2.9}$$

and $u_i(t, k) = \widehat{u}_i(t, k_l^i)$, $k_l^i \le k < k_{l+1}^i$, $l \in \mathbb{Z}_+$. Learning gain matrices $\Gamma_1 \triangleq \text{diag}(\Gamma_1^1, \ldots, \Gamma_1^N)$ and $\Gamma_2 \triangleq \text{diag}(\Gamma_2^1, \ldots, \Gamma_2^N)$ are designed later.

Remark 2.1 In Eqs. (2.8) and (2.9), the key is to choose the appropriate broadcasting iteration sequence $\{k_l^i\}$. It is uneconomical if the interval $[k_l^i, k_{l+1}^i)$ is small. In addition, the consensus of system (2.1) is achieved inefficiently if the interval $[k_l^i, k_{l+1}^i - 1)$ is large. As a result, a feasible alternative for node i is to use the neighbor's states at the k_l^ith iteration until a pre-defined event is triggered at the k_{l+1}^ith iteration. Similarly, the next neighbor's information is updated by the states at k_{l+1}^ith iteration until the event is triggered.

Remark 2.2 In existing research (Tan et al. 2011; Meng et al. 2014, 2015a), the ILC algorithms are requested to be updated in each iteration. Actually, it is uneconomical and unnecessary to update the control strategy in each iteration when small changes in the tracking error occur in some adjacent iteration steps. In comparison to the existing literature, our control strategies have two twofold advantages: 1) ILC algorithms (2.8) and (2.9) are only updated when trigger conditions (2.3) and (2.4) hold. Moreover, different from the traditional method, the event-triggered controllers in this chapter are related to time t and iterative learning step k. 2) Quantization is further considered in event-triggered controllers due to its potential in practical applications such as industrial control systems.

2.4 Convergence Analysis

Some notations are presented in the following for the sake of discussion. For any diagonal matrix $P \triangleq \text{diag}(p_1, \ldots, p_N)$, $\Phi(P) \triangleq \{G = \text{diag}(g_1, \ldots, g_N) : g_i = p_i, or\ g_i = 0,\ \forall\ i \in \{1, \ldots, N\}\}$. Obviously, $\Phi(P)$ is a finite set. In addition, the basic assumption of the learning system (2.1) is given as $x_i(0, k) = x^*(0)$, $x_i(t, 0) = x_{i0}(t)$, and $\|x_{i0}(t)\| \le \varepsilon_i, \forall\, t \in \{0, 1, \ldots, T\}, i = 1, 2, \ldots, N,\ k \in \mathbb{Z}_+$, ε_i is a positive constant. For the sake of simplicity, the desired target trajectory is assumed as $x^*(t) = 0,\ \forall\, t \in \{0, 1, \ldots, T\}$. In the following part, our purpose is to achieve the tracking of the system (2.1) with event-triggered strategy (2.3) or (2.4) by utilizing the iterative learning control strategies (2.8) or (2.9).

In this chapter, each node with dynamic behavior (2.1) broadcasts its state and updates its control protocol based on its own iteration channel (see Fig. 2.2). In Fig. 2.2, asynchronous broadcasting iterations are denoted as a sequence $\{k_l\}$, $l \in \mathbb{Z}_+$, $k_0^1 = k_0^2 = \ldots = k_0^N = k_0 = 0$. Moreover, the controller components may occur

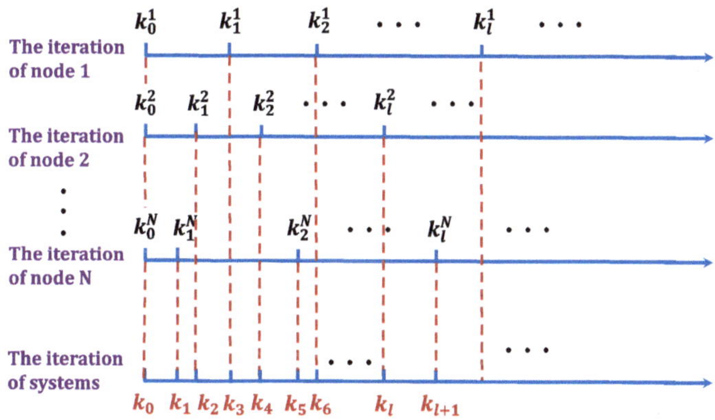

Fig. 2.2 Multi-channel asynchronous broadcasting in the iterations

$k_q^i = k_s^j$ for $\forall\, i,\ j \in \{1, 2, \ldots, N\}, i \neq j,\ q,\ s \in \mathbb{Z}_+$. In this case, we regard k_q^i and k_s^j as two different terms k_l and k_{l+1} for some $l \in \mathbb{Z}_+$.

According to Fig. 2.2, Eqs. (2.8) and (2.9) can be written as

$$\widehat{u}(t, k_{l+1}) = \widehat{u}(t, k_l) + D_l \Gamma_1 L \widehat{x}(t+1, k), \tag{2.10}$$

and

$$\widehat{u}(t, k_{l+1}) = \widehat{u}(t, k_l) + D_l \Gamma_2 L q(\widehat{x}(t+1, k)), \tag{2.11}$$

where $l \in \mathbb{Z}_+$, $\widehat{u}(t, k_l) = (\widehat{u}_1(t, k_l), \ldots, \widehat{u}_N(t, k_l))^{\mathrm{T}}$, $q(\widehat{x}(t, k)) = (q(\widehat{x}_1(t, k)), \ldots, q(\widehat{x}_N(t, k)))^{\mathrm{T}}$. Moreover,

$$u(t, k) = \widehat{u}(t, k_l), \quad k_l \leq k < k_{l+1}. \tag{2.12}$$

About matrices D_l ($l \in \mathbb{Z}_+$), we propose the following property.

Property 2.1 Matrix $D_l \triangleq \mathrm{diag}\big(d_1^l, \ldots, d_N^l\big) \in \Phi(I_N), \forall\, i \in \{1, \ldots, N\}, d_i^l = 1$ or $d_j^l = 0$. Matrix D_l is a matrix with one nonzero diagonal element (i.e., the nonzero diagonal element is 1), and other elements are zero. That is, $\exists\, i \in \{1, \ldots, N\}$ such that $b_i^l = 1$, and $b_j^l = 0,\ \forall\, j \in \{1, \ldots, N\}, j \neq i,\ l \in \mathbb{Z}_+$.

Proof According to Fig. 2.2, $\forall\ l \in \mathbb{Z}_+$, there only exists an $i \in \{1, \ldots, N\}$ and a $s \in \mathbb{Z}_+$ which satisfy that $k_l = k_s^i$. As a result, one has $d_i^l = 1$ and $d_j^l = 0$, $\forall\, j \in \{1, \ldots, N\}, j \neq i, l \in \mathbb{Z}_+$. Therefore, matrix D_l is a matrix with one nonzero diagonal element 1. □

2.4.1 The Consensus Under Event-Triggered Strategy Without Quantization

Let $\delta(x(t, k)) \triangleq (\delta_1(x_1(t, k)), \ldots, \delta_N(x_N(t, k)))^{\mathrm{T}}$, from (2.3), one obtains

$$\delta(x(t, k)) = \gamma \cdot L\widehat{x}(t, k) = \gamma \cdot L(e(t, k) + \delta(x(t, k))), \tag{2.13}$$

where $\widehat{x}(t, k) \triangleq (\widehat{x}_1(t, k), \ldots, \widehat{x}_N(t, k))^{\mathrm{T}} \in \mathbb{R}^N$, $\gamma \triangleq \mathrm{diag}\big(\widehat{\gamma}_1, \ldots, \widehat{\gamma}_N\big)$ with $\widehat{\gamma}_i = \gamma_i$ (or $-\gamma_i$), $i = 1, \ldots, N$. If parameter γ satisfy $\|\gamma\| < \frac{1}{\|L\|}$, one has from (2.13)

$$\delta(x(t, k)) = (I_N - \gamma L)^{-1} \cdot \gamma \cdot Le(t, k). \tag{2.14}$$

Theorem 2.1 *Consider system (2.1) with the learning controller (2.8) triggered by the event condition (2.3). In the finite set $t \in \{0, 1, \ldots, T\}$, the tracking error $e(t, k)$ such that*

$$\lim_{k \to \infty} e(t, k) = \mathbf{0},\ \ \forall\, t \in \{0, 1, \ldots, T\}, \tag{2.15}$$

if the learning parameter γ *satisfy* $\|\gamma\| < \frac{1}{\|L\|}$ *and learning matrix* Γ_1 *such that*

$$\rho\big(\widehat{\Psi}_{l_s}\big) < 1, \tag{2.16}$$

where the matrices

$$\begin{aligned}\widehat{\Psi}_{l_s} &= \Psi_{l_{s+1}-1}\Psi_{l_{s+1}-2}\dots\Psi_{l_s},\ \forall\, s \in \mathbb{Z}_+, l_s \in \mathbb{Z}_+,\ l_0 = 0,\\ \Psi_{l_s} &= I_N + BD_{l_s}\Gamma_1 L(I_N - \gamma L)^{-1}.\end{aligned}$$

Proof In the proof, we discuss the following two cases.
Case 1. We mainly consider that there exists $k \in \mathbb{Z}_+$ such that $k_l \le k,\ k+1 < k_{l+1}$, $l \in \mathbb{Z}_+$. $\forall\, t \in \{0, 1, \dots, T\}$, one has from (2.1), (2.12) and the boundary conditions

$$\begin{aligned}&e(t+1, k+1) - e(t+1, k)\\ &= x(t+1, k+1) - x(t+1, k)\\ &= C\big(x(t, k+1) - x(t, k)\big) + B\big(u(t, k+1) - u(t, k)\big)\\ &= C^{t+1}\big(x(0, k+1) - x(0, k)\big) = \mathbf{0},\end{aligned} \tag{2.17}$$

which means that $e(t, k+1) = e(t, k)$. That is,

$$e(t, k) = e(t, k_l),\ \forall\, k_l \le k < k_{l+1}, \tag{2.18}$$

where $\forall\, t \in \{0, 1, \dots, T\}, l \in \mathbb{Z}_+$.
Case 2. We discuss how $e(t, k)$ changes at the iterations sequence $\{k_l\}$, $l \in \mathbb{Z}_+$. $\forall\, t \in \{0, 1, \dots, T\}$, one has from (2.1), (2.11), (2.12), and (2.18)

$$\begin{aligned}&e(t+1, k_{l+1}) - e(t+1, k_l)\\ &= x(t+1, k_{l+1}) - x(t+1, k_l)\\ &= C\big(x(t, k_{l+1}) - x(t, k_l)\big) + B\big(\widehat{u}(t, k_{l+1}) - \widehat{u}(t, k_l)\big)\\ &= C\big(x(t, k_{l+1}) - x(t, k_l)\big) + BD_l\Gamma_1 L\widehat{x}(t+1, k)\\ &= C\theta(t, k_l) + BD_l\Gamma_1 L\big(\delta(t+1, k_l) + e(t+1, k_l)\big),\end{aligned} \tag{2.19}$$

where $k_l \le k < k_{l+1}$, $\theta(t, k_l) = x(t, k_{l+1}) - x(t, k_l)$. One has from (2.1), (2.8), and (2.19)

$$\begin{aligned}\theta(t, k_l) &= C\theta(t-1, k_l) + BD_l\Gamma_1 L\widehat{x}(t, k_l)\\ &= C\theta(t-1, k_l) + BD_l\Gamma_1 L\big(\delta(t, k_l) + e(t, k_l)\big).\end{aligned} \tag{2.20}$$

According to (2.20) and the boundary conditions, one can obtain that

$$\theta(t, k_l) = \sum_{\iota=0}^{t} C^{\iota} BD_l\Gamma_1 L\big(\delta(t-\iota, k_l) + e(t-\iota, k_l)\big), \tag{2.21}$$

where $\forall\, t \in \{0, 1, \ldots, T\},\ l \in \mathbb{Z}_+$. It follows from (2.19) and (2.21) that

$$\begin{aligned} e(t+1, k_{l+1}) = {} & (I_N + BD_l\Gamma_1 L)e(t+1, k_l) + BD_l\Gamma_1 L\delta(t+1, k_l) \\ & + \sum_{\iota=0}^{t} C^{\iota+1} BD_l\Gamma_1 L\big(\delta(t-\iota, k_l) + e(t-\iota, k_l)\big). \end{aligned} \tag{2.22}$$

Combining (2.14) and (2.22), one has

$$e(t+1, k_{l+1}) = \Psi_l e(t+1, k_l) + \sum_{\iota=0}^{t} \Theta_l^{\iota} e(t-\iota, k_l), \tag{2.23}$$

where $\Psi_l = I_N + BD_l\Gamma_1 L(I_N - \gamma L)^{-1}$, $\Theta_l^{\iota} = C^{\iota+1} BD_l\Gamma_1 L(I_N - \gamma L)^{-1}$, $\forall\ \iota \in \{0, 1, \ldots, t\}$. Let $U(k) = (e^{\mathrm{T}}(0, k), \ldots, e^{\mathrm{T}}(T, k))^{\mathrm{T}}$, then

$$U(k_{l+1}) = \Omega_l U(k_l),\ l \in \mathbb{Z}_+, \tag{2.24}$$

where the matrix

$$\Omega_l = \begin{pmatrix} \Psi_l & O & O & \cdots & O & O & O \\ \Theta_l^1 & \Psi_l & O & \cdots & O & O & O \\ \Theta_l^2 & \Theta_l^1 & \Psi_l & \cdots & O & O & O \\ \vdots & \vdots & \vdots & \cdots & \vdots & \vdots & \vdots \\ \Theta_l^T & \Theta_l^{T-1} & \Theta_l^{T-2} & \cdots & \Theta_l^2 & \Theta_l^1 & \Psi_l \end{pmatrix}.$$

In (2.24), it is not easy to verify that $\rho(\Omega_l) < 1$ since matrix D_l is a matrix with one nonzero diagonal element. To achieve the error convergence, we obtain that from (2.24)

$$U(k_{l_{s+1}}) = \widehat{\Omega}_{l_s} U(k_{l_s}),\ s \in \mathbb{Z}_+, \tag{2.25}$$

where $\widehat{\Omega}_{l_s} = \Omega_{l_{s+1}-1}\Omega_{l_{s+1}-2}\cdots\Omega_{l_s}$, and $l_s \in \mathbb{Z}_+,\ \forall\ s \in \mathbb{Z}_+$, and $l_0 = 0$. For $k_{l_s} \le k < k_{l_{s+1}}$, $s \in \mathbb{Z}_+$, from (2.18) and (2.24), one has

$$U(k) = \begin{cases} U(k_{l_s}), & k_{l_s} \le k < k_{l_s+1}; \\ \Omega_{l_s} U(k_{l_s}), & k_{l_s+1} \le k < k_{l_s+2}; \\ \Omega_{l_s+1}\Omega_{l_s} U(k_{l_s}), & k_{l_s+2} \le k < k_{l_s+3}; \\ \cdots\cdots & \\ \widehat{\Omega}_{l_s} U(k_{l_s}), & k_{l_{s+1}-1} \le k < k_{l_{s+1}}. \end{cases} \tag{2.26}$$

From (2.25), one can immediately get $U(k_{l_{s+1}}) \le \widehat{\Omega}_{l_s}\widehat{\Omega}_{l_{s-1}}\cdots\widehat{\Omega}_{l_0} U(0)$. Note that $\widehat{\Omega}_{l_s}$ is lower-triangular and $U(0)$ is bounded, one has $U(k_{l_s}) \to 0$, as $s \to +\infty$ ($i.e.$, $k \to +\infty$), if $\rho(\widehat{\Omega}_{l_s}) < 1, \forall\, s \in \mathbb{Z}_+$. Moreover, $\rho(\widehat{\Omega}_{l_s}) < 1$ if condition (2.16) holds. Thus, according to the results in Cases 1 and 2, one has $e(t,k) \to \mathbf{0}$, $\forall\, t \in \{0, 1, \ldots, T\}$, as $k \to +\infty$ if the condition (2.16) is satisfied.

Note that the iteration $k \in \mathbb{Z}_+$, which means $k_{l+1} - k_l \geq 1,\ l \in \mathbb{Z}_+$. It further implies that Zeno behaviors will not happen under the event-triggered condition (2.3). □

Remark 2.3 It is worth mentioning that D_l is a matrix with one nonzero diagonal element in Property 2.1. If the inequality $D_{l_s} + D_{l_s+1} + \cdots + D_{s+1-1} \geq I_N$ holds, i.e., $l_{s+1} - l_s \geq N, \forall\, s \in \mathbb{Z}_+$, it is not hard to achieve the condition (2.16) by selecting an appropriate learning gain matrix Γ_1.

2.4.2 The Consensus Under Event-Triggered Strategy with Quantization

In this subsection, we mainly study the consensus problem under event-triggered strategy with quantization. Assume the matrix L is lower-triangular, and define $\Lambda(t,k) \triangleq \widetilde{\Lambda}(t,k) I_N$. From (2.7), one gets

$$q(\widehat{x}(t,k)) = [I_N + \Lambda(t,k)]\widehat{x}(t,k). \tag{2.27}$$

According to (2.7), (2.4), and (2.13), one has

$$\delta(x(t,k)) = \gamma \cdot Lq(\widehat{x}(t,k)) = \gamma \cdot L(I_N + \Lambda(t,k))(e(t,k) + \delta(t,k)). \tag{2.28}$$

Similar to (2.14), if we choose that $\|\gamma\| < \frac{1}{(1+\sigma)\cdot\|L\|}$, from (2.7) and (2.28),

$$\delta(x(t,k)) = [I_N - \gamma L(I_N + \Lambda(t,k))]^{-1} \cdot \gamma \cdot L(I_N + \Lambda(t,k))e(t,k), \tag{2.29}$$

where $t \in \{0, 1, \ldots, T\}, k \in \mathbb{Z}_+$. In the following part, we discuss the learning tracking problem of system (2.1) in line with the following definitions and lemma.

Definition 2.1 A scalar l is called an interval parameter if it lies between two boundaries according to $l \in [\underline{l}, \overline{l}]$, where $\underline{l}$ is the minimum value of l and $\overline{l}$ is the maximum value of l.

Definition 2.2 An interval matrix S^I is defined as a matrix which is a member of the interval plant $\mathfrak{S}^I$ given by

$$\mathfrak{S}^I = \left\{ S^I : S^I = (s_{ij}^I),\ s_{ij}^I \in [\underline{s_{ij}}, \overline{s_{ij}}],\ i, j = 1, 2, \ldots, m \right\},$$

where $\overline{s_{ij}}$ is the maximum extreme value of the ith row and jth column elements of the uncertain plant, and $\underline{s_{ij}}$ is the minimum extreme value of the ith row and jth column elements of the uncertain plant.

Definition 2.3 The upper bound matrix $\overline{S}$ is a matrix whose elements are $\overline{s_{ij}}$. The lower bound matrix $\underline{S}$ is a matrix whose elements are $\underline{s_{ij}}$. The vertex matrices $\mathfrak{S}^v$ are defined by

$$\mathfrak{S}^v = \left\{S^v : S^v = (s_{ij}^v),\ s_{ij}^v \in \{\underline{s_{ij}}, \overline{s_{ij}}\},\ i, j = 1, \ldots, m\right\}.$$

Lemma 2.1 (Han and Lee 1994; Shih et al. 1998) *For given interval matrix S^I, the spectral radius of S^I is bounded by the maximum value of the spectral radii of vertex matrices S^v. S^I and S^v are given in Definitions 2.2 and 2.3.*

According to Lemma 2.1, we propose the following property.

Property 2.2 Matrix L is designed to be lower-triangular. Then, one obtains that the spectral radius of matrix $\widehat{\Phi}(t, k_l)$ is bounded by the maximum value of the spectral radii of matrices $\widehat{\Phi}_l^1$ and $\widehat{\Phi}_l^2$, where the matrices

$$\begin{aligned}
\widehat{\Phi}(t, k_l) &= I_N + BD_l\Gamma_2 L\big(I_N + \Lambda(t, k_l)\big)\big(I_N - \gamma L(I_N + \Lambda(t, k_l))\big)^{-1}, \\
\widehat{\Phi}_l^1 &= I_N + (1-\sigma)BD_l\Gamma_2 L\big(I_N - (1-\sigma)\gamma L\big)^{-1}, \\
\widehat{\Phi}_l^2 &= I_N + (1+\sigma)BD_l\Gamma_2 L\big(I_N - (1+\sigma)\gamma L\big)^{-1}.
\end{aligned}$$

Proof Matrix $\widehat{\Phi}(t, k_l)$ is lower-triangular since matrix L is lower-triangular and matrices $B,\ D_l,\ \Gamma_2$ are all diagonal. Hence, $\rho(\widehat{\Phi}(t, k_l))$ is decided by the diagonal elements

$$\xi_i = 1 + b_i d_i^l \Gamma_2^i \frac{l_{ii}(1 + \widetilde{\Lambda}(t, k_l))}{1 - \widehat{\gamma}_i l_{ii}(1 + \widetilde{\Lambda}(t, k_l))},$$

where $i = 1, \ldots, N$. Then, one has

$$\xi_i = 1 + b_i d_i^l \Gamma_2^i \widehat{\gamma}_i^{-1}\left(\frac{1}{1 - \widehat{\gamma}_i l_{ii}(1 + \widetilde{\Lambda}(t, k_l))} - 1\right).$$

Note that $\widetilde{\Lambda}(t, k_l) \in [-\sigma, \sigma]$, from Lemma 2.1, it is not hard to know the spectral radius of matrix $\widehat{\Phi}(t, k_l)$ is bounded by the maximum value of the spectral radii of matrices $\widehat{\Phi}_l^1$ and $\widehat{\Phi}_l^2$. □

Theorem 2.2 *Consider system (2.1) with the learning controller* (2.9) *triggered by the event condition* (2.4). *In the finite interval $t \in [0, T]$, the tracking error $e(t, k)$ such that*

$$\lim_{k\to\infty} e(t, k) = \mathbf{0},\ \forall\, t \in \{0, 1, \ldots, T\} \tag{2.30}$$

if the learning parameter γ satisfy $\|\gamma\| < \frac{1}{(1+\sigma)\cdot\|L\|}$ and matrix Γ_2 such that

$$\max\left\{\rho(\widetilde{\Phi}_{l_s}^1), \rho(\widetilde{\Phi}_{l_s}^2)\right\} < 1, \tag{2.31}$$

where the matrices

$$\widetilde{\Phi}^i_{l_s} = \widehat{\Phi}^i_{l_{s+1}-1}\widehat{\Phi}^i_{l_{s+1}-2}\cdots\widehat{\Phi}^i_{l_s},\ i=1,2,\ l_s\in\mathbb{Z}_+,\ \forall\, s\in\mathbb{Z}_+,\ l_0=0,$$
$$\widehat{\Phi}^1_{l_s} = I_N + (1-\sigma)BD_{l_s}\Gamma_2 L\big(I_N-(1-\sigma)\gamma L\big)^{-1},$$
$$\widehat{\Phi}^2_{l_s} = I_N + (1+\sigma)BD_{l_s}\Gamma_2 L\big(I_N-(1+\sigma)\gamma L\big)^{-1}.$$

Proof In this proof, we also discuss the following two cases.

Case 1. There exists $k\in\mathbb{Z}_+$ such that $k_l\le k,\ k+1<k_{l+1},\ l\in\mathbb{Z}_+$. Similar to Theorem 2.1, one obtains

$$e(t,k)=e(t,k_l),\ \forall\, k_l\le k<k_{l+1}, \tag{2.32}$$

where $\forall\, t\in\{0,1,\ldots,T\},\ l\in\mathbb{Z}_+$.

Case 2. We further discuss how $e(t,k)$ changes at the iteration sequence $\{k_l\},\ l\in\mathbb{Z}_+$. According to (2.1), (2.11), (2.12), and (2.27), it is not hard to know

$$\begin{aligned}&e(t+1,k_{l+1})-e(t+1,k_l)\\&=C\theta(t,k_l)+BD_l\Gamma_2L(I_N+\Lambda(t,k_l))(\delta(t+1,k_l)+e(t+1,k_l)),\end{aligned} \tag{2.33}$$

where $\theta(t,k_l)=x(t,k_{l+1})-x(t,k_l)$. Similar to (2.21), one gets

$$\theta(t,k_l)=\sum_{\iota=0}^{t}C^{\iota}BD_l\Gamma_2L(I_N+\Lambda(t-\iota,k_l))(\delta(t-\iota,k_l)+e(t-\iota,k_l)), \tag{2.34}$$

where $\forall\, t\in\{0,1,\ldots,T\},\ k\in\mathbb{Z}_+$. Combining (2.33) with (2.34), one obtains

$$\begin{aligned}&e(t+1,k_{l+1})\\&=(I_N+BD_l\Gamma_2L(I_N+\Lambda(t+1,k_l)))e(t+1,k_l)\\&\quad+BD_l\Gamma_2L(I_N+\Lambda(t+1,k_l))\delta(t+1,k_l)\\&\quad+\sum_{\iota=0}^{t}C^{\iota+1}BD_l\Gamma_2L(I_N+\Lambda(t-\iota,k_l))(\delta(t-\iota,k_l)+e(t-\iota,k_l)).\end{aligned} \tag{2.35}$$

In terms of (2.29), the equality (2.35) becomes

$$e(t+1,k_{l+1})=\widehat{\Phi}(t+1,k_l)e(t+1,k_l)+\sum_{\iota=0}^{t}\widehat{\Theta}(t-\iota,k_l)e(t-\iota,k_l), \tag{2.36}$$

where $\widehat{\Phi}(t+1,k_l)=I_N+BD_l\Gamma_2L(I_N+\Lambda(t+1,k_l))(I_N-\gamma L(I_N+\Lambda(t+1,k_l)))^{-1}$, $\widehat{\Theta}(t-\iota,k_l)=C^{\iota+1}BD_l\Gamma_2L(I_N+\Lambda(t-\iota,k_l))(I_N-\gamma L(I_N+\Lambda(t-\iota,k_l)))^{-1}$.

Let $U(k) = (e^{\mathrm{T}}(0, k), \ldots, e^{\mathrm{T}}(T, k))^{\mathrm{T}}$, then one obtains that

$$U(k_{l+1}) = \overline{\Omega}_l U(k_l),\ l \in \mathbb{Z}_+, \tag{2.37}$$

where the matrix

$$\overline{\Omega}_l = \begin{pmatrix} \widehat{\Phi}(0, k_l) & O & O & \cdots & O & O \\ \widehat{\Theta}(0, k_l) & \widehat{\Phi}(1, k_l) & O & \cdots & O & O \\ \widehat{\Theta}(0, k_l) & \widehat{\Theta}(1, k_l) & \widehat{\Phi}(2, k_l) & \cdots & O & O \\ \vdots & \vdots & \vdots & \cdots & \vdots & \vdots \\ \widehat{\Theta}(0, k_l) & \widehat{\Theta}(1, k_l) & \widehat{\Theta}(2, k_l) & \cdots & \widehat{\Theta}(T-1, k_l) & \widehat{\Phi}(T, k_l) \end{pmatrix}.$$

In (2.37), it is also difficult to check that $\rho(\overline{\Omega}_l) < 1$ since matrix D_l is a matrix with one nonzero diagonal element. As a result, we obtain that from (2.37)

$$U(k_{l_{s+1}}) = \widetilde{\Omega}_{l_s} U(k_{l_s}),\ s \in \mathbb{Z}_+, \tag{2.38}$$

where $\widetilde{\Omega}_{l_s} = \overline{\Omega}_{l_{s+1}-1}\overline{\Omega}_{l_{s+1}-2}\cdots\overline{\Omega}_{l_s}$, $l_s \in \mathbb{Z}_+$, $s \in \mathbb{Z}_+$, and $l_0 = 0$.

For $k_{l_s} \leq k < k_{l_{s+1}}$, one obtains from (2.18) and (2.24)

$$U(k) = \begin{cases} U(k_{l_s}), & k_{l_s} \leq k < k_{l_s+1}; \\ \overline{\Omega}_{l_s} U(k_{l_s}), & k_{l_s+1} \leq k < k_{l_s+2}; \\ \overline{\Omega}_{l_s+1}\overline{\Omega}_{l_s} U(k_{l_s}), & k_{l_s+2} \leq k < k_{l_s+3}; \\ \cdots\cdots \\ \widetilde{\Omega}_{l_s} U(k_{l_s}), & k_{l_{s+1}-1} \leq k < k_{l_{s+1}}. \end{cases} \tag{2.39}$$

In terms of the equality (2.38), one has $U(k_{l_{s+1}}) \leq \widetilde{\Omega}_{l_s}\widetilde{\Omega}_{l_{s-1}}\cdots\widetilde{\Omega}_{l_0}U(0)$. $\widetilde{\Omega}_{l_s}$ is lower-triangular and $U(0)$ is bounded, one immediately gets $U(k_{l_s}) \to 0$, as $s \to +\infty$ if $\rho(\widetilde{\Omega}_{l_s}) < 1, \forall s \in \mathbb{Z}_+$. From (2.37) and (2.38), one has $\rho(\widetilde{\Omega}_{l_s}) < 1$ if $\rho(\widetilde{\Phi}_{l_s}(t, k_l)) < 1$, where $\widetilde{\Phi}_{l_s}(t, k_l) = \widehat{\Phi}_{l_{s+1}-1}(t, k_l)\widehat{\Phi}_{l_{s+1}-2}(t, k_l)\cdots\widehat{\Phi}_{l_s}(t, k_l)$, $\forall\ t \in \{0, 1, \ldots, T\}$, $\forall\ s \in \mathbb{Z}_+$. According to Property 2.2, one knows that $\rho(\widetilde{\Phi}_{l_s}(t, k_l)) < 1$ if condition (2.31) holds. Thus, according to (2.32) and (2.39), one knows $e(t, k) \to \mathbf{0}$, $\forall\, t \in \{0, 1, \ldots, T\}$ as $k \to +\infty$ if condition (2.31) is satisfied.

In addition, the iteration $k \in \mathbb{Z}_+$, it means that $k_{l+1} - k_l \geq 1,\ l \in \mathbb{Z}_+$. As a result, Zeno behaviors will not happen under the designed event condition (2.4). □

Remark 2.4 Generally speaking, the system state converges to a set but not a trajectory when quantization is considered in the controller (Zhang et al. 2015). However, by choosing parameter to satisfy $\|\gamma\| < \frac{1}{(1+\sigma)\cdot\|L\|}$ and using the interval matrix properties, the system state can converge to the desired trajectory when the quantized controller (2.9) is considered in this chapter.

Remark 2.5 In comparison with Theorem 4.2 of Zhang et al. (2015), the advantage of Theorem 2.2 in this chapter is that one can obtain $\lim_{k\to\infty} e(t, k) = \mathbf{0}$, $\forall\ t \in$

$\{0, 1, \ldots, T\}$, by designing the quantized controller to be related to time t and iterative learning step k. Note that in literature (Zhang et al. 2015), all nodes converge to a set in Theorem 4.2.

Remark 2.6 Compared with Meng et al. (2015b), the advantages of our results in Sect. 2.4.2 are: (1) The ILC algorithm (2.9) in this chapter is only updated when the event-triggering condition (2.4) held. However, the ILC strategies (5) in Meng et al. (2015b) are updated in each iteration k. It is unnecessary and costly to update the ILC algorithm in each iteration k when the iterative controller changes a little in some contiguous learning steps. (2) In Theorem 4.1 of Meng et al. (2015b), the communication network is required to have a spanning tree, which is removed in Theorem 2.2.

Remark 2.7 In this chapter, the desired trajectory is assumed as $x^*(t) = 0,\ \forall\ t \in \{0, 1, \ldots, T\}$. In fact, this assumption can be removed if $\delta_i(x_i(t,k))$ is designed to be $\delta_i(x_i(t,k)) \triangleq \widehat{e}_i(t,k) - e_i(t,k)$, and one changes $\widehat{x}_j(t,k)$ as $\widehat{e}_j(t,k)$ in (2.3), (2.4), (2.8), and (2.9). Furthermore, $\widehat{e}_i(t,k) = \widehat{x}_i(t,k) - x^*(t),\ \forall\ i \in \{1, 2, \ldots, N\},\ t \in \{0, 1, \ldots, T\},\ \forall\ k \in \mathbb{Z}_+$. Then, one can obtain similar results to Theorems 2.1 and 2.2.

2.5 Simulation Example

In this section, a numerical example is presented to demonstrate the validity of the results in Theorems 2.1 and 2.2.

Example 2.1 Consider a discrete-time system (2.1) with three nodes. Let the terminal time $T = 10$, $x(t,k) = (x_1(t,k), x_2(t,k), x_3(t,k))^{\mathrm{T}} \in \mathbb{R}^3$, $\sigma = 0.5$, and the matrices

$$L = \begin{pmatrix} 0.8 & 0 & 0 \\ 0.6 & 1 & 0 \\ 0.8 & 1 & 0.5 \end{pmatrix}, \quad C = \begin{pmatrix} 1 & 0 & 0 \\ 0 & 1 & 0 \\ 0 & 0 & 1.5 \end{pmatrix}, \quad B = \begin{pmatrix} 0.8 & 0 & 0 \\ 0 & -1 & 0 \\ 0 & 0 & 0.6 \end{pmatrix}.$$

$D_{l_s} \in \{D_1, D_2, D_3\}$ with $D_1 = \mathrm{diag}(1,0,0)$, $D_2 = \mathrm{diag}(0,1,0)$, $D_3 = \mathrm{diag}(0,0,1)$, $\forall\ s \in \mathbb{Z}_+$. The desired trajectory is defined as $x^*(t) = 0,\ x_i(t,0) = x_{i0}(t) = \sin(t), \forall\, t \in \{0, 1, \ldots, 10\}, i = 1, 2, 3$. According to $\|L\| = 1.8338$, we set that $\gamma = \mathrm{diag}(0.25, 0.25, 0.25)$, and $\Gamma_1 = \mathrm{diag}(-1, 0.5, -0.4)$. Hence, one knows $\rho(\Psi_1\Psi_2\Psi_3) < 1$, where $\Psi_i = I_N + BD_i\Gamma_1 L(I_N - \gamma L)^{-1},\ i = 1, 2, 3$.

As mentioned in Remark 2.3, condition (2.16) in Theorem 2.1 holds if one chooses $D_{l_s} + D_{l_s+1} + D_{l_s+2} + \cdots + D_{s+1-1} \geq I_N,\ \forall\ s \in \mathbb{Z}_+$. Hence, under the iterative controller (2.8) with event-triggered strategy (2.3), the tracking error $e(t,k) \to \mathbf{0}$ *when* $k \to +\infty$.

Figure 2.3 shows the state trajectories of three nodes in the system (2.1) with the learning controller (2.8) and event-triggered strategy (2.3) at different iteration

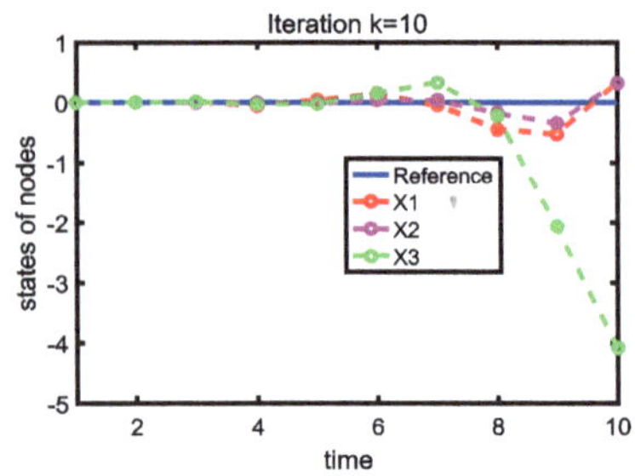

(a) State trajectories at the 10th iteration

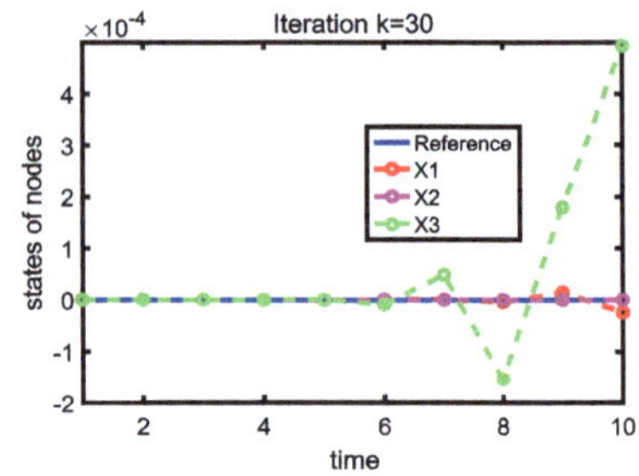

(b) State trajectories at the 30th iteration

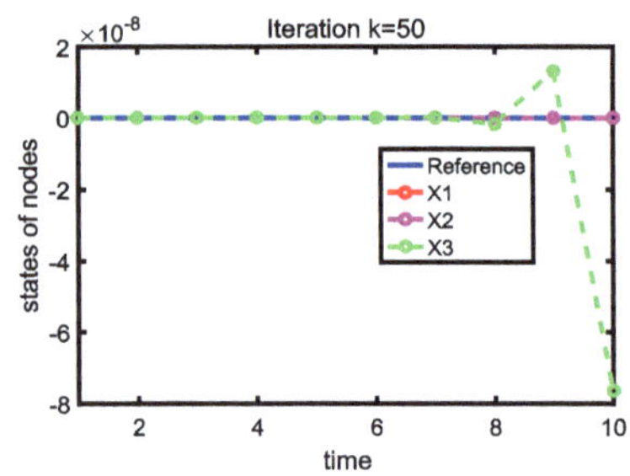

(c) State trajectories at the 50th iteration

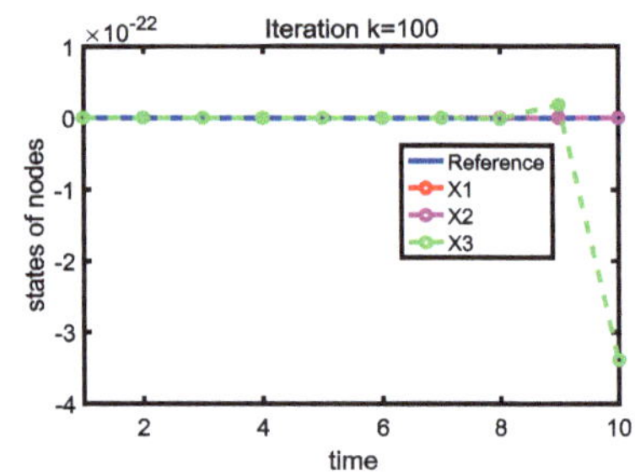

(d) State trajectories at the 100th iteration

Fig. 2.3 The trajectories of state $x(t, k)$ in system (2.1) with the learning controller (2.8) and event-triggered strategy

step k. According to Fig. 2.3, one knows that the state of each node asymptotically achieves the desired trajectory as the iteration step k increases.

In Theorem 2.2, we choose $\Gamma_2 = \text{diag}(-0.5, 0.5, -1)$, one has $\rho\big(\widehat{\Phi}_1^1\widehat{\Phi}_2^1\widehat{\Phi}_3^1\big) < 1$ and $\rho\big(\widehat{\Phi}_1^2\widehat{\Phi}_2^2\widehat{\Phi}_3^2\big) < 1$ with $\widehat{\Phi}_i^1 = I_N + (1-\sigma)BD_i\Gamma_2 L\big(I_N + (1-\sigma)(I_N - (1-\sigma)\gamma L)^{-1}\gamma L\big)$, $\widehat{\Phi}_i^2 = I_N + (1+\sigma)BD_i\Gamma_2 L\big(I_N + (1+\sigma)(I_N - (1+\sigma)\gamma L)^{-1}\gamma L\big)$, $i = 1, 2, 3$. Meanwhile, as mentioned in Remark 2.3, condition (2.31) in Theorem 2.2 was satisfied if $D_{l_s} + D_{l_s+1} + D_{l_s+2} + \cdots + D_{s+1-1} \geq I_N, \forall\, s \in \mathbb{Z}_+$. As a result, under the learning controller (2.9) with event-triggered strategy (2.4), the tracking error $e(t, k) \to \mathbf{0}$ when $k \to +\infty$.

Figure 2.4 shows the state trajectories of three nodes in system (2.1) with the learning controller (2.9) and event-triggered strategy (2.4) at different iteration step k. From Figs. 2.4, one knows that the state of each node asymptotically achieves the desired reference as the iteration step k increases.

Remark 2.8 In conditions (2.16) and (2.31), the learning parameters can be chosen to satisfy the two conditions. For example, in our numerical simulation, for each D_i $(i = 1, 2, 3)$, the learning matrices Γ_1 and Γ_2 are appropriately diagonal matrices to satisfy $\Psi_i^{ii} < 1$, $\widehat{\Phi}_i^{1ii} < 1$, and $\widehat{\Phi}_i^{2ii} < 1$, where Ψ_i^{ii}, $\widehat{\Phi}_i^{1ii}$, and $\widehat{\Phi}_i^{2ii}$ are the ith elements of the diagonal lines of matrices Ψ_i, $\widehat{\Phi}_i^1$, and $\widehat{\Phi}_i^2$, respectively.

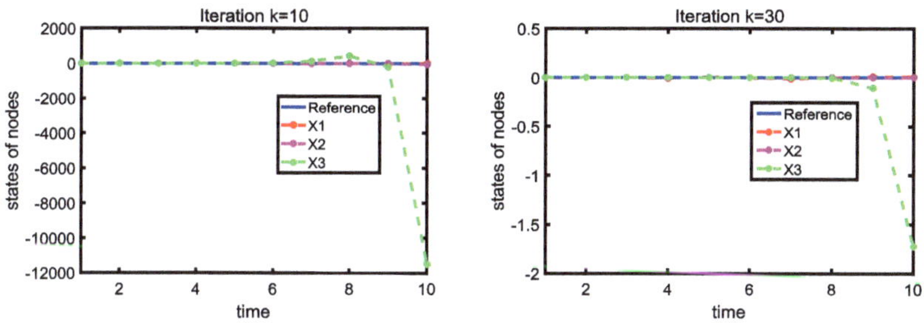

(a) State trajectories at the 10th iteration(b) State trajectories at the 30th iteration

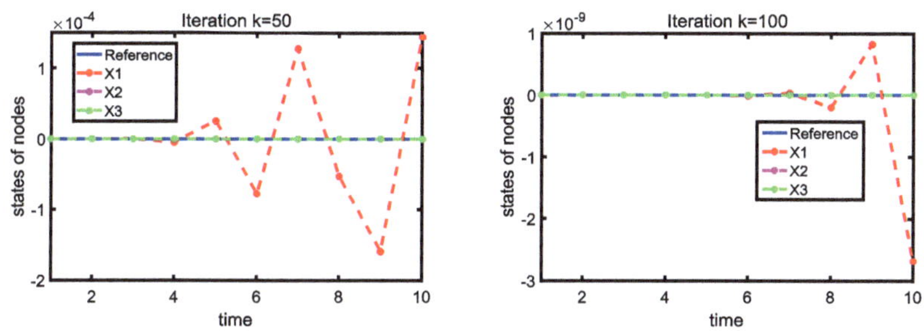

(c) State trajectories at the 50th iteration(d) State trajectories at the 100th iteration

Fig. 2.4 The trajectories of state $x(t, k)$ in system (2.1) with the learning controller (2.9) and event-triggered strategy (2.4)

2.6 Summary

In this chapter, we have discussed the consensus problem for discrete-time systems with event-triggered mechanisms and quantization. The iterative learning control strategies have been designed to achieve the target. The event-triggered mechanism in our iterative learning controllers has been used to reduce the number of iterative steps to be updated. Moreover, a quantization strategy has been introduced into the event-triggered controller to address communication bandwidth constraints. Some relaxed conditions have been proposed to achieve the tracking by using the interval matrix properties. Finally, a numerical example has been provided to illustrate the effectiveness of the main results.

References

Abidi K, Xu JX (2011) Iterative learning control for sampled-data systems: from theory to practice. IEEE Trans Ind Electron 58(7):3002–3015

Arimoto S, Kawamura S, Miyazaki F (1984) Bettering operation of robots by learning. J Robotic Syst 1(2):123–140

Delchamps D (1990) Stabilizing a linear system with quantized state feedback. IEEE Trans Automat Control 35(8):916–924

Fan Y, Feng G, Wang Y, Song C (2013) Distributed event-triggered control of multi-agent systems with combinational measurements. Automatica 49(2):671–675

Fu MY, Xie LH (2005) The sector bound approach to quantized feedback control. IEEE Trans Automat Control 50(11):1698–1711

Han HS, Lee JG (1994) Necessary and sufficient conditions for stability of time-varying discrete interval matrices. Inter J Control 59(4):1021–1029

Hu WF, Liu L, Feng G (2016) Consensus of linear multi-agent systems by distributed event-triggered strategy. IEEE Trans Cybern 46(1):148–157

Huang DQ, Xu JX, Li XF, Xu C, Yu M (2013) D-type anticipatory iterative learning control for a class of inhomogeneous heat equations. Automatica 49(8):2397–2408

Li JS, Li JM (2014) Adaptive iterative learning control for coordination of second-order multi-agent systems. Inter J Robust Nonlin Control 24(18):3282–3299

Liu ZW, Guan ZH, Li T, Zhang XH, Xiao JW (2012) Quantized consensus of multi-agent systems via broadcast gossip algorithms. Asian J Control 14(6):1634–1642

Mazo M, Tabuada P (2011) Decentralized event-triggered control over wireless sensor/actuator networks. IEEE Trans Automat Control 56(10):2456–2461

Meng DY, Moore KL (2016) Learning to cooperate: networks of formation agents with switching topologies. Automatica 64:278–293

Meng DY, Jia YM, Du JP, Yu FS (2013) Tracking algorithms for multiagent systems. IEEE Trans Neural Netw Learn Syst 24(10):1660–1676

Meng DY, Jia YM, Du JP, Zhang J (2014) On iterative learning algorithms for the formation control of nonlinear multi-agent systems. Automatica 50(1):291–295

Meng DY, Jia YM, Du JP (2015) Robust consensus tracking control for multiagent systems with initial state shifts, disturbances, and switching topologies. IEEE Trans Neural Netw Learn Syst 26(4):809–824

Meng DY, Jia YM, Du JP (2015) Robust iterative learning protocols for finite-time consensus of multi-agent systems with interval uncertain topologies. Inter J Syst Sci 46(5):857–871

Mezghani M, Roux G, Cabassud M, Lann M, Dahhou B, Casamatta G (2002) Application of iterative learning control to an exothermic semibatch chemical reactor. IEEE Trans Control Syst Tech 10(6):822–834

Peng C, Han QL (2013) A novel event-triggered transmission scheme and control co-design for sampled-data control systems. IEEE Trans Automat Control 58(10):2620–2626

Peng C, Han QL, Yue D (2013) To transmit or not to transmit: a discrete event-triggered communication scheme for networked takagi-sugeno fuzzy systems. IEEE Trans Fuzzy Syst 21(1):164–170

Shih MH, Lur YY, Pang CT (1998) An inequality for the spectral radius of an interval matrix. Linear Algebra Appl 274(1):27–36

Tan Y, Xu JX, Norrlöf M, Freeman C (2011) On reference governor in iterative learning control for dynamic systems with input saturation. Automatica 47(11):2412–2419

Wang ZD, Shen B, Shu HS, Wei GL (2012) Quantized H_∞ control for nonlinear stochastic time-delay systems with missing measurements. IEEE Trans Automat Control 57(6):1431–1444

Xiong WJ, Ho DWC, Yu XH (2016) Saturated finite interval iterative learning for tracking of dynamic systems with hnn-structural out-put. IEEE Trans Neural Netw Learn Syst 27(7):1578–1584

Xu JX, Tan Y, Lee T (2004) Iterative learning control design based on composite energy function with input saturation. Automatica 40(8):1371–1377

Yue D, Tian EG, Han QL (2013) A delay system method for designing event-triggered controllers of networked control systems. IEEE Trans Automat Control 58(2):475–481

Zhang ZQ, Hao F, Zhang L, Wang L (2014) Consensus of linear multi-agent systems via event-triggered control. Inter J Control 87(6):1243–1251

Zhang ZQ, Zhang L, Hao F, Wang L (2015) Distributed event-triggered consensus for multi-agent systems with quantisation. Inter J Control 88(6):1112–1122

Zhang RK, Hou ZS, Chi RH, Ji HH (2015) Adaptive iterative learning control for nonlinearly parameterised systems with unknown time-varying delays and input saturations. Inter J Control 88(6):1133–1141

Zou L, Wang ZD, Gao HJ, Liu XH (2015) Event-triggered state estimation for complex networks with mixed time delays via sampled data information: the continuous-time case. IEEE Trans Cybern 45(12):2804–2815

Chapter 3
Consensus Under Limited Information Communication

3.1 Introduction

In this chapter, we shall discuss the problem of quantized iterative learning for digital networks consisting of many packet radio terminals sent to a single hub station. Digital networks include packet data networks, cellular and micro-cellular networks, and mobile satellite networks. In general, a digital network is a complex circuit network composed of large assemblies of logic gates. Digital networks are more useful than analog circuit networks because they make it easier for an electronic device to switch to one of several known states rather than precisely reproducing a continuous range of values.

Unlike other learning-type control strategies, such as adaptive control, neural networks, and repetitive control, ILC aims to improve performance by incorporating error information into the control for subsequent iterations. By learning from previous executions and performing the same task multiple times, ILC has the advantage of achieving high performance with low instantaneous tracking error despite of large model uncertainty and repetitive disturbances. ILC has received widespread attention (Huang et al. 2014; Kurek and Zaremba 1993; Li and Li 2014; Liu and Jia 2012; Meng and Jia 2012; Meng et al. 2012; Wang et al. 2004; Xu and Tan 2003) in recent years due to its broad applications in industries, such as robotic manipulators, hard disk drives, rapid thermal processing, and chemical polymerization/crystallization. In Wang et al. (2004), based on a new output-recurrent fuzzy neural network, a direct adaptive iterative learning control is proposed for a class of repeatable nonlinear systems with unknown nonlinearities and variable initial resetting errors. In Meng and Jia (2012), the problem of finite-time output consistency for multi-agent systems is studied using the ILC methods. It is shown in Liu and Jia (2012) that the formation problem is equivalently transformed into a stability control problem over finite-time intervals by means of iterative learning approach.

Most of the existing ILC works on networks require the fixed network topology. As we all know, in networks, due to various reasons such as link failure and the

W. Xiong et al., *Iterative Learning Control for Network Systems Under Constrained Information Communication*, Intelligent Control and Learning Systems 12,
https://doi.org/10.1007/978-981-97-0926-7_3

change of environment, the communication topology is often time-varying. Hence, in Meng et al. (2014), deal with the problems of formation control for multi-agent systems with nonlinear dynamics and switching network topologies. However, the union of the switching graphs in each iteration needs to have spanning trees from beginning to end. Also, it is shown in Meng et al. (2015) that these uncertainties vary dynamically not only along the time axis but also on the iteration axis. However, the obtained conditions (12) and (17) in Meng et al. (2015) have to be satisfied for all iterations, which is hard to check when the iteration k tends to infinity. Therefore, one of our goals in this chapter is to consider time-varying topologies on digital networks and put forward some loose standards for the learning control of the time-varying networks.

In recent years, due to the limited bandwidth, quantized communications have gained great importance from both theoretic and engineering points of view (Carli et al. 2010; Liu et al. 2010, 2012a, b; Yang et al. 2011, 2012). A quantized system is discussed in Xia et al. (2013) with finite-level quantized input computed from quantized measurements. In Li and Xie (2011), consider discrete-time distributed average consensus with finite communication data rate and time-varying communication topologies. Kar and Moura in Dong et al. (2012a, b) discuss quantized average consensus with stochastic link failures by static infinite-level and finite-level uniform quantizers. In Wang et al. (2012), the quantized H_∞ control problem for a class of nonlinear random time-delay network-based systems with probabilistic data loss is investigated. As far as the authors know, there has been little work on the ILC with limited bandwidth. Therefore, another purpose of this chapter is to discuss the consistent tracking problem of digital networks in a limited interval by designing an adaptive program to select the number of quantization levels. The number of quantization levels of each quantizer is adjusted online according to whether the associated channel is valid or not in the last step.

3.2 Problem Formulation

Consider the following network consisting of N nodes (N is a positive integer). Let $x_i(t) \in \mathbb{R}$ and $u_i(t) \in \mathbb{R}$ be the state and control input of the ith node, the network model can be written as Li and Xie (2011); Li et al. (2014)

$$x_i(t+1) = x_i(t) + u_i(t), \tag{3.1}$$

where $i = 1, 2, \ldots, N,\ t \in \{0, T\}$ is the fixed time interval with $T > 1$.

As the introduction mentions, we will consider iterative learning in the model (3.1). In other words, each node has to deal with two independent dynamic processes: the first process shows the system's dynamics about time t and the second process describes the system's dynamics about the iterative learning k. Therefore, the ith node performing repetitive tasks over the fixed time interval $t \in [0, T]$ has the following dynamics over $k \in \mathbb{Z}_+$

$$x_i(t+1,k) = x_i(t,k) + u_i(t,k), \tag{3.2}$$

where $u_i(t,k)$ is the iterative learning controller, and $t \in \{0,\ldots,T\}$, $k \in \mathbb{Z}_+$, $i = 1,2,\ldots,N$.

3.2.1 Quantizer Design

Similar to Li and Xie (2011), we assume that only quantized data can be transmitted between agents. For each communication channel, the state value of the sender is firstly encoded to quantized data and then transmitted. After the data is received, a decoder is used by the receiver to get an estimate of the sender's state. For $\forall\, i,\ j \in \{1,2,\ldots,N\}$, if $j \in \Pi_i$, the encoder $\varphi_{ij}(t,k)$ associated with i for channel (i,j) is designed as

$$\begin{cases} \zeta_{ij}(t,0) = 0, \\ \zeta_{ij}(t,k+1) = b(t,k)\alpha_{ij}(t)s_{ij}(t,k+1) + \zeta_{ij}(t,k), \\ s_{ij}(t,k+1) = q_\nu\left(\frac{x_i(t,k+1)-\zeta_{ij}(t,k)}{b(t,k)}\right), \end{cases} \tag{3.3}$$

where $t = 0,1,2,\ldots,T,\ k \in \mathbb{Z}_+$, $x_i(t,k)$ and $s_{ij}(t,k)$ are the input and the output of $\varphi_{ij}(t,k)$, respectively. $\zeta_{ij}(t,k)$ is the internal state of $\varphi_{ij}(t,k)$, $q_\nu(\cdot)$ is a quantizer, and ν changes with time t and iterative learning k, $b(t,k) > 0$ is a scaling function. If node $l \notin \Pi_i$, the channel (i,l) will never be active, and it is unnecessary to design an encoder for (i,l). The main role of the scaling function $b(t,k)$ is to avoid the saturation of the finite-level quantizer. The decoder $\psi_{ij}(t,k)$ associated with i for channel (i,j) is defined as

$$\begin{cases} \widehat{x}_{ij}(t,0) = 0, \\ \widehat{x}_{ij}(t,k+1) = b(t,k)\alpha_{ij}(t)s_{ij}(t,k+1) + \widehat{x}_{ij}(t,k), \end{cases} \tag{3.4}$$

where $t = 0,1,2,\ldots,T,\ \ k \in \mathbb{Z}_+,\ \widehat{x}_{ij}(t,k)$ is the output of $\psi_{ij}(t,k)$. Note that, node j receives $s_{ij}(t,k)$ from encoder $\varphi_{ij}(t,k)$ in channel (i,j). Then, the estimate $\widehat{x}_{ij}(t,k)$ of $x_i(t,k)$ is obtained from decoder $\psi_{ij}(t,k)$ in channel (i,j).

In (3.3), the function $q_\nu(\cdot) : \mathbb{R} \to \Omega$ is a finite-level uniform symmetric quantizer that maps from $\mathbb{R}$ to a discrete set of quantized levels and takes the form:

$$q_\nu(m) = \begin{cases} 0, & if\ -\frac{1}{2} < m < \frac{1}{2}, \\ i, & if\ \frac{2i-1}{2} < m < \frac{2i+1}{2}, \\ \nu, & if\ m \geq \frac{2\nu-1}{2}, \\ -q_\nu(-m), & if\ m < -\frac{1}{2}, \end{cases} \tag{3.5}$$

where $i = 1, 2, \ldots, \nu - 1$. Clearly, the set of quantization levels is $\Omega = \{0, \pm 1, \ldots, \pm \nu\}$ and the number of quantization levels is $2\nu + 1$.

3.2.2 ILC Scheme Design

Let $x^*(t)$ be the desired consensus trajectory of $x_i(t, k)$ in (3.2). Define $\theta_i(t, k) = x_i(t, k+1) - x_i(t, k)$ and the tracking error $e_i(t, k) = x^*(t) - x_i(t, k)$. In this chapter, we aim to design a distributed ILC $u_i(t, k)$ with a finite-level uniform quantization scheme to achieve $e_i(t, k) \to 0$, $i = 1, 2, \ldots, N$, $\forall\, t \in [1, T]$, when $k \to +\infty$. The distributed ILC $u_i(t, k)$ with a finite-level uniform quantization scheme is designed as

$$u_i(t, k+1) = u_i(t, k) + \Delta u_i(t, k),\ i = 1, 2, \ldots, N, \tag{3.6}$$

and

$$\Delta u_i(t-1, k) = r \sum_{j \in \Pi_i} \alpha_{ij}(t)\big(\widehat{x}_{ji}(t, k) - \zeta_{ij}(t, k)\big) + \sum_{j=1}^{N} g_{ij}(t) e_j(t, k), \tag{3.7}$$

where $t \in [1, T]$, $i = 1, 2, \ldots, N$, $j \in \Pi_i$, and $k \in \mathbb{Z}_+$. $r > 0$ is the learning gain and $G(t) = (g_{ij}(t))_{N \times N}$ is the learning gain matrix, which will be designed together in the following section. Moreover, we define

$$\Lambda_{ij}(t, k) = s_{ij}(t, k+1) - \frac{x_i(t, k+1) - \zeta_{ij}(t, k)}{b(t, k)}. \tag{3.8}$$

Remark 3.1 In this chapter, the communication topologies of digital networks are thought to be time-varying, which generalizes the fixed communication topologies in Li and Li (2014); Meng and Jia (2012); Meng et al. (2012). Furthermore, the iterative learning controller $u_i(t, k)$ is considered to receive the information with quantized communication. As a matter of fact, in a digital network, due to the limited capability of the transmission bandwidth of information channels, it is actually possible for the controller $u_i(t)$ to transmit its feedback information with quantized communication.

Throughout the whole chapter, we need the following assumptions and lemma.

Assumption 3.1 The boundary conditions for system (3.2) are given as $x_i(0, k) = x_{i0}$, where x_{i0} is a constant, for $i = 1, 2, \ldots, N$, $t \in [0, T]$, $\forall\, k \in \mathbb{Z}_+$.

Assumption 3.2 The cardinal number $d_i(t)$ satisfies that $d^* = \max\limits_{0 \le t \le T} d_i(t)$, where $d^* > 0$ is a constant.

Lemma 3.1 (Li and Xie 2011) *Suppose that*

$$y(k+1) = \big(1 - \varsigma(k)\big)y(k) + \eta(k), \tag{3.9}$$

where $k \in \mathbb{Z}_+,\ 0 < \varsigma(k) \le 1,\ \eta(k) \ge 0$, *one has*

$$y(k+1) \le \left[\prod_{t=0}^{k}(1-\varsigma(t))\right]y(0) + \sup_{0\le i\le k}\frac{\eta(i)}{\varsigma(i)}. \tag{3.10}$$

Moreover, if $\sum\limits_{k=1}^{\infty}\varsigma(k) = \infty$, *one has* $\lim\limits_{k\to\infty}\sup y(k) \le \lim\limits_{k\to\infty}\sup\frac{\eta(k)}{\varsigma(k)}$.

Lemma 3.2 (Horn and Johnson 2012) *For a matrix* $A \in \mathbb{R}^{N\times N}$, *there must exist a matrix norm* $\|\cdot\|$ *such that*

$$\|A\| \le \rho(A) + \varepsilon, \tag{3.11}$$

where $\forall\ \varepsilon > 0,\ \rho(A)$ *is the spectral radius of matrix* A.

Proof According to Jordan decomposition theorem, there exists an invertible matrix P such that

$$P^{-1}AP = \begin{pmatrix} \lambda_1 & \delta_1 & 0 & \cdots & 0 & 0 \\ 0 & \lambda_2 & \delta_2 & \cdots & 0 & 0 \\ \vdots & \vdots & \vdots & \cdots & \vdots & \vdots \\ 0 & 0 & 0 & \cdots & \lambda_{N-1} & \delta_{N-1} \\ 0 & 0 & 0 & \cdots & 0 & \lambda_N \end{pmatrix},$$

where $\lambda_i\ (i = 1, 2, \ldots, N)$ are the eigenvalues of matrix A and $\delta_j = 1$ or $0,\ j = 1, 2, \ldots, N-1$.

Let $D = \mathrm{diag}(1, \varepsilon, \varepsilon^2, \ldots, \varepsilon^{N-1})$, one has

$$D^{-1}P^{-1}APD = D^{-1}JD = \begin{pmatrix} \lambda_1 & \varepsilon\delta_1 & 0 & \cdots & 0 & 0 \\ 0 & \lambda_2 & \varepsilon\delta_2 & \cdots & 0 & 0 \\ \vdots & \vdots & \vdots & \cdots & \vdots & \vdots \\ 0 & 0 & 0 & \cdots & \lambda_{N-1} & \varepsilon\delta_{N-1} \\ 0 & 0 & 0 & \cdots & 0 & \lambda_N \end{pmatrix}.$$

Then, define the matrix norm

$$\|A\| = \left\|D^{-1}P^{-1}APD\right\|_{\infty}, \tag{3.12}$$

one can get $\|D^{-1}P^{-1}APD\|_{\infty} \le \max_i(|\lambda_i| + \varepsilon) = \rho(A) + \varepsilon$. □

Lemma 3.3 (Horn and Johnson 2012) *For a matrix $A \in \mathbb{R}^{N\times N}$, there must exist a vector norm $\|x\|_*$ such that*

$$\|Ax\|_* \leq \|A\| \cdot \|x\|_*, \tag{3.13}$$

where $\|A\|$ is an arbitrary matrix norm of matrix A, and $\forall\, x = (x_1, \ldots, x_N)^T \in \mathbb{R}^N$ is a vector.

Proof For any vector $\chi \neq 0 \in \mathbb{R}^N$, define the vector norm

$$\|x\|_* = \left\|x\chi^{\mathrm{T}}\right\|, \tag{3.14}$$

where $\|\cdot\|$ is the matrix norm in (3.12). Then, one has $\|Ax\|_* = \|Ax\chi^{\mathrm{T}}\| \leq \|A\| \cdot \|x\chi^{\mathrm{T}}\| = \|A\| \cdot \|x\|_*$. □

Remark 3.2 In this chapter, the matrix norm and the vector norm will be given in (3.12) and (3.14). Especially, in (3.14), for $\forall\, x = (x_1, \ldots, x_N)^{\mathrm{T}} \in \mathbb{R}^N$, if $|x_i| = \max_{1\leq j\leq N} |x_j|$, $i \in \{1, 2, \ldots, N\}$, we will choose $\chi = (\chi_1, \ldots, \chi_N)^{\mathrm{T}}$ with $\chi_i = 1$ and $\chi_j = 0,\ j = 1, 2, \ldots, N,\ j \neq i$. If we define matrix $A = \mathrm{diag}(a_{11}, \ldots, a_{nn})$ with $a_{ii} = 1$ and $a_{jj} = 0$, $j = 1, 2, \ldots, N,\ j \neq i$, one has that $\|Ax\|_* = |x_i|$ and $\|A\| = 1$. That is, $|x_i| \leq \|x\|_*$, which means $|x_j| \leq \|x\|_*$, $\forall\, j = 1, 2, \ldots, N$.

3.3 Main Results

From (3.3) and (3.4), one has that $\widehat{x}_{ji}(t, k) = \zeta_{ji}(t, k)$, for $t \in [1, T]$, $i = 1, 2, \ldots, N,\ j \in \Pi_i$, and $k \in \mathbb{Z}_+$. Then, one obtains from (3.7)

$$\begin{aligned}
&\Delta u_i(t-1, k) \\
&= r \sum_{j\in\Pi_i} \alpha_{ij}(t)\big[\widehat{x}_{ji}(t, k) - x_j(t, k) + x_j(t, k) \\
&\quad - x_i(t, k) + x_i(t, k) - \zeta_{ij}(t, k)\big] + \sum_{j=1}^{N} g_{ij}(t)e_j(t, k) \\
&= r \sum_{j\in\Pi_i} \alpha_{ij}(t)\big[\delta_{ji}(t, k) - \delta_{ij}(t, k)\big] + r \sum_{j\in\Pi_i} \alpha_{ij}(t)\big[x_j(t, k) - x_i(t, k)\big] \\
&\quad + \sum_{j=1}^{N} g_{ij}(t)e_j(t, k),
\end{aligned} \tag{3.15}$$

where $\delta_{ji}(t, k) = \zeta_{ji}(t, k) - x_j(t, k)$. Since $e_i(t, k) = x^*(t) - x_i(t, k)$, one has

$$
\begin{aligned}
& e_i(t+1,k+1) - e_i(t+1,k) \\
&= x_i^*(t+1) - x_i(t+1,k+1) - x_i^*(t+1) + x_i(t+1,k) \\
&= -\big[x_i(t,k+1) + u_i(t,k+1) - x_i(t,k) - u_i(t,k)\big] \\
&= -\theta_i(t,k) - \Delta u_i(t,k),
\end{aligned} \tag{3.16}
$$

where $\theta_i(t,k) = x_i(t,k+1) - x_i(t,k)$. Here, $\theta_i(t,k) = x_i(t-1,k+1) + u_i(t-1,k+1) - x_i(t-1,k) - u_i(t-1,k) = \theta_i(t-1,k) + \Delta u_i(t-1,k)$. Under Assumption 3.1, one can obtain that

$$
\theta_i(t,k) = \sum_{j=0}^{t-1} \Delta u_i(j,k), \tag{3.17}
$$

for $t \in [1,T], i = 1,2,\ldots,N,\ j \in \Pi_i$, and $k \in \mathbb{Z}_+$. According to (3.3), (3.7), and (3.8), one can obtain that

$$
\alpha_{ij}(t)\delta_{ij}(t,k+1) = \alpha_{ij}(t)b(t,k)\Lambda_{ij}(t,k), \tag{3.18}
$$

where $t \in [1,T], i = 1,2,\ldots,N,\ j \in \Pi_i$, and $k \in \mathbb{Z}_+$.

Denote $e(t,k) = (e_1(t,k),\ldots,e_N(t,k))^{\mathrm{T}}$, $\theta(t,k) = (\theta_1(t,k),\ldots,\theta_N(t,k))^{\mathrm{T}}$, $\Delta u(t,k) = (\Delta u_1(t,k),\ldots,\Delta u_N(t,k))^T$, $\alpha_i(t) = (\alpha_{1i}(t),\ldots,\alpha_{Ni}(t))^T$, $\beta_i(t) = (\alpha_{i1}(t),\ldots,\alpha_{iN}(t))^T$, $\Lambda^1(t,k) = (\Lambda_{11}(t,k),\ldots,\Lambda_{N1}(t,k),\Lambda_{12}(t,k),\Lambda_{22}(t,k),\ldots,\Lambda_{NN}(t,k))^{\mathrm{T}}$, $\Lambda^2(t,k) = (\Lambda_{11}(t,k),\Lambda_{12}(t,k),\ldots,\Lambda_{1N}(t,k),\Lambda_{21}(t,k),\Lambda_{22}(t,k),\ldots,\Lambda_{NN}(t,k))^{\mathrm{T}}$, $\Xi_1(t) = \mathrm{diag}(\alpha_1^{\mathrm{T}}(t),\ldots,\alpha_N^{\mathrm{T}}(t))$, $\Xi_2(t) = \mathrm{diag}(\beta_1^{\mathrm{T}}(t),\ldots,\beta_N^{\mathrm{T}}(t))$, $r(t) = (r_{ij}(t))_{N\times N}$. From (3.15), (3.16), (3.17), and (3.18), it follows that

$$
\begin{aligned}
& e(t+1,k+1) \\
&= -rb(t+1,k-1)\Big[\Xi_1(t+1)\Lambda^1(t+1,k-1) - \Xi_2(t+1)\Lambda^2(t+1,k-1)\Big] \\
&+ \big(I - rL(t+1) - G(t+1)\big)e(t+1,k) - \sum_{j=0}^{t-1}\Delta u(j,k),
\end{aligned} \tag{3.19}
$$

where $G(t) = (g_{ij}(t))_{N\times N},\ t \in [1,T], k \in \mathbb{Z}_+$, and $k > 0$. Moreover,

$$
\begin{aligned}
e(1,k+1) &= \big[I - rL(1) - G(1)\big]e(1,k) \\
&\quad - rb(1,k-1)\big[\Xi_1(1)\Lambda^1(1,k-1) - \Xi_2(1)\Lambda^2(1,k-1)\big].
\end{aligned}
$$

Remark 3.3 In the following, the objective is to design the appropriate ILC $\Delta u_i(t,k)$ to achieve that $e_i(t,k) \to 0$ over the finite interval $t \in [1,T]$, as $k \to \infty$. From (3.2), (3.3), (3.4), and (3.5), one can see that $\Delta u_i(t,k)$ mainly depends on the three parameters: the scaling function $b(t,k)$, the learning gain r, and the number of quantization levels $\nu_{ij}(t,k),\ t \in [1,T], k \in \mathbb{Z}_+, i = 1,2,\ldots,N,\ j \in \Pi_i$. One of the key roles of the three parameters is to avoid the saturation of the finite-level quantizers.

Theorem 3.1 *Consider system (3.2). Under Assumptions 3.1 and 3.2, in the finite interval* $[1, T]$, *the tracking error* $e(t, k)$ *such that*

$$\lim_{k\to\infty} \frac{\|e(t,k)\|_*}{b_M(k)} \le \frac{\tau + C_M^1}{b_M(1)} + \frac{r(T+1)\mu^2 d^*}{1 - \|Q\|\mu}, \tag{3.20}$$

where $\|\cdot\|_*$ *is defined in (3.14), and* $t \in [1, T]$. *Moreover, when* $\lim_{k\to\infty} b_M(k) \to 0$, *one can get*

$$\lim_{k\to\infty} \|e(t,k)\|_* \to 0, \tag{3.21}$$

if the numbers of quantization levels $v_{i,j}^m(\cdot)$ $(i = 1, 2, \ldots, N,\ j \in \Pi_i)$ *satisfy*

$$v_{i,j}^m(1) \ge \frac{C_M^1 + \max\limits_{0\le t\le T} |x^*(t)|}{b_m(0)} - \frac{1}{2}, \tag{3.22}$$

$$v_{i,j}^m(2) \ge \begin{cases} \gamma_1, & if\ \alpha_{ij} = 1, \\ \gamma_2, & if\ \alpha_{ij} = 0, \end{cases} \tag{3.23}$$

and

$$v_{i,j}^m(k+1) \ge \begin{cases} \gamma_3, & if\ \alpha_{ij} = 1, \\ \gamma_4, & if\ \alpha_{ij} = 0, \end{cases} \tag{3.24}$$

where the parameters are as

$$\gamma_1 = \Big[\frac{1}{2} + rd^*T\Big]\frac{b_M(0)}{b_m(1)} + \big[r(d^*+1) + Ng\big]TC_M^1 \cdot \frac{1}{b_m(1)} - \frac{1}{2},$$
$$\gamma_2 = \frac{b_M(0)}{b_m(1)}\Big[v_{i,j}^m(1) + \frac{1}{2} + rd^*T\Big] + \big[r(d^*+1) + Ng\big]TC_M^1 \cdot \frac{1}{b_m(1)} - \frac{1}{2},$$
$$\gamma_3 = \frac{\big[\frac{1}{2} + rd^*T\big]b_M(k-1)}{b_m(k)} + \frac{\Big(\frac{\tau + C_M^1}{b_M(1)} + \frac{r(T+1)\mu^2 d^*}{1-\|Q\|\mu}\Big)[r(d^*+1) + Ng]Tb_M(k)}{b_m(k)} - \frac{1}{2},$$
$$\gamma_4 = \frac{\big[v_{i,j}^m(k) + \frac{1}{2} + rd^*T\big]b_M(k-1)}{b_m(k)} + \frac{\Big(\frac{\tau + C_M^1}{b_M(1)} + \frac{r(T+1)\mu^2 d^*}{1-\|Q\|\mu}\Big)[r(d^*+1) + Ng]Tb_M(k)}{b_m(k)} - \frac{1}{2},$$

$$v_{i,j}^m(k) = \min_{0\le t\le T} v_{i,j}(t,k),\ k \in \mathbb{Z}_+,\ \ \mu = \sup_{k\ge 0} \frac{b_M(k)}{b_M(k+1)} \le \frac{9}{10}\frac{1}{\|Q\|},$$
$$C_M^1 = \max_{1\le i\le N}\max_{0\le t\le T} \big|e_i(t,1)\big|,\ \ g = \max_{1\le i,j\le N}\max_{0\le t\le T} \big|g_{ij}(t)\big|,\ \ \tau = (T+N-2)\varepsilon,$$
$$0 < b_m(k) = \min_{0\le t\le T} b(t,k),\ \ 0 < b_M(k) = \max_{0\le t\le T} b(t,k),\ \ k = 0, 1, 2, \ldots.$$

And matrix Q is given as follows:

$$Q = \begin{pmatrix} \|I-H(1)\| & 0 & 0 & \cdots & 0 & 0 \\ \|H(1)\| & \|I-H(2)\| & 0 & \cdots & 0 & 0 \\ \|H(1)\| & \|H(2)\| & \|I-H(3)\| & \cdots & 0 & 0 \\ \vdots & \vdots & \vdots & \cdots & \vdots & \vdots \\ \|H(1)\| & \|H(2)\| & \|H(3)\| & \cdots & \|H(T-1)\| & \|I-H(T)\| \end{pmatrix}, \tag{3.25}$$

with $H(t) = rL(t) + G(t),\ t = 1, 2, \ldots, T$.

Proof We discuss two cases in the proof.

(1) Consider $\alpha_{ij} = 1$. Equations (3.3) and (3.8) lead to

$$\zeta_{ij}(t, k+1) = x_i(t, k) + b(t, k-1)\Lambda_{ij}(t, k-1), \tag{3.26}$$

where $t \in [1, T],\ k \in \mathbb{Z}_+,\ i = 1, 2, \ldots, N,\ j \in \Pi_i$. We shall discuss the following two cases.

(a) When $k = 0$, one has from (3.3) and (3.22)

$$\left|\frac{x_i(t,1) - \zeta_{ij}(t,0)}{b(t,0)}\right| = \left|\frac{x_i(t,1)}{b(t,0)}\right| \le \frac{C_M^1 + \max\limits_{0\le t\le T}|x^*(t)|}{b_m(0)} \le \nu_{i,j}^m(1) + \frac{1}{2}, \tag{3.27}$$

where $t \in [1, T],\ i = 1, 2, \ldots, N,\ j \in \Pi_i$. Combining (3.5) with (3.27) leads to

$$\max_{i,j} \left|\Lambda_{ij}(t, 0)\right| \le \frac{1}{2}. \tag{3.28}$$

(b) When $k > 0$, with (3.26), one can obtain that $\delta_{ij}(t, k) = \zeta_{ij}(t, k) - x_i(t, k) = b(t, k-1)\Lambda_{ij}(t, k-1)$. Then, one has from (3.17) and (3.26)

$$\begin{aligned} \left|\frac{x_i(t,k+1) - \zeta_{ij}(t,k)}{b(t,k)}\right| &= \left|\frac{x_i(t,k+1) - x_i(t,k) - b(t,k-1)\Lambda_{ij}(t,k-1)}{b(t,k)}\right| \\ &\le \frac{\left|\theta_i(t,k)\right|}{b(t,k)} + \frac{b(t,k-1)}{b(t,k)}\left|\Lambda_{ij}(t,k-1)\right| \\ &\le \frac{\sum\limits_{j=0}^{t-1}\left|\Delta u_i(j,k)\right|}{b(t,k)} + \frac{b(t,k-1)}{b(t,k)}\left|\Lambda_{ij}(t,k-1)\right|. \end{aligned} \tag{3.29}$$

Let $k = 1$, combining (3.15), (3.23), (3.28), and (3.29), one has

$$
\begin{aligned}
\left|\frac{x_i(t,2)-\zeta_{ij}(t,1)}{b(t,1)}\right| &\le \frac{\sum_{j=0}^{t-1}\left|\Delta u_i(j,1)\right|}{b(t,1)}+\frac{1}{2}\cdot\frac{b(t,0)}{b(t,1)}\\
&\le \left[\frac{1}{2}+rd^*T\right]\frac{b_M(0)}{b_m(1)}+\left[r(d^*+1)+Ng\right]TC_M^1\cdot\frac{1}{b_m(1)}\\
&\le \nu_{i,j}^m(2)+\frac{1}{2},
\end{aligned}
\tag{3.30}
$$

where $t\in[1,T],\ i=1,2,\ldots,N,\ j\in\Pi_i$. It follows that

$$
\max_{i,j}\left|\Lambda_{ij}(t,1)\right|\le\frac{1}{2}. \tag{3.31}
$$

According to (3.15) and (3.19), one has

$$
\begin{aligned}
\|e(t+1,k+1)\|_* &\le \|I-rL(t+1)-G(t+1)\|\cdot\|e(t+1,k)\|_*\\
&\quad+\sum_{j=1}^{t}\|rL(j)+G(j)\|\cdot\|e(j,k)\|_*\\
&\quad+2r(T+1)b_M(k-1)d^*\max_{i,j}|\Lambda_{ij}(t,k-1)|,
\end{aligned}
\tag{3.32}
$$

where $t\in[1,T],\ i=1,2,\ldots,N,\ j\in\Pi_i,\ k=1,2,\ldots$. Moreover, $\|e(1,k+1)\|_*\le\|I-rL(1)-G(1)\|\cdot\|e(1,k)\|_*+2rb_M(k-1)d^*\max_{i,j}|\Lambda_{ij}(t,k-1)|$. Denote $\omega(k)=(\|e(1,k)\|_*,\ldots,\|e(T,k)\|_*)^{\mathrm{T}}$, one knows

$$
\left\|\omega(k+1)\right\|_*\le\left\|Q\right\|\left\|\omega(k)\right\|_*+2r(T+1)b_M(k-1)d^*\max_{i,j}\left|\Lambda_{ij}(t,k-1)\right|,
\tag{3.33}
$$

where Q has been given in (3.25). Let $h(k)=1-\|Q\|\frac{b_M(k)}{b_M(k+1)}$ and $z(k)=\frac{\omega(k)}{b_M(k)}$, then

$$
\begin{aligned}
\|z(k+1)\|_* &= \frac{\|\omega(k+1)\|_*}{b_M(k+1)}\\
&\le (1-h(k))\|z(k)\|_*+2r(T+1)\frac{b_M(k-1)}{b_M(k+1)}d^*\max_{i,j}|\Lambda_{ij}(t,k-1)|,
\end{aligned}
\tag{3.34}
$$

where $k=1,2,\ldots$. Based on (3.11) and Remark 3.2, one has $\|\omega(1)\|_*\le(T-1)\varepsilon+\max_{1\le t\le T}\|e(t,1)\|_*$ and $\|e(t,1)\|_*\le(N-1)\varepsilon+\max_{1\le i\le N}|e_i(t,1)|$. That is,

$$
\|\omega(1)\|_*\le(T+N-2)\varepsilon+\max_{1\le t\le T}\max_{1\le i\le N}|e_i(t,1)|\le\tau+C_M^1, \tag{3.35}
$$

where $\tau \triangleq (T+N-2)\varepsilon$. According to $\mu = \sup\limits_{k\geq 0} \frac{b_M(k)}{b_M(k+1)} \leq \frac{9}{10} \cdot \frac{1}{\|Q\|}$, it follows that $\frac{1}{10} \leq h(k) \leq 1, \ \forall\, k \in \mathbb{Z}_+$. From Lemma 3.1 and (3.35), one has

$$\|z(k+1)\|_* \leq \frac{\tau + C_M^1}{b_M(1)} + \frac{2r(T+1)\mu^2 d^*}{1-\|Q\|\mu} \sup_{1\leq t\leq T,\ 0\leq j\leq k-1} \max_{i,j} |\Lambda_{ij}(t, j)|, \tag{3.36}$$

where $0 < k \in \mathbb{Z}_+$.

Now, we prove that if $\max\limits_{ij} |\Lambda_{ij}(t,l)| \leq \frac{1}{2}$ for $t \in [1, T],\ l = 0, 1, 2, k-1$, then $\max\limits_{ij} |\Lambda_{ij}(t,k)| \leq \frac{1}{2}$ for $t \in [1, T]$, $0 < k \in \mathbb{Z}_+$. Suppose $\max\limits_{0\leq l\leq k-1} \max\limits_{ij} |\Lambda_{ij}(t,l)| \leq \frac{1}{2}$. Note that $\max\limits_{i,j} |x_j(t,k) - x_i(t,k)| \leq \max\limits_{i} |e_i(t,k)| + \max\limits_{j} |e_j(t,k)|$, from (3.15), (3.18), (3.24), (3.29), and the results in Remark 3.2, one has

$$\left| \frac{x_i(t, k+1) - \zeta_{ij}(t,k)}{b(t,k)} \right| \leq \nu_{i,j}^m(k+1) + \frac{1}{2}, \tag{3.37}$$

where $t \in [1, T],\ i = 1, 2, \ldots, N,\ j \in \Pi_i$. It follows that

$$\max_{i,j} \left|\Lambda_{ij}(t,k)\right| \leq \frac{1}{2}. \tag{3.38}$$

Consider $a_{ij} = 0$. (3.3) leads to

$$\zeta_{ij}(t, k+1) = \zeta_{ij}(t,k), \tag{3.39}$$

where $t \in [1, T],\ k \in \mathbb{Z}_+,\ i = 1, 2, \ldots, N,\ j \in \Pi_i$. We also discuss the following two cases.

(a) When $k = 0$, one can have the same results with (3.27) and (3.28). The details are omitted here.

(b) When $k > 0$, with (3.39), it follows that

$$\begin{aligned}
&\left| \frac{x_i(t, k+1) - \zeta_{ij}(t,k)}{b(t,k)} \right| \\
&= \left| \frac{x_i(t, k+1) - \zeta_{ij}(t,k-1)}{b(t,k)} \right| \\
&\leq \frac{\left|\theta_i(t,k)\right|}{b(t,k)} + \frac{\left|x_i(t,k) - \zeta_{ij}(t,k-1)\right|}{b(t,k-1)} \cdot \frac{b(t,k-1)}{b(t,k)} \\
&\leq \frac{\sum\limits_{j=0}^{t-1} \left|\Delta u_i(j,k)\right|}{b(t,k)} + \frac{\left|x_i(t,k) - \zeta_{ij}(t,k-1)\right|}{b(t,k-1)} \cdot \frac{b(t,k-1)}{b(t,k)}.
\end{aligned} \tag{3.40}$$

Let $k = 1$, combining (3.15), (3.23), (3.27), (3.28), and (3.40), one has

$$\begin{aligned}
&\left|\frac{x_i(t,2)-\zeta_{ij}(t,1)}{b(t,1)}\right| \\
&\le \frac{\sum_{j=0}^{t-1}\left|\Delta u_i(j,1)\right|}{b(t,1)} + \left[v_{i,j}^m(1)+\frac{1}{2}\right]\cdot\frac{b(t,0)}{b(t,1)} \\
&\le \left[v_{i,j}^m(1)+\frac{1}{2}+rd^*T\right]\frac{b_M(0)}{b_m(1)} + \left[r(d^*+1)+Ng\right]TC_M^1\cdot\frac{1}{b_m(1)} \\
&\le v_{i,j}^m(2)+\frac{1}{2},
\end{aligned} \tag{3.41}$$

where $t \in [1, T],\ i = 1, 2, \ldots, N,\ j \in \Pi_i$. It follows that

$$\max_{i,j}\left|\Lambda_{ij}(t,1)\right| \le \frac{1}{2}. \tag{3.42}$$

Similar to the proof of part (1), according to (3.24) and (3.40), one also has

$$\max_{i,j}\left|\Lambda_{ij}(t,k)\right| \le \frac{1}{2}, \tag{3.43}$$

where $t \in [1, T],\ i = 1, 2, \ldots, N,\ j \in \Pi_i,\ k \in \mathbb{Z}_+$.

Hence, from (3.36), (3.38), and (3.43), one can obtain (3.20). Moreover, (3.20) leads to that $\lim_{k\to\infty} \|e(t,k)\|_* \to 0$, if $\lim_{k\to\infty} b_M(k) \to 0$. Based on Remark 3.2, one has that $\lim_{k\to\infty} |e_i(t,k)| \to 0$, if $\lim_{k\to\infty} b_M(k) \to 0$, $i = 1, 2, \ldots, N$. □

Remark 3.4 In Theorem 3.1, condition $\mu = \sup_{k\ge 0} \frac{b_M(k)}{b_M(k+1)} \le \frac{9}{10}\cdot\frac{1}{\|Q\|}$ could be as $\mu = \sup_{k\ge 0} \frac{b_M(k)}{b_M(k+1)} \le \iota\cdot\frac{1}{\|Q\|}$. Here, the coefficient ι could be changed as any constant with $0 < \iota < 1$, which guarantees that the conditions in Lemma 3.1 could be satisfied.

Remark 3.5 From (3.21), the tracking error $\|e(t,k)\|_* \to 0$, if $\lim_{k\to\infty} b_M(k) \to 0$. Here, $\lim_{k\to\infty} b_M(k) \to 0$ is not difficult to be achieved. For example, denote the scaling function $b(t,k) = t\cdot(\frac{1}{2})^k$, $t \in [1, T]$, $k \in \mathbb{Z}_+$, one has $\lim_{k\to\infty} b_M(k) \to 0$. Moreover, the tracking problem can be achieved exponentially fast.

Remark 3.6 Theorem 3.1 gives a tracking goal which can be faultlessly achieved over a finite-time interval. It is essentially different from the typical tracking control which usually pays close attention to the system performance when t tends to infinity (see Li and Xie 2011; Li et al. 2014). For a sort of discrete-time digital networks, the high-precision tracking performance in a limited interval can be reached through using a repetitive type of ILC. In consequence, designing the ILCs is feasible in some practical applications.

Remark 3.7 In Meng et al. (2014, 2015), the iterative learning tracking has been discussed for multi-agent systems with switching network topologies. Compared with Meng et al. (2014, 2015), there are two differences in our chapter: (1) The iterative learning controller $u_i(t, k)$ is considered to be quantized. In a digital network, due to the finite capability of transmission bandwidth of information channels, etc., it is feasible for the controller $u_i(t)$ to transmit its feedback information with quantized communication. (2) There is no requirement for the structure of the network graphs. However, in Meng et al. (2014), the union of the switching graphs in each iteration always requires spanning trees.

3.4 Simulation Example

In this section, a numerical example is provided to prove the usefulness of the result in Theorem 3.1.

Example 3.1 Consider a simple digital computer network with three computers. All three computers are viewed as nodes, and they transfer the data from the nodes over a point-to-point communication channel. In this case, the point-to-point communication channels are variable with time t. For simplicity, we only consider the variable communication channels in Fig. 3.1. It is possible to describe the mathematical model of the digital computer network in the system (3.1). In this example, the goal is to implement the ILC method in the (3.1) system (see system (3.2)) in order to discuss the consensus tracking of the set of nodes.

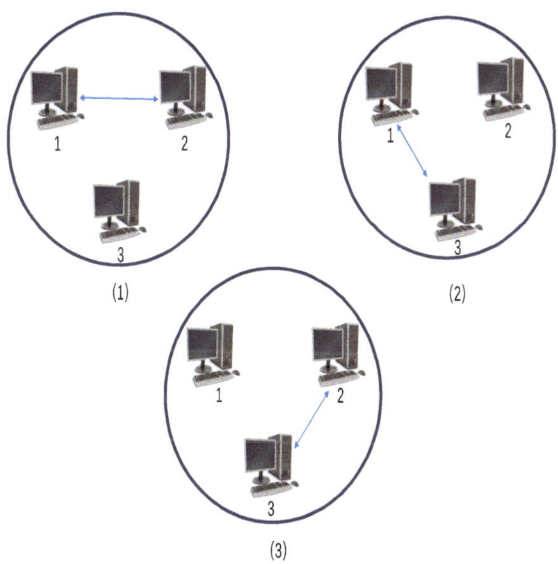

Fig. 3.1 A simple computer network with variable communication channels

Here, denote $T = 10$, the learning gain $r = 0.15$, $\varepsilon = 0.01$ in (3.12), $x(t,k) = (x_1(t,k), x_2(t,k), x_3(t,k))^{\mathrm{T}} \in \mathbb{R}^3$, the scaling function $b(t,k) = 3\cos(\pi t) \cdot (\frac{4}{5})^k$, $t \in \{1, \ldots, 10\}$, $k \in \mathbb{Z}_+, \mathcal{A}(t) = [(\alpha_{ij}(t))_{3\times 3}]$, and the learning gain matrices $G(t) = [(g_{ij}(t))_{3\times 3}]$ satisfy

$$\mathcal{A}(3s) = \begin{pmatrix} 0 & 1 & 0 \\ 1 & 0 & 0 \\ 0 & 0 & 0 \end{pmatrix}, \ \mathcal{A}(3s+1) = \begin{pmatrix} 0 & 0 & 1 \\ 0 & 0 & 0 \\ 1 & 0 & 0 \end{pmatrix}, \ \mathcal{A}(3s+2) = \begin{pmatrix} 0 & 0 & 0 \\ 0 & 0 & 1 \\ 0 & 1 & 0 \end{pmatrix},$$

and

$$G(3s) = \begin{pmatrix} 0.3 & 0 & -0.1 \\ 0 & 0.3 & -0.1 \\ -0.1 & -0.1 & 0.3 \end{pmatrix}, \ G(3s+1) = \begin{pmatrix} 0.3 & -0.1 & 0 \\ -0.1 & 0.3 & -0.1 \\ 0 & -0.1 & 0.3 \end{pmatrix},$$
$$G(3s+2) = \begin{pmatrix} 0.3 & -0.1 & -0.1 \\ -0.1 & 0.3 & 0 \\ -0.1 & 0 & 0.3 \end{pmatrix},$$

where $s = 0, 1, 2, 3$. Then, one has $d^* = 1$, $g = 0.3$, $\mu = \sup_{k\geq 0} \frac{b_M(k)}{b_M(k+1)} = 1.11$, and $\|Q\| = 0.68$. As a result, one has $\mu < 1.3043 = \frac{9}{10} \cdot \frac{1}{\|Q\|}$. Define the desired trajectory $x^*(t) = 0.2\sin(t) + 0.2t$ in the finite-time interval $\{1, \ldots, 10\}$.

According to Theorem 3.1, in the finite interval $\{1, \ldots, 10\}$, the numbers of quantization levels $\nu_{i,j}^m(\cdot)$ $(i = 1, \ldots, N,\ j \in \Pi_i)$ are selected as follows: $\nu_{i,j}^m(1) = 0.60$ and $\nu_{i,j}^m(2) = 19.25$, if $a_{ij} = 1$, or $\nu_{i,j}^m(2) = 20.5$, if $a_{ij} = 0$, and

$$\nu_{i,j}^m(k+1) = \begin{cases} 108.54, & if\ \alpha_{ij} = 1, \\ \frac{10}{9}\nu_{i,j}^m(k) + 108.54, & if\ \alpha_{ij} = 0. \end{cases} \tag{3.44}$$

In the case of arbitrary boundary conditions, according to Theorems 3.1 and Remark 3.2, the ILC scheme (3.15) can guarantee that the tracking error $e_i(t,k)$ has $|e_i(t,k)| \to 0$ $(i = 1, 2, \ldots, N)$ as $k \to +\infty$. Figure 3.2a–d shows the state trajectories of three nodes in the system (3.2) at different iteration k. Figure 3.3 shows the tracking error with respect to iteration step k and time t. From Figs. 3.2 and 3.3, it is clear that the tracking error of each node asymptotically goes to zero on the finite interval $\{1, \ldots, 10\}$ as the iteration step k increases.

Furthermore, the corresponding simulation results of the number of quantization levels $\nu_{12}(t,k)$ at different iteration k can also be observed in Fig. 3.4. Also, Fig. 3.5 shows the number of quantization levels $\nu_{12}(t,k)$ with respect to iteration k and time t. According to Fig. 3.4a–d, it can be seen that the variation of $\nu_{12}(t,k)$ is not large when it changes with t and k.

Remark 3.8 Compared with Meng et al. (2014, 2015), our approach has no requirement on the structure of the time-varying networks by designing an appropriate quantized ILC scheme. Meanwhile, a scaling function has to be applied to avoid the

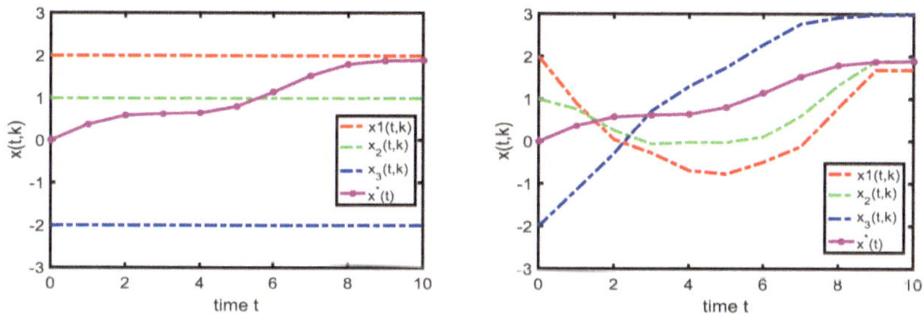

(a) The tracking state with respect to time t (k=0)

(b) The tracking state with respect to time t (k=2)

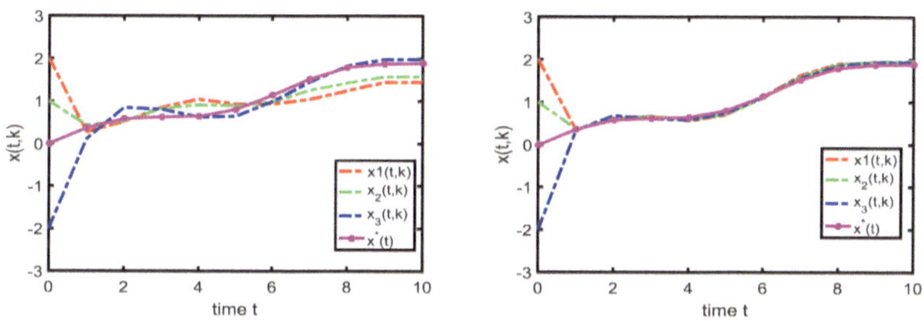

(c) The tracking state with respect to time t (k=10)

(d) The tracking state with respect to time t (k=20)

Fig. 3.2 The tracking error with respect to time t

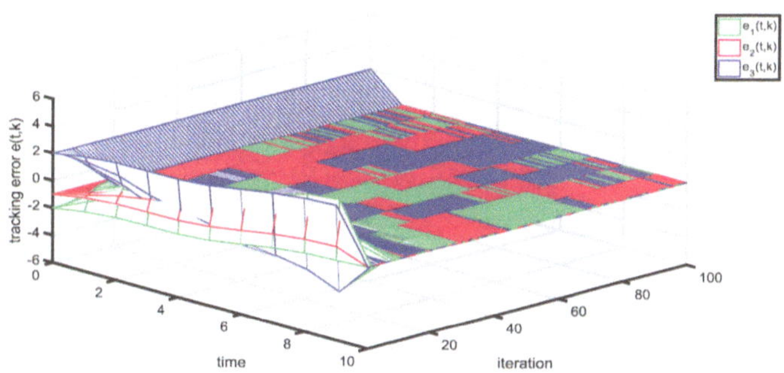

Fig. 3.3 The trajectories of tracking error $e(t, k)$ in the system (3.1)

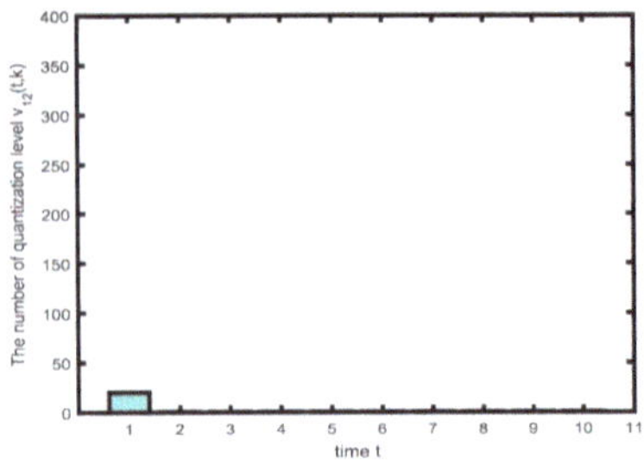

(a) The number of quantization levels $\nu_{12}(t,k)$ with respect to time $t(k=0)$

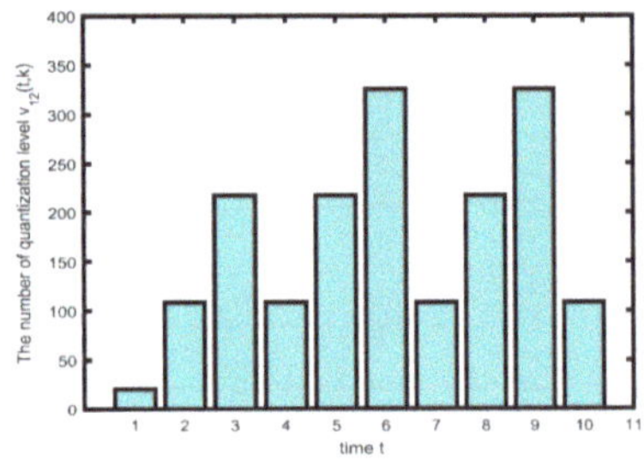

(b) The number of quantization levels $\nu_{12}(t,k)$ with respect to time $t(k=2)$

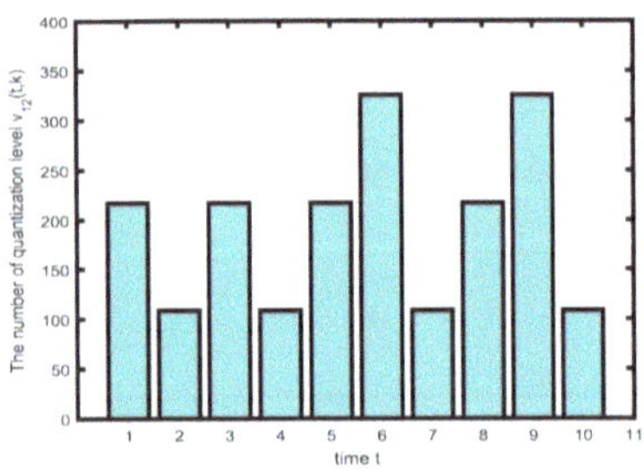

(c) The number of quantization levels $\nu_{12}(t,k)$ with respect to time $t(k=5)$

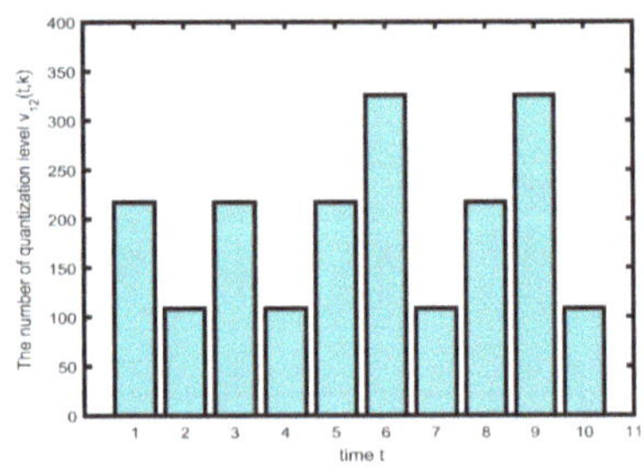

(d) The number of quantization levels $\nu_{12}(t,k)$ with respect to time $t(k=20)$

Fig. 3.4 The number of quantization levels $\nu_{12}(t,k)$ with respect to time t

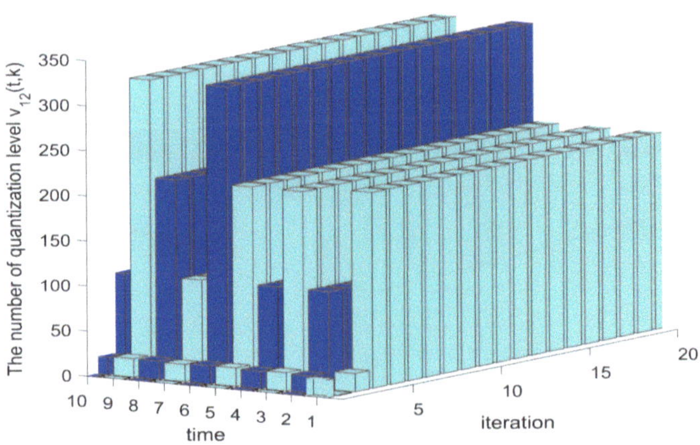

Fig. 3.5 The number of quantization levels $\nu_{12}(t,k)$ with respect to iteration k and time t

saturation of the finite-level quantizers. Compared with Carli et al. (2010), Yang et al. (2011), this chapter designs finite-level quantizers but no infinite-level ones due to communication channels with limited capacity. Moreover, the tracking objective can be achieved over a finite-time interval by using quantized ILC method.

3.5 Summary

The quantized iterative learning problem of digital networks with time-varying topologies has been discussed in this chapter. The quantized iterative learning controllers have been designed to illustrate the limited information communication in the process of encoding and decoding. Then, a relaxed condition has been proposed to achieve the tracking in a finite interval by using the quantized iterative learning controllers. Finally, numerical examples with the simulations have been provided to illustrate the usefulness of the obtained criteria.

References

Carli R, Fagnani F, Frasca P, Zampieri S (2010) Gossip consensus algorithms via quantized communication. Automatica 46(1):70–80

Dong HL, Wang ZD, Gao HJ (2012) Distributed filtering for a class of time-varying systems over sensor networks with quantization errors and successive packet dropouts. IEEE Trans Signal Process 60(6):3164–3173

Dong HL, Wang ZD, Gao HJ (2012) On design of quantized fault detection filters with randomly occurring nonlinearities and mixed time-delays. Signal Process 92(4):1117–1125

Horn RA, Johnson CR (2012) Matrix Analysis. Cambridge University Press, London

Huang DQ, Xu JX, Venkataramanan V, Huynh T (2014) High-performance tracking of piezoelectric positioning stage using current-cycle iterative learning control with gain scheduling. IEEE Trans Ind Electron 61(2):1085–1098

Kurek JE, Zaremba MB (1993) Iterative learning control synthesis based on 2-D system theory. IEEE Trans Automat Control 38(1):121–125

Li JS, Li JM (2014) Adaptive iterative learning control for coordination of second-order multi-agent systems. Int J Robust Nonlin Control 24(18):3282–3299

Li T, Xie LH (2011) Distributed consensus over digital networks with limited bandwidth and time-varying topologies. Automatica 47(9):2006–2015

Li DQ, Liu QP, Wang XF, Yin ZX (2014) Quantized consensus over directed networks with switching topologies. Syst Control Lett 65:13–22

Liu Y, Jia YM (2012) An iterative learning approach to formation control of multi-agent systems. Syst Control Lett 61(1):148–154

Liu M, Ho DWC, Niu YG (2010) Robust filtering design for stochastic system with mode-dependent output quantization. IEEE Trans Signal Process 58(12):6410–6416

Liu ZW, Guan ZH, Li T, Zhang XH, Xiao JW (2012a) Quantized consensus of multi-agent systems via broadcast gossip algorithms. Asian J Control 14(6):1634–1642

Liu TF, Jiang ZP, Hill DJ (2012b) Small-gain based output-feedback controller design for a class of nonlinear systems with actuator dynamic quantization. IEEE Trans Autom Control 57(5):1326–1332

Meng DY, Jia YM (2012) Iterative learning approaches to design finite-time consensus protocols for multi-agent systems. Syst Control Lett 61(1):187–194

Meng DY, Jia YM, Du JP, Yu FS (2012) Tracking control over a finite interval for multi-agent systems with a time-varying reference trajectory. Syst Control Lett 61(7):807–818

Meng DY, Jia YM, Du JP, Zhang J (2014) On iterative learning algorithms for the formation control of nonlinear multi-agent systems. Automatica 50(1):291–295

Meng DY, Jia YM, Du JP (2015) Robust consensus tracking control for multiagent systems with initial state shifts, disturbances, and switching topologies. IEEE Trans Neural Netw Learn Syst 26(4):809–824

Wang YC, Chien CJ, Teng CC (2004) Direct adaptive iterative learning control of nonlinear systems using an output-recurrent fuzzy neural network. IEEE Trans Syst Man Cybern 34(3):1348–1359

Wang ZD, Shen B, Shu HS, Wei GL (2012) Quantized control for nonlinear stochastic time-delay systems with missing measurements. IEEE Trans Autom Control 57(6):1431–1444

Xia YQ, Yan JJ, Shi P, Fu MY (2013) Stability analysis of discrete-time systems with quantized feedback and measurements. IEEE Trans Ind Informat 9(1):313–324

Xu JX, Tan Y (2003) Linear and Nonlinear Iterative Learning Control. Springer, Berlin

Yang GN, Shi P, Liu GP, Gao HJ (2011) Network-based feedback control for systems with mixed delays based on quantization and dropout compensation. Automatica 47(12):2805–2809

Yang W, Liu M, Shi P (2012) H_∞ filtering for nonlinear stochastic systems with sensor saturation, quantization and random packet losses. Signal Process 92(6):1387–1396

Chapter 4
Consensus Under Switching Topology and Observer Information

4.1 Introduction

In the past few decades, more attention has been attracted to the research of complex networks, such as Boolean networks (Huang et al. 2020), neural networks (Huang et al. 2018), and so on. However, running a large piece of equipment with a network structure is expensive, so multi-agent systems are used as a replacement for costly systems. The fundamental function of multi-agent systems is to solve complex practical problems effectively, which is beyond the ability of a single agent. Nowadays, multi-agent systems have been widely applied in system coordination (Cao et al. 2013), intelligent transportation systems (Adler et al. 2005), cooperative control (He et al. 2020; Meng 2018), etc. Some meaningful theoretical results have been presented according to the Lyapunov approach and linear matrix inequality (He et al. 2017; Hosseini et al. 2019). In real applications, it is only sometimes effective for control techniques mentioned in the above literatures when the systems require high precision in a finite time.

ILC is an effective control method for the tracking problem. Uchiyama proposed it in 1978, and became an important branch of intelligent control. The main idea is to achieve tracking by updating the control input repeatedly, where the update information comes from the tracking information of previous iterations. Since it can deal with the dynamic system with uncertainty in a simple way, this control technique has been widely applied in many fields, such as chemical engineering (Oh and Lee 2016), robot control (Bouakrif et al. 2013), freeway traffic (Chi et al. 2013), and initial state drift (Chen et al. 1999; Chi et al. 2015; Wei and Li 2017), etc. Furthermore, works are contributing to ILC consensus tracking problems of multi-agent systems, see references therein (Xiong et al. 2016a, b; Yang et al. 2016), etc. However, it is worth mentioning the initial state drift of multi-agent systems is not fully considered, which needs to be further studied.

On the one hand, partial information of some agents may be unmeasurable because of technical limitations. Thus, output measurements of agents are observed for the

W. Xiong et al., *Iterative Learning Control for Network Systems Under Constrained Information Communication*, Intelligent Control and Learning Systems 12,
https://doi.org/10.1007/978-981-97-0926-7_4

consensus tracking problems, and various observer-based learning control strategies have been developed (Chen et al. 2010; Huang et al. 2015; Li et al. 2017). In Huang et al. (2015), two state observers were designed to estimate the unavailable control system and reference state information. In Li et al. (2017), the consensus problem for linear multi-agent systems was investigated from a two-dimensional system framework of iterative learning control with unmeasurable states. Since malware attacks may change the structure of the communication networks between agents (Kwon and Hwang 2016; Liu et al. 2017; Wan et al. 2017), it is difficult to achieve consensus tracking in this situation. Motivated by the above facts, it is necessary to design appropriate observers to improve the tracking performance. Unlike the work mentioned above, the observers are designed in line with switching topologies, which is general and realistic in the application.

On the other hand, consensus tracking is an important issue for iterative learning processes, especially when switching topologies are involved (Meng and Moore 2016; Meng et al. 2015). In some existing work, only one type of switching topologies are considered (Ai et al. 2018; Dai and Guo 2017; Su and Lin 2018; Wen et al. 2016). In Dai and Guo (2017), the leader-following consensus problem was investigated for nonlinear multi-agent systems with semi-Markov switching topologies. Moreover, a disturbance observer was designed to estimate the disturbances suffered by the followers, and the consensus tracking problem was considered with external disturbances and switching topologies in Ai et al. (2018). Later, independent switching topologies were introduced in Wan et al. (2017), Wen et al. (2017), meaning observers and controllers sometimes have their communication networks, and cyber-attacks can change the two communication networks independently. Thus, it inspires us to consider the consensus tracking problem with independent switching topologies in the control and observation networks. Furthermore, observer-based design strategies are proposed in terms of separate switching topologies.

4.2 Problem Formulation

Achieving consensus is one of the most important issues in network system research. Nevertheless, it is challenging when the real state is unobtainable and the communication topology changes. As a result, this problem will be analyzed in the rest part.

In this chapter, the dynamic behavior of each node i $(i = 1, \ldots, M)$ is described as follows:

$$\begin{cases} x_i(t+1,k) = Ax_i(t,k) + Bu_i(t,k), \\ y_i(t+1,k) = Cx_i(t+1,k), \end{cases} \tag{4.1}$$

where the index k denotes the kth iteration, $x_i(t,k) \in \mathbb{R}^n$, $y_i(t,k) \in \mathbb{R}^m$ and $u_i(t,k) \in \mathbb{R}^p$ denote the state, output, and input of node i at time t, respectively. $A \in \mathbb{R}^{n\times n}$,

$B \in \mathbb{R}^{n\times p}$ and $C \in \mathbb{R}^{m\times n}$. Furthermore, assume that (A, B) is stabilizable and (A, C) is detectable.

Since the full state measurements may be unavailable due to technical limitations, the tracking protocols have to be designed based on the related partial state measurements. Considering the switching topologies, the observer can be described as follows:

$$\begin{cases} z_i(t+1,k) = Az_i(t,k) + Bu_i(t,k) \\ \qquad\qquad + \sum_{j=1}^{M} \breve{\alpha}_{i,j}^{\breve{o}(k)} F\big[\varepsilon_j(t+1,k-1) - \varepsilon_i(t+1,k-1)\big], \\ \omega_i(t+1,k) = Cz_i(t+1,k), \end{cases} \tag{4.2}$$

where $z_i(t,k)$ is the kth observer's state of node i, and $\varepsilon_i(t,k) = \omega_i(t,k) - y_i(t,k) = C[z_i(t,k) - x_i(t,k)]$ represents the output error. $\breve{\alpha}_{i,j}^{\breve{o}(k)}$ is an element of adjacency matrix with switching topology, $\breve{o}(k)$ denotes the switching signal of observation output errors, which is presented in a more detailed description in the following. It means that node j can receive information from node i in the sense of switching topology. Observer gain matrix $F \in \mathbb{R}^{n\times m}$.

Let $x^*(t)$ denote the desired state, $e_i(t,k) = x_i(k,t) - x^*(t)$ denotes the tracking error, and $\hat{e}_i(k,t) = x_i(k,t) - z_i(k,t)$ represents the observation error. $\omega_i(t+1,k)$ is the output of the observer embedded in follower i. According to (4.1) and (4.2), the error system is described as follows:

$$\begin{cases} e_i(t+1,k) = Ax_i(t,k) + Bu_i(t,k) - x^*(t+1), \\ \hat{e}_i(t+1,k) = A\hat{e}_i(t,k) - \sum_{j=1}^{M} \breve{\alpha}_{i,j}^{\breve{o}(k)} F[\varepsilon_j(t+1,k-1) - \varepsilon_i(t+1,k-1)]. \end{cases} \tag{4.3}$$

According to the previous discussion, the task is to design appropriate learning control strategies such that the tracking error $e_i(t,k)$ and observation error $\hat{e}_i(t,k)$ can asymptotically converge to zero, i.e.,

$$\lim_{k\to\infty} e_i(t,k) = \mathbf{0}, \quad i = 1, \ldots, M,$$

and

$$\lim_{k\to\infty} \hat{e}_i(t,k) = \mathbf{0}, \quad i = 1, \ldots, M.$$

4.3 ILC Scheme

The appropriate control strategy is first designed in this section and the initial state learning scheme is proposed later.

4.3.1 Controller Design

In this chapter, the communication switching topologies of the controllers are assumed to be unrelated to the switching topologies of the observers. In other words, two types of switching topologies are independent. To describe the changes in network communication structure, the directed switching graphs are used to indicate communication topologies among M nodes.

Assume that a given graph set $\hat{\mathbb{F}}_1 = \{\mathbb{F}_1^1, \ldots, \mathbb{F}_1^{\upsilon_1}\}$ is used to describe the switching topologies for observation output errors, a graph set $\hat{\mathbb{F}}_2 = \{\mathbb{F}_2^1, \ldots, \mathbb{F}_2^{\upsilon_2}\}$ describes the switching topologies for observation states. Switching signals are described by piecewise constant function $\breve{o}(k) : \mathbb{Z}_+ \to \{1, \ldots, \upsilon_1\}$ and $\acute{o}(k) : \mathbb{Z}_+ \to \{1, \ldots, \upsilon_2\}$, i.e., $\mathbb{F}_1^{\breve{o}(k)} \in \hat{\mathbb{F}}_1$ and $\mathbb{F}_2^{\acute{o}(k)} \in \hat{\mathbb{F}}_2$.

Since the information of relative partial state measurements can be used, then the control strategy can be designed as follows:

$$\begin{cases} u_i(t,k) \; is \; any \; initial \; input, \; k = 1, \\ u_i(t,k) = u_i(t,k-1) + \sum_{j=1}^{M} \acute{\alpha}_{i,j}^{\acute{o}(k)} P_1\big[z_j(t+1,k-1) - z_i(t+1,k-1)\big] \\ \qquad - q_i P_1 e_i(t+1,k-1), \;\; k > 1, \end{cases} \tag{4.4}$$

where $P_1 \in \mathbb{R}^{p\times n}$ denotes the learning gain matrix.

According to (4.3) and (4.4), it is not difficult to obtain

$$\begin{cases} e_i(t+1,k) = A\big[e_i(t,k) - e_i(t,k-1)\big] + \big(I_n - q_i B P_1\big)e_i(t+1,k-1) \\ \qquad + \sum_{j=1}^{M} \acute{\alpha}_{i,j}^{\acute{o}(k)} B P_1\big[z_j(t+1,k-1) - z_i(t+1,k-1)\big], \\ \hat{e}_i(t+1,k) = A\hat{e}_i(t,k) - \sum_{j=1}^{M} \breve{\alpha}_{i,j}^{\breve{o}(k)} F\big[\varepsilon_j(t+1,k-1) - \varepsilon_i(t+1,k-1)\big]. \end{cases} \tag{4.5}$$

Note that $z_j(t+1,k-1) - z_i(t+1,k-1) = -[\hat{e}_j(t+1,k-1) - \hat{e}_i(t+1,k-1)] + [e_j(t+1,k-1) - e_i(t+1,k-1)]$, and $\varepsilon_i(t,k) = -C\hat{e}_i(t,k)$. Denote $\mathbf{e}(t,k) = (e_1(t,k)^{\mathrm{T}}, \ldots, e_M(t,k)^{\mathrm{T}})^{\mathrm{T}}$ and $\hat{\mathbf{e}}(t,k) = (\hat{e}_1(t,k)^{\mathrm{T}}, \ldots, \hat{e}_M(t,k)^{\mathrm{T}})^{\mathrm{T}}$. Then one obtains

$$\begin{aligned} \mathbf{e}(t+1,k) &= \big(I_M \otimes A\big)\big[\mathbf{e}(t,k) - \mathbf{e}(t,k-1)\big] \\ &\quad + \big[I_M \otimes I_n - (Q_M + \mathcal{L}^{\acute{o}(k)}) \otimes BP_1\big]\mathbf{e}(t+1,k-1) \\ &\quad + \big(\mathcal{L}^{\acute{o}(k)} \otimes BP_1\big)\hat{\mathbf{e}}(t+1,k-1), \\ \hat{\mathbf{e}}(t+1,k) &= I_M \otimes A\hat{\mathbf{e}}(t,k) - \mathcal{L}^{\breve{o}(k)} \otimes FC\hat{\mathbf{e}}(t+1,k-1), \end{aligned}$$

where $Q_M = \mathrm{diag}(q_1, \ldots, q_M)$. From the above inequalities, it is easy to get

$$\tilde{\mathbf{e}}(t+1,k) = \Omega_1 \tilde{\mathbf{e}}(t,k) + \Omega_2 \tilde{\mathbf{e}}(t,k-1) + \Omega_k^{\check{o},\acute{o}} \tilde{\mathbf{e}}(t+1,k-1), \tag{4.6}$$

where $k \geq 2$, $\theta \geq 0$, and $\tilde{\mathbf{e}}(t,k) = (\mathbf{e}(t,k)^{\mathrm{T}}, \hat{\mathbf{e}}(t,k)^{\mathrm{T}})^{\mathrm{T}}$ denotes the integrated error. In addition, the matrices

$$\Omega_1 = \begin{pmatrix} I_M \otimes A & O \\ O & I_M \otimes A \end{pmatrix}, \Omega_2 = \begin{pmatrix} -I_M \otimes A & O \\ O & O \end{pmatrix},$$
$$\Omega_k^{\check{o},\acute{o}} = \begin{pmatrix} I_M \otimes I_n - (Q_M + \mathcal{L}^{\acute{o}(k)}) \otimes BP_1 & \mathcal{L}^{\acute{o}(k)} \otimes BP_1 \\ O & -\mathcal{L}^{\check{o}(k)} \otimes FC \end{pmatrix}.$$

4.3.2 Initial State Learning Scheme

It is worth mentioning that initial state drift is an important issue, and the inconsistency of the initial state leads to robustness. Then, different schemes are proposed to solve this problem, such as the initial state learning law in Chen et al. (1999), Chi et al. (2015), Wei and Li (2017). Inspired by the above literatures, one can design an initial state learning strategy as follows:

$$\begin{cases} x_i(0,k) = x_i(0,k-1) - q_i P_2 e_i(0,k-1) \\ \qquad\qquad + \sum_{j=1}^{M} \acute{\alpha}_{i,j}^{\acute{o}(k)} P_2 \big[x_j(0,k-1) - x_i(0,k-1)\big], \\ z_i(0,k) = z_i(0,k-1) + q_i P_2 \hat{e}_i(0,k-1) \\ \qquad\qquad + \sum_{j=1}^{M} \acute{\alpha}_{i,j}^{\acute{o}(k)} P_2 \big[z_j(0,k-1) - z_i(0,k-1)\big], \end{cases} \tag{4.7}$$

where $P_2 \in \mathbb{R}^{n \times n}$ is the initial learning matrix. Then,

$$\begin{cases} e_i(0,k) = (I_n - q_i P_2) e_i(0,k-1) \\ \qquad\qquad + \sum_{j=1}^{M} \acute{\alpha}_{i,j}^{\acute{o}(k)} P_2 \big[e_j(0,k-1) - e_i(0,k-1)\big], \\ \hat{e}_i(0,k) = (I_n - q_i P_2) \hat{e}_i(0,k-1) - q_i P_2 e_i(0,k-1) \\ \qquad\qquad + \sum_{j=1}^{M} \acute{\alpha}_{i,j}^{\acute{o}(k)} P_2 \big[\hat{e}_j(0,k-1) - \hat{e}_i(0,k-1)\big]. \end{cases} \tag{4.8}$$

It is not hard to get

$$\mathbf{e}(0,k) = \big[I_M \otimes I_n - (Q_M + \mathcal{L}^{\acute{o}(k)}) \otimes P_2\big] \mathbf{e}(0,k-1),$$
$$\hat{\mathbf{e}}(0,k) = \big[I_M \otimes I_n - (Q_M + \mathbb{L}^{\acute{o}(k)}) \otimes P_2\big] \hat{\mathbf{e}}(0,k-1) - (Q_M \otimes P_2) \mathbf{e}(0,k-1). \tag{4.9}$$

According to (4.9), we obtain the following equation

$$\tilde{\mathbf{e}}(0,k)=\Phi_k^{\acute{o}}\tilde{\mathbf{e}}(0,k-1),$$

where the matrix

$$\Phi_r^{\acute{o}}=\begin{pmatrix} I_M\otimes I_n-(Q_M+\mathcal{L}^{\acute{o}(k)})\otimes P_2 & O \\ -Q_M\otimes P_2 & I_M\otimes I_n-(Q_M+\mathcal{L}^{\acute{o}(k)})\otimes P_2 \end{pmatrix}. \quad (4.10)$$

4.4 Convergence Analysis

In this section, the main results are presented. First, the following hypothesis is necessary to derive the main results.

Assumption 4.1 For $\check{o}(k)\in\{1,\ldots,v_1\}$ and $\acute{o}(k)\in\{1,=\cdots,v_2\}$, both $\mathbb{F}_1^{\check{o}(k)}$ and $\mathbb{F}_2^{\acute{o}(k)}$ contain a directed spanning tree, which takes the desired state (virtual leader) as the root node.

Theorem 4.1 *Suppose Assumption 4.1 holds. By using learning control strategies* (4.4) *and* (4.7), *the integrated error* $\tilde{\mathbf{e}}(t,k)$ *of the discrete-time error system* (4.5) *can asymptotically converge to zero with independent switching topologies, i.e.,* $\lim_{k\to\infty}\tilde{\mathbf{e}}(t,k)=\mathbf{0},\ \forall\, t\in\mathcal{S}$ *if there exist an observation matrix* F*, a learning gain matrix* P_1 *and an initial learning gains matrix* P_2*, such that*

$$\varrho_1=\max_{\check{o}(k),\acute{o}(k)}\left\{\left\|\Omega_k^{\check{o},\acute{o}}\right\|\right\}<1,\ \ \varrho_2=\max_{\acute{o}(k)}\left\{\left\|\Phi_k^{\acute{o}}\right\|\right\}<1.$$

Proof According to (4.6) and (4.10), by calculation, one has

$$\begin{aligned}
\tilde{\mathbf{e}}(0,k)&=\Phi_r^{\acute{o}}\tilde{\mathbf{e}}(0,k-1),\\
\tilde{\mathbf{e}}(1,k)&=\left[\Omega_1\Phi_k^{\acute{o}}+\Omega_2\right]\tilde{\mathbf{e}}(0,k-1)+\Omega_k^{\check{o},\acute{o}}\tilde{\mathbf{e}}(1,k-1),\\
\tilde{\mathbf{e}}(2,k)&=\left[\Omega_1^2\Phi_k^{\acute{o}}+\Omega_1\Omega_2\right]\tilde{\mathbf{e}}(0,k-1)+\left[\Omega_1\Omega_k^{\check{o},\acute{o}}+\Omega_2\right]\tilde{\mathbf{e}}(1,k-1)\\
&\quad+\Omega_k^{\check{o},\acute{o}}\tilde{\mathbf{e}}(2,k-1),\\
\tilde{\mathbf{e}}(3,k)&=\left[\Omega_1^3\Phi_k^{\acute{o}}+\Omega_1^2\Omega_2\right]\tilde{\mathbf{e}}(0,k-1)+\left[\Omega_1^2\Omega_k^{\check{o},\acute{o}}+\Omega_1\Omega_2\right]\tilde{\mathbf{e}}(1,k-1)\\
&\quad+\left[\Omega_1\Omega_k^{\check{o},\acute{o}}+\Omega_2\right]\tilde{\mathbf{e}}(2,k-1)+\Omega_k^{\check{o},\acute{o}}\tilde{\mathbf{e}}(3,k-1),\\
&\quad\cdots\cdots\cdots\cdots
\end{aligned}$$

$$\tilde{\mathbf{e}}(t+1,k) = \Big[\Omega_1^{t+1}\Phi_k^{\acute{o}} + \Omega_1^t\Omega_2\Big]\tilde{\mathbf{e}}(0,k-1) + \sum_{l=0}^{t-1}\Big[\Omega_1^{t-l}\Omega_k^{\check{o},\acute{o}} + \Omega_1^{t-1-l}\Omega_2\Big]\tilde{\mathbf{e}}(l+1,k)$$
$$+\Omega_k^{\check{o},\acute{o}}\tilde{\mathbf{e}}(t+1,k-1),\ t \geq 1.$$

Then, it is not difficult to get

$$\underbrace{\begin{pmatrix} \tilde{\mathbf{e}}(0,k) \\ \tilde{\mathbf{e}}(1,k) \\ \tilde{\mathbf{e}}(2,k) \\ \tilde{\mathbf{e}}(3,k) \\ \vdots \\ \tilde{\mathbf{e}}(T,k) \end{pmatrix}}_{\mathbf{H}(k)} = \underbrace{\begin{pmatrix} \Phi_k^{\acute{o}} & O & O & \cdots & O \\ \Omega_1\Phi_k^{\acute{o}} + \Omega_2 & \Omega_k^{\check{o},\acute{o}} & O & \cdots & O \\ \Omega_1(\Omega_1\Phi_k^{\acute{o}} + \Omega_2) & \Omega_1\Omega_k^{\check{o},\acute{o}} + \Omega_2 & \Omega_k^{\check{o},\acute{o}} & \cdots & O \\ \vdots & \vdots & \vdots & \cdots & \vdots \\ \Omega_1^{T-1}(\Omega_1\Phi_k^{\acute{o}} + \Omega_2) & \Omega_1^{T-2}(\Omega_1\Omega_k^{\check{o},\acute{o}} + \Omega_2) & \cdots & \cdots & \Omega_k^{\check{o},\acute{o}} \end{pmatrix}}_{\Lambda(\Phi_k^{\acute{o}},\Omega_k^{\check{o},\acute{o}})}\mathbf{H}(k-1).$$

From the above analysis, one obtains that $\mathbf{H}(k) = \prod_{\kappa=0}^{k-1}\Lambda(\Phi_{k-\kappa}^{\acute{o}}, \Omega_{k-\kappa}^{\check{o},\acute{o}})\mathbf{H}(1)$. Furthermore,

$$\prod_{\kappa=0}^{k-1}\Lambda(\Phi_{k-\kappa}^{\acute{o}}, \Omega_{k-\kappa}^{\check{o},\acute{o}}) = \begin{pmatrix} \prod_{\kappa=0}^{k-1}\Phi_{k-\kappa}^{\acute{o}} & O & \cdots & O \\ \star & \prod_{\kappa=0}^{k-1}\Omega_{k-\kappa}^{\check{o},\acute{o}} & \cdots & O \\ \vdots & \vdots & \cdots & \vdots \\ \star & \star & \cdots & \prod_{\kappa=0}^{k-1}\Omega_{k-\kappa}^{\check{o},\acute{o}} \end{pmatrix}, \tag{4.11}$$

where $\star$ denotes the block matrix.

Next, it needs to prove that $\prod_{\kappa=0}^{k-1}\Lambda(\Phi_{k-\kappa}^{\acute{o}}, \Omega_{k-\kappa}^{\check{o},\acute{o}})$ converge to the zero matrix. That is to say, one needs to prove that every block matrix tends to the zero matrix. In fact, taking norm on each block matrix, it is not hard to get that

$$\left\|\prod_{\kappa=0}^{k-1}\Phi_{k-\kappa}^{\acute{o}}\right\| \leq \prod_{\kappa=0}^{k-1}\left\|\Phi_{k-\kappa}^{\acute{o}}\right\|,\quad \left\|\prod_{\kappa=0}^{k-1}\Omega_{k-\kappa}^{\check{o},\acute{o}}\right\| \leq \prod_{\kappa=0}^{k-1}\left\|\Omega_{k-\kappa}^{\check{o},\acute{o}}\right\|. \tag{4.12}$$

According to Theorem 4.1, one knows $\lim_{k\to\infty}\prod_{\kappa=0}^{k-1}\left\|\Phi_{k-\kappa}^{\acute{o}}\right\| = 0$ and $\lim_{k\to\infty}\prod_{\kappa=0}^{k-1}\left\|\Omega_{k-\kappa}^{\check{o},\acute{o}}\right\| = 0$. By calculation, it is not difficult to get that the norm of each block matrix "$\star$" converges to zero as $r \to \infty$. Thus, one can know that each block matrix of $\prod_{\kappa=0}^{k-1} \Lambda(\Phi_{k-\kappa}^{\acute{o}}, \Omega_{k-\kappa}^{\check{o},\acute{o}})$ converges to the zero matrix as $k \to \infty$. And it also implies that $\lim_{k\to\infty} \mathbf{H}(k) = \mathbf{0}$, then it is obvious that $\lim_{k\to\infty} \tilde{\mathbf{e}}(t,k) = \mathbf{0}$, $t \in \mathcal{S}$. □

Corollary 4.1 *Suppose Assumption* 4.1 *and* $\tilde{\mathbf{e}}(0,k) \equiv \mathbf{0}$ *are satisfied. By using learning control strategy* (4.4), *then the integrated error* $\tilde{\mathbf{e}}(t,k)$ *of the discrete-time system can converge to zero asymptotically in the sense of independent switching topologies, i.e.,* $\lim_{k\to\infty} \tilde{\mathbf{e}}(t,k) = \mathbf{0}$, $\forall\, t \in \mathcal{S}$ *if there exist an observation matrix* F *and a learning gain* P_1 *such that*

$$\varrho_1 = \max_{\check{o}(k),\acute{o}(k)} \left\{ \|\Omega_k^{\check{o},\acute{o}}\| \right\} < 1.$$

Proof In this situation, one knows $\Phi_k^{\acute{o}} = I$ and $\tilde{\mathbf{e}}(0,k) = \mathbf{0}$, it is not difficult to get

$$\underbrace{\begin{pmatrix} \tilde{\mathbf{e}}(1,k) \\ \tilde{\mathbf{e}}(2,k) \\ \tilde{\mathbf{e}}(3,k) \\ \vdots \\ \tilde{\mathbf{e}}(T,k) \end{pmatrix}}_{\tilde{\mathbf{H}}(k)} = \underbrace{\begin{pmatrix} \Omega_k^{\check{o},\acute{o}} & O & \cdots & O \\ \Omega_1\Omega_k^{\check{o},\acute{o}} + \Omega_2 & \Omega_r^{\check{o},\acute{o}} & \cdots & O \\ \vdots & \vdots & \cdots & \vdots \\ \Omega_1^{T-2}(\Omega_1\Omega_k^{\check{o},\acute{o}} + \Omega_2) & \cdots & \cdots & \Omega_k^{\check{o},\acute{o}} \end{pmatrix}}_{\tilde{\Lambda}(\Omega_k^{\check{o},\acute{o}})} \tilde{\mathbf{H}}(k-1).$$

Since $\tilde{\mathbf{e}}(0,k) \equiv \mathbf{0}$, it means that the first column of $\Lambda(\Phi_k^{\acute{o}}, \Omega_k^{\check{o},\acute{o}})$ does not work on the iterative correction of $\tilde{\mathbf{e}}(0,k)$. Thus, $\Lambda(\Phi_k^{\acute{o}}, \Omega_k^{\check{o},\acute{o}})$ becomes $\tilde{\Lambda}(\Omega_k^{\check{o},\acute{o}})$. Similar as the proof of Theorem 4.1, one can immediately obtain the result. □

Corollary 4.2 *Suppose Assumption* 4.1 *hold, and there are no switching topologies. By using learning control strategies* (4.4) *and* (4.7), *the integrated error* $\tilde{\mathbf{e}}(t,k)$ *can asymptotically converge to zero , i.e.,* $\lim_{k\to\infty} \tilde{\mathbf{e}}(t,k) = \mathbf{0}$, $\forall\, t \in \mathcal{S}$ *if there exist an observation matrix* F, *a learning gain matrix* P_1, *and the initial learning gains matrix* P_2 *satisfying the following inequalities*

$$\begin{aligned} \rho\big(I_M \otimes I_n - (Q_M + \mathcal{L}) \otimes BP_1\big) &< 1, \\ \rho\big(-\mathcal{L} \otimes FC\big) &< 1, \\ \rho\big(I_M \otimes I_n - (Q_M + \mathcal{L}) \otimes P_2\big) &< 1. \end{aligned}$$

Proof If there are no switching topologies, Corollary 4.2 is a special case of Theorem 4.1. Similar to the proof of Theorem 4.1, $\Lambda(\Phi_k^{\acute{o}}, \Omega_k^{\check{o},\acute{o}})$ becomes

$$\Lambda(\Phi,\Omega) = \begin{pmatrix} \Phi & O & \cdots O \\ \Omega_1\Phi + \Omega_2 & \Omega & \cdots O \\ \vdots & \vdots & \cdots \vdots \\ \Omega_1^{T-1}(\Omega_1\Phi + \Omega_2) & \Omega_1^{T-2}(\Omega_1\Omega + \Omega_2) & \cdots\ \Omega \end{pmatrix},$$

where the matrices

$$\Omega = \begin{pmatrix} I_M \otimes I_n - (Q_M + \mathcal{L}) \otimes BP_1 & \mathcal{L} \otimes BP_1 \\ O & -\mathcal{L} \otimes FC \end{pmatrix},$$

and

$$\Phi = \begin{pmatrix} I_M \otimes I_n - (Q_M + \mathcal{L}) \otimes P_2 & O \\ -Q_M \otimes P_2 & I_M \otimes I_n - (Q_M + \mathcal{L}) \otimes P_2 \end{pmatrix}.$$

From the knowledge of matrix analysis, only $\varrho(\Lambda(\Phi, \Omega)) < 1$ is required to ensure the tracking error convergence. In line with the definition of spectral radius, $\varrho(\Lambda(\Phi, \Omega)) < 1$ is equivalent to $\varrho(I_M \otimes I_n - (Q_M + \mathcal{L}) \otimes BP_1) < 1$, $\varrho(-\mathcal{L} \otimes FC) < 1$, and $\varrho(I_M \otimes I_n - (Q_M + \mathcal{L}) \otimes P_2) < 1$. Then, one obtains the result. □

Remark 4.1 Compared with Chen et al. (2010), the advantage is that switching signals between nodes are considered in this chapter. Moreover, control strategies are designed according to distributed observers and partial state measurement. In comparison with Meng et al. (2015), the difference is that our learning control strategies are designed in line with distributed observation and two independent switching topologies. Moreover, our controller is designed to track the desired state instead of the leader's output trajectories.

Remark 4.2 Compared with Wen et al. (2017), Wan et al. (2017), our control strategies are designed according to distributed observations and learning strategies, then full tracking can be achieved under the independent switching topologies. Moreover, there is no limitation about desired states in this chapter, meaning learning control strategies are applicable. Compared with Li et al. (2017), the control strategy in this chapter is not adaptive, and it is based on distributed observers and switching topologies, which may be appropriate for the complex situation.

Remark 4.3 Compared with Chen et al. (1999), the initial state learning strategy is designed according to the distributed observers in this paper. Moreover, the hypothesis about the directed graph is relatively weak. Our controller is designed without activation functions and neural network weights, and it is more convenient in real-life applications.

Corollary 4.3 *Suppose Assumption* 4.1 *hold, and the initial strategy* (4.7) *is not utilized, i.e.,* $\tilde{\mathbf{e}}(0, k) = \tilde{\mathbf{e}}(0, k-1)$. *By using the ILC strategy* (4.4), *the integrated error* $\tilde{\mathbf{e}}(t, k)$ *of the discrete-time system is bounded if there exists an observation matrix* F *and a learning gain matrix* P_1 *make matrices such that* $\varrho_1 = \max_{\check{o}(k), \acute{o}(k)} \left\{ \left\| \Omega_k^{\check{o}, \acute{o}} \right\| \right\} < 1$.

Proof From Theorem 4.1, according to (4.6) and $\tilde{\mathbf{e}}(0, k) = \tilde{\mathbf{e}}(0, k-1)$, it is not hard to know

$$\mathbf{H}(k) = \begin{pmatrix} I & O & O & \cdots & O \\ \Omega_1 + \Omega_2 & \Omega_k^{\check{o},\acute{o}} & O & \cdots & O \\ \Omega_1(\Omega_1 + \Omega_2) & \Omega_1 \Omega_k^{\check{o},\acute{o}} + \Omega_2 & \Omega_k^{\check{o},\acute{o}} & \cdots & O \\ \vdots & \vdots & \vdots & \cdots & \vdots \\ \Omega_1^{T-1}(\Omega_1 + \Omega_2) & \Omega_1^{T-2}(\Omega_1 \Omega_k^{\check{o},\acute{o}} + \Omega_2) & \cdots & \cdots & \Omega_k^{\check{o},\acute{o}} \end{pmatrix} \mathbf{H}(k-1). \quad (4.13)$$

Similar to the proof of Theorem 4.1, it is easy to know $\mathbf{H}(k) = \prod_{l=0}^{k-1} \Lambda(I, \Omega_{k-\kappa}^{\check{o},\acute{o}})\mathbf{H}(1)$. By matrix multiplication and induction, then the Eq. (4.13) becomes

$$\mathbf{H}(k) = \begin{pmatrix} I & O & \cdots & O \\ * & \prod_{\kappa=0}^{r} \Omega_{r-\kappa}^{\check{o},\acute{o}} & \cdots & O \\ \vdots & \vdots & \cdots & \vdots \\ * & * & \cdots & \prod_{\kappa=0}^{k-1} \Omega_{k-\kappa}^{\check{o},\acute{o}} \end{pmatrix} \mathbf{H}(1). \quad (4.14)$$

From the condition of Theorem 4.3, one can derive $\prod_{\kappa=0}^{k} \Omega_{k-\kappa}^{\check{o},\acute{o}} \to O$ as $k \to \infty$. Meanwhile, since

$$\left\| \sum_{l=1}^{k-1} \left(\prod_{\kappa=0}^{l-1} \Omega_{k-\kappa}^{\check{o},\acute{o}} \right) \right\| \leq \sum_{l=1}^{r-1} \prod_{\kappa=0}^{l-1} \left\| \Omega_{r-\kappa}^{\check{o},\acute{o}} \right\| \leq \left[\varrho_1 + \varrho_1^2 + \cdots + \varrho_1^{k-1} \right] \leq \frac{\varrho_1}{1 - \varrho_1},$$

one knows that $\sum_{l=1}^{k-1} \left(\prod_{\kappa=0}^{l-1} \Omega_{k-\kappa}^{\check{o},\acute{o}} \right)$ is bounded as $r \to \infty$. Thus, $\left[I + \sum_{l=1}^{k-1} (\prod_{\kappa=0}^{l-1} \Omega_{k-\kappa}^{\check{o},\acute{o}}) \right] [\Omega_1 + \Omega_2]$ is also bounded. For the same reason, it can also be derived that each matrix "$*$" is bounded by calculation. According to the above analysis, the result is obtained immediately. □

Remark 4.4 From Theorem 4.3, one knows that the iterative learning strategy (4.4) cannot guarantee the integrated error $\tilde{\mathbf{e}}(t, k)$ converges to zero. Therefore, it is necessary to design initial state learning strategies. In addition, the convergence of $\tilde{\mathbf{e}}(t, k)$ is also related to $\Phi_k^{\acute{o}}$ according to Theorem 4.1, which means that the initial state learning strategy plays an important role in the convergence analysis.

Corollary 4.4 *It is assumed that the two kinds of switching signals are periodic, i.e.,* $\check{o}(k) = \check{o}(k + \delta_1)$, $\delta_1 \geq v_1$ *and* $\acute{o}(k) = \acute{o}(k + \delta_2)$, $\delta_2 \geq v_2$. *Meanwhile, suppose Assumption* 4.1 *holds, and the initial strategy* (4.7) *is not utilized, i.e.,* $\tilde{\mathbf{e}}(0, k) = \tilde{\mathbf{e}}_(0, k - 1)$. *By using the ILC strategy* (4.4), *the integrated error* $\tilde{\mathbf{e}}(t, k)$ *of the discrete-time system is bounded if there exist an observation matrix* F *and a learning gain matrix* P_1 *satisfy the following inequalities*

$$\rho\bigg(\prod_{l=1}^{\delta_2}\Big[I_M \otimes I_n - (Q_M + \mathcal{L}^{\acute{o}(l)}) \otimes BP_1\Big]\bigg) < 1,$$

$$\rho\bigg(\prod_{l=1}^{\delta_1}\Big[- \mathcal{L}^{\check{o}(l)} \otimes FC\Big]\bigg) < 1.$$

Proof In this situation, for $\theta \in \mathcal{S} \setminus \{0\}$, it is not hard to know that (4.13) becomes

$$\tilde{\mathbf{H}}(k) = \tilde{\Lambda}(\Omega_k^{\check{o},\acute{o}})\tilde{\mathbf{H}}(k-1) + \underbrace{\begin{pmatrix} \Omega_1+\Omega_2 & O & \cdots & O \\ O & \Omega_1(\Omega_1+\Omega_2) & \cdots & O \\ \vdots & \vdots & \cdots & \vdots \\ O & O & \cdots & \Omega_1^{T-1}(\Omega_1+\Omega_2) \end{pmatrix}}_{\tilde{\Lambda}(\Omega_1,\Omega_2)} \begin{pmatrix} \tilde{\mathbf{e}}(0,k) \\ \tilde{\mathbf{e}}(0,k) \\ \vdots \\ \tilde{\mathbf{e}}(0,k) \end{pmatrix},$$

where $\tilde{\Lambda}(\Omega_k^{\check{o},\acute{o}})$ is the same as that in Corollary 4.1. According to above recurrence relation, it is not difficult to get

$$\begin{aligned}\tilde{\mathbf{H}}(k) &= \tilde{\Lambda}(\Omega_k^{\check{o},\acute{o}})\tilde{\mathbf{H}}(k-1) + \tilde{\Lambda}(\Omega_1,\Omega_2)\big(\mathbf{1}\otimes\tilde{\mathbf{e}}(0,k)\big) \\ &= \tilde{\Lambda}(\Omega_k^{\check{o},\acute{o}})\tilde{\Lambda}(\Omega_{k-1}^{\check{o},\acute{o}})\tilde{\mathbf{H}}(k-2) + \tilde{\Lambda}(\Omega_k^{\check{o},\acute{o}})\tilde{\Lambda}(\Omega_1,\Omega_2)\big(\mathbf{1}\otimes\tilde{\mathbf{e}}(0,k-1)\big) \\ &\quad +\tilde{\Lambda}(\Omega_1,\Omega_2)\big(\mathbf{1}\otimes\tilde{\mathbf{e}}(0,k)\big) \\ &= \cdots\cdots\end{aligned}$$

Note $\tilde{\mathbf{e}}(0,k) = \tilde{\mathbf{e}}(0,k-1)$ and using induction, the above equality becomes

$$\begin{aligned}\tilde{\mathbf{H}}(k) &= \prod_{\kappa=0}^{k-2}\tilde{\Lambda}(\Omega_{k-\kappa}^{\check{o},\acute{o}})\tilde{\mathbf{H}}(1) + \sum_{\kappa=0}^{k-3}\bigg(\prod_{l=0}^{\kappa}\tilde{\Lambda}(\Omega_{k-l}^{\check{o},\acute{o}})\bigg)\tilde{\Lambda}(\Omega_1,\Omega_2)(\mathbf{1}\otimes\tilde{\mathbf{e}}(0,k)) \\ &\quad +\tilde{\Lambda}(\Omega_1,\Omega_2)\big(\mathbf{1}\otimes\tilde{\mathbf{e}}(0,k)\big).\end{aligned} \tag{4.15}$$

Further, one can get that $\prod_{\kappa=0}^{k-2}\tilde{\Lambda}(\Omega_{k-\kappa}^{\check{o},\acute{o}})\tilde{\mathbf{H}}(1) = \mathbf{0}$ as $k \to \infty$. Note $\tilde{\Lambda}(\Omega_k^{\check{o},\acute{o}}) = \tilde{\Lambda}(\Omega_{k-\lambda}^{\check{o},\acute{o}})$, where $\lambda = [\delta_1, \delta_2]$ denotes the least common multiple of δ_1 and δ_2. Then, for a sufficient large k,

$$\begin{aligned}&\sum_{\kappa=0}^{r-3}\bigg(\prod_{l=0}^{\kappa}\tilde{\Lambda}(\Omega_{r-l}^{\check{o},\acute{o}})\bigg) \\ &= \bigg(\tilde{\Lambda}(\Omega_r^{\check{o},\acute{o}}) + \cdots + \prod_{l=0}^{\lambda-1}\tilde{\Lambda}(\Omega_{r-l}^{\check{o},\acute{o}})\bigg) + W_\lambda\bigg(\tilde{\Lambda}(\Omega_r^{\check{o},\acute{o}}) + \cdots + \prod_{l=0}^{\lambda-1}\tilde{\Lambda}(\Omega_{r-l}^{\check{o},\acute{o}})\bigg) \\ &\quad +W_\lambda^2\bigg(\tilde{\Lambda}(\Omega_r^{\check{o},\acute{o}}) + \cdots + \prod_{l=0}^{\lambda-1}\tilde{\Lambda}(\Omega_{r-l}^{\check{o},\acute{o}})\bigg) + \cdots\end{aligned}$$

$$= \sum_{p=0}^{(r-3)/\lambda} W_\lambda^p \left(\tilde{\Lambda}(\Omega_r^{\breve{o},\acute{o}}) + \tilde{\Lambda}(\Omega_r^{\breve{o},\acute{o}})\tilde{\Lambda}(\Omega_{r-1}^{\breve{o},\acute{o}}) + \cdots + \prod_{l=0}^{\lambda-1} \tilde{\Lambda}(\Omega_{r-l}^{\breve{o},\acute{o}}) \right),$$

where $W_\lambda = \prod_{l=0}^{\lambda-1} \tilde{\Lambda}(\Omega_{k-l}^{\breve{o},\acute{o}})$. Note $\rho(W_\lambda) = \rho\Big(\prod_{l=0}^{\lambda-1} \tilde{\Lambda}(\Omega_{k-l}^{\breve{o},\acute{o}})\Big) = \rho\Big(\prod_{l=0}^{\lambda-1} \Omega_{k-l}^{\breve{o},\acute{o}}\Big) < 1$, and $\sum_{p=0}^{(r-3)/\lambda} W_\lambda^p$ is absolute convergence. As a result, $\sum_{\kappa=0}^{k-3} \Big(\prod_{l=0}^{\kappa} \tilde{\Lambda}(\Omega_{k-l}^{\breve{o},\acute{o}})\Big)$ is absolute convergence. According to $\tilde{\mathbf{e}}(0, k) = \tilde{\mathbf{e}}(0, k-1)$, one can know (4.15) is bounded as $k \to \infty$. □

Remark 4.5 From Corollary 4.4, the sufficient condition is weaker than Corollary 4.3. However, the result requires more harsh conditions for switching topologies. Analogously, the sufficient condition of Corollary 4.2 is weaker than Theorem 4.1, but the conclusion is obtained without switching topologies. This condition shows malicious attacks bring undesirable impacts on the system. It is meaningful to study the tracking problem under switching topologies.

4.5 Simulation Example

In this section, a numerical example is presented to demonstrate the validity of the results in this chapter.

Example 4.1 Considering system (4.1), the observer state of node i is described by (4.2) and $u_i(t, k)$ is designed as (4.4). Particularly, set $\mathcal{S} = \{0, 1, \ldots, 30\}$, $i \in \{1, \ldots, 6\}$, i.e., $M = 6$. The matrices are as follows:

$$A = \begin{pmatrix} 0.1 & 1 \\ -0.8 & 0.2 \end{pmatrix}, \quad B = \begin{pmatrix} 0.1 & 0.05 \\ 0.05 & 0.2 \end{pmatrix}, \quad C = \begin{pmatrix} -0.1, & 0.5 \end{pmatrix},$$

and the desired state $x^*(t) = \big(\cos(0.2\pi t) - 0.5, \ \sin(0.2\pi t) - 0.5\big)^{\mathrm{T}}$.

In Fig. 4.1, the weight values of edges are 1 if $(j, i) \in \mathcal{E}$ but 0 otherwise. The red lines represent the communication links among all the followers. The blue lines represent the communication links from the virtual node to the followers. In addition, $\hat{\mathbb{F}}_1 = \{\hat{\mathbb{F}}_1^1, \hat{\mathbb{F}}_1^2\}$ and $\hat{\mathbb{F}}_2 = \{\hat{\mathbb{F}}_2^1, \hat{\mathbb{F}}_2^2\}$. Specifically, $\breve{o}(k) = 1$ when $k = 2n - 1$, and $\breve{o}(k) = 2$ when $k = 2n$. Similarly, $\acute{o}(k) = 1$ when $k = 2n - 1$, and $\acute{o}(k) = 2$ when $k = 2n$. Thus, $\delta_1 = \delta_2 = 2$. In order to describe the error convergence conveniently, the total observation error and total tracking error at the kth iteration are defined as $\sum_{t=0}^{T} \|\hat{e}(t, k)\|_1$ and $\sum_{t=0}^{T} \|\mathbf{e}(t, k)\|_1$, respectively.

Case 1. An initial learning strategy is applied

ILC strategy (4.4) and initial learning strategy (4.7) are utilized. Specifically, the observation matrix and the learning gain matrices are as follows:

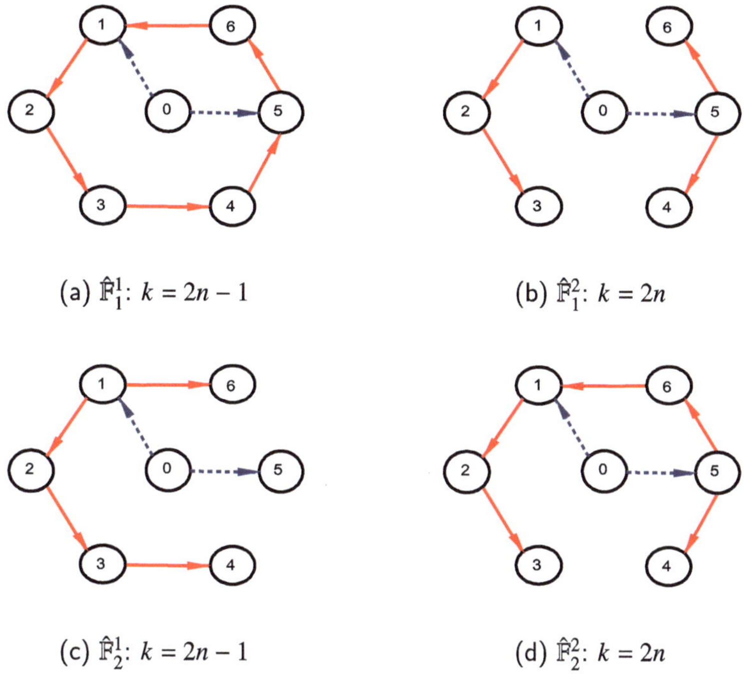

(a) $\hat{\mathbb{F}}_1^1$: $k = 2n - 1$ (b) $\hat{\mathbb{F}}_1^2$: $k = 2n$

(c) $\hat{\mathbb{F}}_2^1$: $k = 2n - 1$ (d) $\hat{\mathbb{F}}_2^2$: $k = 2n$

Fig. 4.1 Two kinds of switching topologies $\hat{\mathbb{F}}_1$ and $\hat{\mathbb{F}}_2$

$$F = \begin{pmatrix} 0.3 \\ -0.1 \end{pmatrix}, \quad P_1 = \begin{pmatrix} 0.5 & -0.1 \\ -0.1 & 0.5 \end{pmatrix}, \quad P_2 = \begin{pmatrix} 0.005 & 0 \\ 0 & 0.005 \end{pmatrix}.$$

Furthermore, let $u_i(t, 1) = \mathbf{0}$, $x_i(0, 1) = \mathbf{0}$ and $z_i(0, 1) = \mathbf{1}$. By calculation, one can get $\varrho_1 = 0.9951 < 1$, $\varrho_2 < 1$, $\rho\Big(\prod_{l=1}^{2}[I_6 \otimes I_2 - (Q_6 + \mathcal{L}^{\hat{\sigma}(l)}) \otimes BP_1]\Big) = 0.9411 < 1$, and $\rho\Big(\prod_{l=1}^{2}[-\mathcal{L}^{\check{\sigma}(l)} \otimes FC]\Big) = 0.0163 < 1$. This implies the conditions of the Theorem 4.1, Corollaries 4.3, and 4.4 hold. Figure 4.2 shows the nodes' state and observation state at the 80th and 400th iteration, respectively.

Figure 4.3a and b show the initial state learning of each node at each iteration. Figure 4.3c and d show the nodes' tracking and observation errors at each iteration. One could conclude that all nodes' states asymptotically converge to the desired states with the increasing of k. Compared with Li et al. (2017), ILC is designed with independent switching topologies suitable for complex situations. The structure of the controller is relatively simple. Compared with Xiong et al. (2016a), the initial state-leaning strategy is considered, and it helps to achieve full tracking.

Case 2. The initial learning strategy is not applied.
Only ILC strategy (4.4) is used. Considering the 6 linear dynamic nodes, the communications among all the nodes are depicted in Fig. 4.1. Observation matrix F and

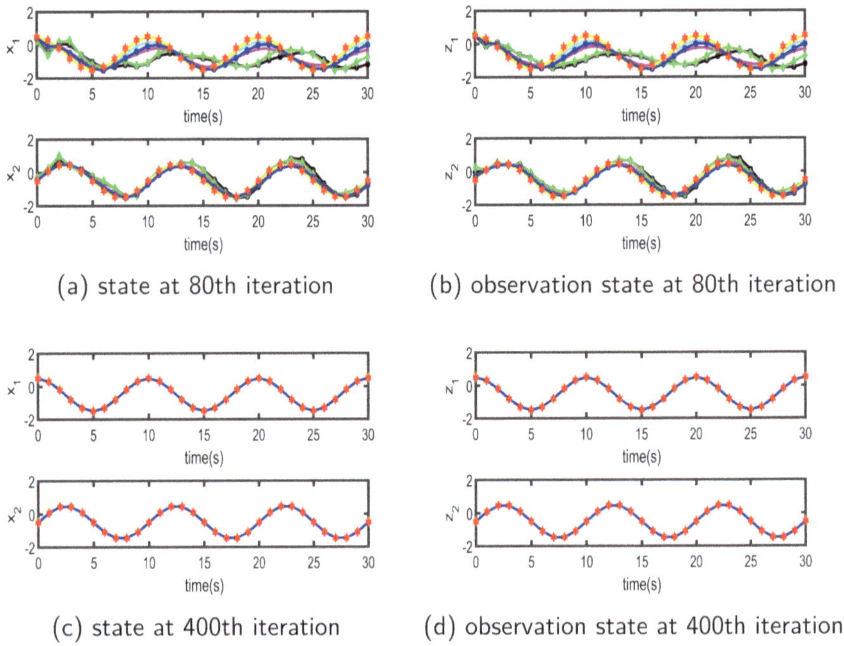

Fig. 4.2 State and observation state of each node at 50th and 150th iteration

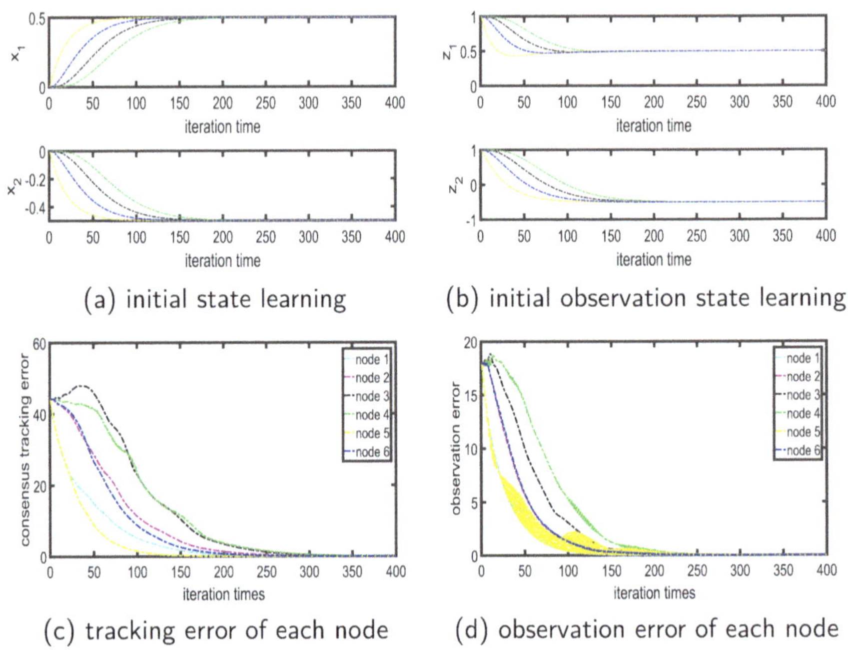

Fig. 4.3 Initial state learning, tracking error, and observation error of each node

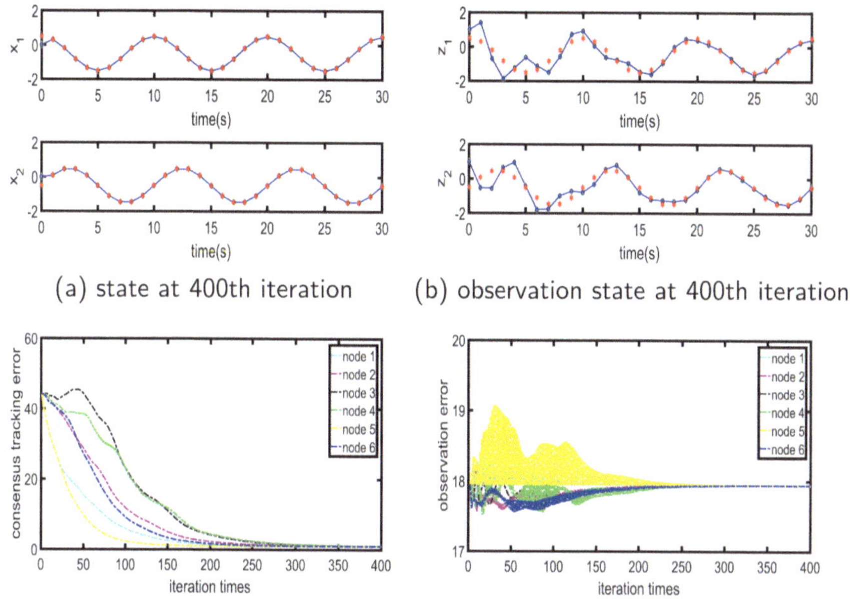

(a) state at 400th iteration (b) observation state at 400th iteration

(c) tracking error of each node (d) observation error of each node

Fig. 4.4 State and error of each node without initial state learning strategy

learning gain matrix P_1 are the same as in case 1. Similarly, one can know that the conditions of Corollaries 4.3 and 4.4 hold.

Figure 4.4a and b show the nodes' state and observation state at the 150th iteration. Figure 4.4c and d show nodes' tracking and observation errors without initial state learning strategy at each iteration. Then, one could conclude that the two types of errors are bounded with the iteration step k increasing. Particularly, observation error cannot be reduced with the increase of k. In addition, it is hard to achieve the consensus under two independent switching topologies because initial state drift exists. The general learning control scheme cannot be achieved. Thus, it is necessary to design initial state learning strategies.

Remark 4.6 In the example, a second-order multi-agent system is presented to illustrate the effectiveness of the theoretical results. The main results are the sufficient conditions for error convergence, which are not based on the Lyapunov method and linear matrix inequalities, and it is easy to check them. Further, the ILC strategy is still effective for high-dimensional multi-agent system, matrices F, P_1 and P_2 can be chosen to satisfy $\varrho_1 < 1$ and $\varrho_2 < 1$ according to Theorem 4.1.

4.6 Summary

In this chapter, the observer-based consensus problem for linear discrete-time systems has been discussed by utilizing the learning control schemes. Unlike traditional ILC schemes, our ILC schemes have been designed in light of distributed observers and two independent switching topologies. By using the control strategies, sufficient convergent conditions have been derived in the presence of two independent switching topologies. Also, it is interesting to know that the initial learning strategies have played a significant role in the theoretical analysis.

References

Adler J, Satapathy G, Manikonda V, Bowles B, Blue VJ (2005) A multi-agent approach to cooperative traffic management and route guidance. Transport Res Part B 39(4):297–318
Ai XL, Yu JQ, Jia ZY, Yang D, Xu X, Shen YC (2018) Disturbance observer-based consensus tracking for nonlinear multiagent systems with switching topologies. Inter J Robust Nonlin Contr 28(6):2144–2160
Bouakrif F, Boukhetala D, Boudjema F (2013) Velocity observer-based iterative learning control for robot manipulators. Inter J Syst Sci 44(2):214–222
Cao YC, Yu WW, Ren W, Chen GR (2013) An overview of recent progress in the study of distributed multi-agent coordination. IEEE Trans Ind Inform 9(1):427–438
Chen Y, Wen C, Gong Z, Sun M (1999) An iterative learning controller with initial state learning. IEEE Trans Automat Control 44(2):371–376
Chen WS, Li RH, Li J (2010) Observer-based adaptive iterative learning control for nonlinear systems with time-varying delays. Inter J Auto Comput 7(4):438–446
Chi RH, Hou ZS, Jin ST, Wang DW, Hao JE (2013) A data-driven iterative feedback tuning approach of ALINEA for freeway traffic ramp metering with PARAMICS simulations. IEEE Trans Ind Inform 9(4):2310–2317
Chi RH, Hou ZS, Jin ST (2015) A data-driven adaptive ILC for a class of nonlinear discrete-time systems with random initial states and iteration-varying target trajectory. J Franklin Ins 352(6):2407–2424
Dai JT, Guo Z (2017) Exponential consensus of non-linear multi-agent systems with semi-markov switching topologies. IET Contr Theory Appl 11(18):3363–3371
He WL, Chen GR, Han QL, Du WL, Cao JD (2017) Multiagent systems on multilayer networks: synchronization analysis and network design. IEEE Trans Syst Man Cyber Syst 47(7):1655–1667
He WL, Xu B, Han QL, Qian F (2020) Adaptive consensus control of linear multiagent systems with dynamic event-triggered strategies. IEEE Trans Cyber 50(7):2996–3008
Hosseini I, Khayatian A, Karimaghaee P, Fiacchini M, Navarro MAD (2019) LMI-based reset unknown input observer for state estimation of linear uncertain systems. IET Control Theory Appl 13(12):1872–1881
Huang DQ, Xu JX, Yang SP, Jin X (2015) Observer based repetitive learning control for a class of nonlinear systems with non-parametric uncertainties. Inter J Robust Nonlin Contr 25(8):1214–1229
Huang C, Wang W, Cao JD, Lu JQ (2018) Synchronization-based passivity of partially coupled neural networks with event-triggered communication. Neurocomputing 319(30):134–143
Huang C, Lu JQ, Ho DWC, Zhai GS, Cao JD (2020) Stabilization of probabilistic boolean networks via pinning control strategy. Inform Sci 510:205–217

Kwon C, Hwang I (2016) Cyber attack mitigation for cyber-physical systems: hybrid system approach to controller design. IET Control Theory Appl 10(7):731–741
Li JS, Liu SY, Li JM (2017) Observer-based distributed adaptive iterative learning control for linear multi-agent systems. Inter J Syst Sci 48(14):2948–2955
Liu CS, Wu J, Long CN, Wang YB (2017) Dynamic state recovery for cyber-physical systems under switching location attacks. IEEE Trans Contr Net Syst 4(1):14–22
Meng DY (2018) Dynamic distributed control for networks with cooperative-antagonistic interactions. IEEE Trans Automat Control 63(5):2311–2326
Meng DY, Moore KL (2016) Learning to cooperate: networks of formation agents with switching topologies. Automatica 64:278–293
Meng DY, Jia YM, Du JP (2015) Robust consensus tracking control for multiagent systems with initial state shifts, disturbances, and switching topologies. IEEE Trans Neur Net Lear Syst 26(4):809–824
Oh SK, Lee JM (2016) Iterative learning model predictive control for constrained multivariable control of batch processes. Comput Chem Eng 93:284–292
Su SZ, Lin ZL (2018) A multiple Lyapunov function approach to distributed synchronization control of multi-agent systems with switching directed communication topologies and unknown nonlinearities. IEEE Trans Contr Net Syst 5(1):23–33
Wan Y, Cao JD, Chen GR, Huang W (2017) Distributed observer-based cyber-security control of complex dynamical networks. IEEE Trans Circuits Syst I Reg Papers 64(11):2966–2975
Wei YS, Li XD (2017) Robust higher-order ILC for non-linear discrete-time systems with varying trail lengths and random initial state shifts. IET Control Theory Appl 11(15):2440–2447
Wen GG, Yu YG, Peng ZX, Rahmani A (2016) Consensus tracking for second-order nonlinear multi-agent systems with switching topologies and a time-varying reference state. Inter J Contr 89(10):2096–2106
Wen GH, Yu WW, Xia YQ, Yu XH, Hu JQ (2017) Distributed tracking of nonlinear multiagent systems under directed switching topology: an observer-based protocol. IEEE Trans Syst Man Cyber Syst 47(5):869–881
Xiong WJ, Yu XH, Patel R, Yu WW (2016a) Iterative learning control for discrete-time systems with event-triggered transmission strategy and quantization. Automatica 72:84–91
Xiong WJ, Yu XH, Chen Y, Gao J (2016b) Quantized iterative learning consensus tracking of digital networks with limited information communication. IEEE Trans Neur Net Lear Syst 28(6):1473–1480
Yang SP, Xu JX, Li XF (2016) Iterative learning control with input sharing for multi-agent consensus tracking. Syst Cont Lett 94:97–106

Chapter 5
Tracking Under Measurable and Unmeasurable State Information

5.1 Introduction

ILC is a useful control mechanism for enhancing dynamic systems' tracking performance by incorporating mistake knowledge into the control for successive iterations. Because of its simplicity and efficacy, ILC has been discovered to be a good alternative in various fields and applications, such as robotic manipulators (Tayebi 2004; Sun et al. 2006), hard disk drives (Wu and Tomizuka 2010), data-driven control (Chi et al. 2015, 2016), rapid thermal processing (Yang et al. 2003), and multi-agent systems (Meng et al. 2014, 2015; Li and Li 2014). For instance, in Chien and Yao (2004), an iterative learning controller is designed for a class of repeatable nonlinear systems with uncertain parameters and initial output resetting error using a model reference adaptive control technique. The relative formation between agents approaches the target formation exponentially, as demonstrated in Meng and Moore (2016). When the union of interaction graphs with spanning trees occurs frequently along the iteration axis, a good performance in formation is observed. In addition, the iterative learning problem for discrete-time systems with an event-triggered scheme and quantization has been studied in Xiong et al. (2016).

The system's structure being discussed is often fixed in most of the ILC-related material mentioned above. In other words, a connection between the nodes is expected to either always exist or never exist. This presumption is obviously illogical and unjust. Most physical systems are nonlinear in many real-world situations. The processes may constantly be vulnerable to various system architecture changes or be affected by real-world factors like bandwidth restrictions, temperature variations, and disruptions from the outside environment. It will thus cause significant problems with system analysis and design. In recent years, it has become clear that the Takagi–Sugeno (T–S) fuzzy model (Takagi and Sugeno 1985) is a useful tool for describing a variety of complex nonlinear real systems, including multiple nonlinear systems (Zheng et al. 2002; Choi 2008; Qiu et al. 2009), switched nonlinear hybrid systems (Su et al. 2015; Niu et al. 2016; Niu and Li 2017), etc. The switching fuzzy

W. Xiong et al., *Iterative Learning Control for Network Systems Under Constrained Information Communication*, Intelligent Control and Learning Systems 12,
https://doi.org/10.1007/978-981-97-0926-7_5

model, for instance, is switched based on the predicated variables, such as states, measurable external variables, and/or time, in Tanaka et al. (2000). The authors of Choi and Park (2004) introduce a discrete-time switching fuzzy system that has properties shared by switched hybrid systems and T–S fuzzy systems. Fuzzy logic theory has since been extensively used in the control domain (Liu et al. 2013; Xiong et al. 2014; Huang et al. 2015; Zhu et al. 2016). The initial goal is to apply ILC in a particular type of T–S fuzzy systems to analyze the control problem of dynamic systems with changing structure, which is motivated by the properties of the T–S fuzzy model.

Applying digital control approaches for ILC design is more practicable from an economical and practical standpoint (Chien 2000; Sun and Wang 2001; Abidi and Xu 2011). The controllers for sampled data are always built to update their data simultaneously. It is obviously nearly impossible for the controller to update their feedback information concurrently in the real application due to the unreliability of information channels and the capacity of transmission bandwidth of networks. Furthermore, due to external and/or internal disturbances, such as wind gusts and electrical sounds, some system states cannot always be measured. The fact that the iterative learning controllers will be created to be both synchronous and asynchronous in the information update serves as one important incentive for this research. In this chapter, the efficacy of each strategy will be demonstrated. Additionally, we take into account situations in which controllers can measure and not measure system states.

Developing synchronous and asynchronous iterative learning controllers makes it highly desirable to explore the tracking problem of T–S fuzzy systems with measurable and unmeasurable state information based on the discussion above. This chapter has two targets: (1) develop T–S fuzzy models to explain dynamic systems with variable structure and create iterative learning controllers to talk about the tracking issue with T–S fuzzy systems; (2) synchronous and asynchronous controllers will be built within the ILC. Additionally, in the controllers, system states will be regarded as measurably and immeasurably.

5.2 Problem Formulation

In Takagi and Sugeno (1985), a fuzzy system is described by a set of IF-THEN rules, which exhibit the property of linear input/output relations. Inspired by Takagi and Sugeno (1985), in order to characterize dynamic systems with variable structure, T–S fuzzy models will be built. We will also consider iterative learning strategies for the T–S fuzzy model. In other words, each node deals with two separate dynamic processes: the first process depicts the dynamics of the system with respect to time t, while the second process portrays the dynamics of the system with respect to iterative learning k. Additionally, the ith rule is written as follows:

Plant Rule i:
IF $x_1(t,k)$ is $\theta_{i1}(t)$, $x_2(t,k)$ is $\theta_{i2}(t)$, ..., and $x_n(t,k)$ is $\theta_{in}(t)$,
THEN

$$\begin{cases} x(t+1,k) = A_i x(t,k) + B_i u(t,k), \\ y(t,k) = C_i x(t,k), \ i = 1, 2, \ldots, r, \end{cases} \tag{5.1}$$

where $x(t,k) = \left(x_1(t,k), \ldots, x_n(t,k)\right)^{\mathrm{T}} \in \mathbb{R}^n$ is the state vector and n is the number of nodes, $\theta_{i1}(t), \ldots, \theta_{in}(t)$ are the fuzzy sets, r is the number of IF-THEN rules. Here, one can see that Rule i is only related to time t but independent to iteration k. $A_i \in \mathbb{R}^{n\times n}$, $B_i \in \mathbb{R}^{n\times n_u}$, $C_i \in \mathbb{R}^{n_o\times n}$ are known constant matrices, for $i = 1, \ldots, r$, n_u and n_o are appropriate positive integers. $y(t,k) = \left(y_1(t,k), \ldots, y_{n_o}(t,k)\right)^{\mathrm{T}} \in \mathbb{R}^{n_o}$ is the output vector. $t \in \{0, 1, \ldots, T\}$, $k \in \mathbb{Z}_+$, and $u(t,k) = \left(u_1(t,k), \ldots, u_{n_u}(t,k)\right)^{\mathrm{T}} \in \mathbb{R}^{n_u}$ is the iterative learning controller, which will be designed later.

The defuzzified output of the T–S fuzzy system (5.1) is represented as follows:

$$\begin{cases} x(t+1,k) = \sum\limits_{i=1}^{r}\sum\limits_{j=1}^{r} h_i(x(t))h_j(x(t))\left[A_i x(t,k) + B_j u(t,k)\right], \\ y(t,k) = \sum\limits_{i=1}^{r} h_i(x(t))C_i x(t,k), \end{cases} \tag{5.2}$$

where $i = 1, \ldots, r$, $h_i(x(t))$ is the simplified form of $h_i(x(t,k))$ because $h_i(x(t,k))$ is only related to time t. $h_i(x(t,k)) = v_i(x(t,k))/\sum_{i=1}^{r} v_i(x(t,k))$, $v_i(x(t,k)) = \prod\limits_{j=1}^{n} \theta_{ij}(x_j(t,k))$, and $\theta_{ij}(x_j(t,k))$ is the membership function of $x_j(t,k)$ in $\theta_{ij}(t)$ and is only related to time t. The property of $v_i(x(t,k))$ is that

$$v_i(x(t,k)) \geq 0, \ \sum_{j=1}^{r} v_j(x(t,k)) > 0, \ i = 1, \ldots, r, \tag{5.3}$$

and hence,

$$h_i(x(t)) \geq 0, \ \sum_{j=1}^{r} h_i(x(t)) = 1, \ i = 1, \ldots, r, \tag{5.4}$$

for $t \in \{0, 1, \ldots, T\}$, $k \in \mathbb{Z}_+$.

For system (5.1), the tracking of a desired trajectory is achieved if

$$\lim_{k\to+\infty} y_i(t,k) = y_i^*(t), \ t \in \{1, \ldots, T\}, \tag{5.5}$$

where $y^*(t) = \left(y_1^*(t), \ldots, y_{n_o}^*(t)\right)^{\mathrm{T}}$ and $y^*(t) = \sum_{i=1}^{r} h_i(x(t))C_i x^*(t) \in \mathbb{R}^{n_o}$ is the desired output status and $x^*(t) = \left(x_1^*(t), \ldots, x_n^*(t)\right)^{\mathrm{T}}$ is the desired trajectory of

$x(t,k)$. Let $e_i(t,k) = y_i(t,k) - y_i^*(t)$ be the tracking error of node i, and $e(t,k) = (e_1(t,k), \ldots, e_{n_o}(t,k))^{\mathrm{T}} \in \mathbb{R}^{n_o}$. Note that the tracking objective (5.5) holds if and only if $\lim_{k\to+\infty} e(t,k) = \mathbf{0}, \forall\, t \in \{1, \ldots, T\}$.

5.3 Sampling Protocols and ILC Schemes

5.3.1 *Two Types of Sampling Protocols*

This chapter aims to construct suitable sampled-data iterative learning controllers to produce the needed reference trajectory for the T–S fuzzy system (5.1). In literature Chien (2000), Sun and Wang (2001), Abidi and Xu (2011), there are inherent constraints since the sampled-data controllers are always built to update their data simultaneously. Due to the instability of information channels and the network's ability to transmit data at a high rate, updating the controller's feedback information simultaneously is practically impossible in many applications. The following will suggest two sample techniques to represent a real-world scenario more accurately.

(1) *Synchronous iterative learning sampling protocol:* All controller components $u_i(\cdot,\cdot)$ $(i = 1, \ldots, n_u)$ update their sampled-data feedback information based on the same iteration, i.e., $k_s^i = k_s^j$, $\forall\, i, j \in \{0, 1, \ldots, n_u\}, s \in \mathbb{Z}_+$. The corresponding synchronous protocol is shown in Fig. 5.1. Here, the synchronous iterative learning sampling protocol is denoted by $\{k_s\}$, $\forall\, s \in \mathbb{Z}_+$.

(2) *Asynchronous iterative learning sampling protocol:* Each controller component $u_i(\cdot,\cdot)$ $(i = 1, 2, \ldots, n)$ updates its control protocol based on its own iteration channel, i.e., $k_s^i \neq k_s^j$, $\exists\, i, j \in \{1, 2, \ldots, n_u\}$, $i \neq j$, $\forall\, s \in \mathbb{Z}_+$. The corresponding asynchronous protocol is shown in Fig. 5.2.

Remark 5.1 The synchronous protocol is compared with the asynchronous one as follows:

(1) **Difference:** Each asynchronous controller uses its own iteration channel to update the feedback data. In other words, the asynchronous protocol may update the feedback

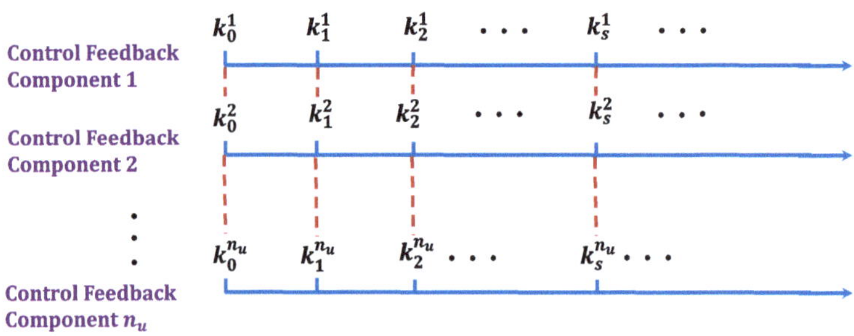

Fig. 5.1 Synchronous transmission of multi-channel sampling protocol

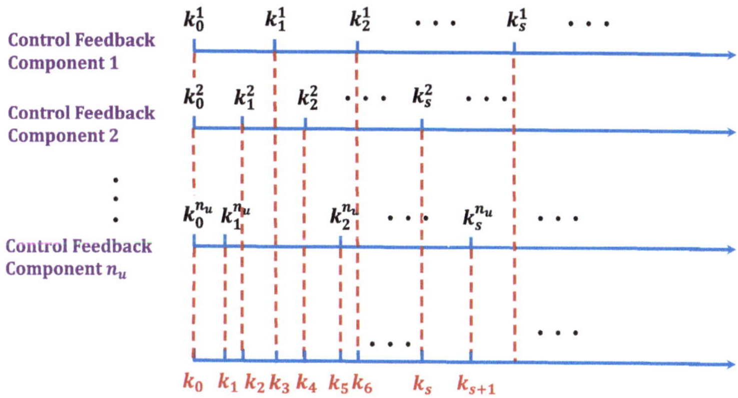

Fig. 5.2 Asynchronous transmission of multi-channel sampling protocol

data at various iteration stages. However, the synchronous protocol's feedback data is constantly updated within the same iteration.
(2) **Difficulty:** The inability to convert asynchronous controllers in the same manner as synchronous ones makes it challenging to design numerous conditions that must be checked in applications.

5.3.2 Two Types of ILC Design

Since the state $x(t, k)$ of model (5.1) may not be measurable, and thus two general cases on the measurability of iterative learning controllers will also be considered in this chapter.

Case 1 When the state $x(t, k)$ of model (5.1) is measurable, the fuzzy synchronous sampled-data ILCs are designed in the following.

Plant Rule i:
IF $x_1(t, k)$ is $\theta_{i1}(t)$, $x_2(t, k)$ is $\theta_{i2}(t)$, ..., and $x_n(t, k)$ is $\theta_{in}(t)$,
THEN

$$u_\omega(t, k_{s+1}) = u_\omega(t, k_s) + \Gamma_i^\omega \sum_{j=1}^{n_o} q_{\omega j}^i \widetilde{e}_j(t+1, k), \tag{5.6}$$

and

$$u_\omega(t, k) = u_\omega(t, k_s), \quad \forall\, k_s \le k < k_{s+1}, \tag{5.7}$$

where $t \in \{0, 1, \dots, T\}$, $\forall\, s \in \mathbb{Z}_+$, $i = 1, 2, \dots, r$, $\omega = 1, 2, \dots, n_u$, $u(t, k)$ is the synchronous controller for rule i, $\widetilde{e}(t, k) = (\widetilde{e}_1(t, k), \dots, \widetilde{e}_{n_o}(t, k))^{\mathrm{T}} \in \mathbb{R}^{n_o}$. $\Gamma_i =$

$\mathrm{diag}(\Gamma_i^1, \ldots, \Gamma_i^{n_u}) \in \mathbb{R}^{n_u \times n_u}$ and $Q_i = (q_{\omega j}^i)_{n_u \times n_o} \in \mathbb{R}^{n_u \times n_o}$ are the learning gain matrices, which will be designed later. For the synchronous controllers, we define

$$\widetilde{e}_j(t, k) = e_j(t, k_s), \quad \forall\, k_s \le k < k_{s+1}, \tag{5.8}$$

where $s \in \mathbb{Z}_+$, $j = 1, \ldots, n_o$. Similarly, for the fuzzy asynchronous control strategies, one has

Plant Rule i:

IF $x_1(t, k)$ is $\theta_{i1}(t)$, $x_2(t, k)$ is $\theta_{i2}(t)$, ..., and $x_n(t, k)$ is $\theta_{in}(t)$,

THEN

$$u_\omega(t, k_{s+1}^\omega) = u_\omega(t, k_s^\omega) + \Gamma_i^\omega \sum_{j=1}^{n_o} q_{\omega j}^i \widetilde{e}_j(t+1, k), \tag{5.9}$$

$$u_\omega(t, k) = u_\omega(t, k_s^\omega), \quad \forall\, k_s^\omega \le k < k_{s+1}^\omega, \tag{5.10}$$

and

$$\widetilde{e}_j(k, k) = e_j(t, k_s^j), \quad \forall\, k_s^j \le k < k_{s+1}^j, \tag{5.11}$$

where $s \in \mathbb{Z}_+$, $\omega = 1, \ldots, n_u$, $j = 1, \ldots, n_o$.

Based on (5.7), (5.8), (5.10), and (5.11), the synchronous and asynchronous controllers in (5.6) and (5.9) for rule i can be written, respectively, as

$$u(t, k_{s+1}) = u(t, k_s) + \Gamma_i Q_i \widetilde{e}(t+1, k), \tag{5.12}$$

and

$$u(t, k_{s+1}^\omega) = u(t, k_s^\omega) + \Gamma_i Q_i \widetilde{e}(t+1, k), \tag{5.13}$$

where $u(t, k_s^\omega) \triangleq (u_1(t, k_s^1), \ldots, u_{n_u}(t, k_s^{n_u}))^{\mathrm{T}}$. According to (5.4), the overall fuzzy synchronous and asynchronous controllers of (5.6) and (5.9) with (5.8) and (5.11) are

$$u(t, k_{s+1}) = u(t, k_s) + \sum_{i=1}^{r} h_i(x(t)) \Gamma_i Q_i \widetilde{e}(t+1, k), \tag{5.14}$$

$$u(t, k_{s+1}^\omega) = u(t, k_s^\omega) + \sum_{i=1}^{r} h_i(x(t)) \Gamma_i Q_i \widetilde{e}(t+1, k). \tag{5.15}$$

Case 2 When the state $x(t, k)$ of model (5.1) is not measurable, let $\widehat{x}(t, k)$ be the estimation of $x(t, k)$ and $\widehat{y}(t, k)$ is the observer output. Similar to that in Yoneyama

(2009), Li et al. (2009), the premise variable is the estimate of the state. Then, one can obtain

$$\begin{cases} \widehat{x}(t+1,k) = \sum_{i=1}^{r}\sum_{j=1}^{r} h_i(\widehat{x}(t))h_j(\widehat{x}(t))\big[A_i\widehat{x}(t,k) + B_j u(t,k)\big], \\ \widehat{y}(t,k) = \sum_{i=1}^{r} h_i(\widehat{x}(t))C_i\widehat{x}(t,k), \end{cases} \tag{5.16}$$

where $i = 1, \ldots, r$.

Here, if u is the overall fuzzy synchronous controller, one has (5.7) and

$$u(t, k_{s+1}) = u(t, k_s) + \sum_{i=1}^{r} h_i(\widehat{x}(t))\Upsilon_i\Lambda_i\widetilde{\widehat{e}}(t+1,k), \tag{5.17}$$

where $\Upsilon_i = \mathrm{diag}(\Upsilon_i^1, \Upsilon_i^{n_u}) \in \mathbb{R}^{n_u \times n_u}$ and $\Lambda_i = (\Lambda_{\omega j}^i)_{n_u \times n_o} \in \mathbb{R}^{n_u \times n_o}$ are also the learning gain matrices, which will be designed later. $\widetilde{\widehat{e}}(t,k) = (\widetilde{\widehat{e}}_1(t,k), \widetilde{\widehat{e}}_2(t,k), \ldots, \widetilde{\widehat{e}}_{n_o}(t,k))^{\mathrm{T}} \in \mathbb{R}^{n_o}$ is the error vector with

$$\widetilde{\widehat{e}}_j(t,k) = \widehat{e}_j(t,k_s), \quad \forall\, k \in [k_s, k_{s+1}), \tag{5.18}$$

where $\widehat{e}_j(t,k) = \widehat{y}_j(t,k) - y_j^*(t)$ is the observer error of node j, $j = 1, 2, \ldots, n_o$, and $\widehat{e}(t,k) = (\widehat{e}_1(t,k), \ldots, \widehat{e}_{n_o}(t,k))^{\mathrm{T}} \in \mathbb{R}^{n_o}$. Similarly, for the overall fuzzy asynchronous controllers, one can obtain from (5.14) and (5.17)

$$u(t, k_{s+1}^j) = u(t, k_s^j) + \sum_{i=1}^{r} h_i(\widehat{x}(t))\Upsilon_i\Lambda_i\widetilde{\widehat{e}}(t+1,k), \tag{5.19}$$

and

$$\widetilde{\widehat{e}}_j(t,k) = \widehat{e}_j(t,k_s^j), \quad k_s^j \le k < k_{s+1}^j, \tag{5.20}$$

where $j = 1, \ldots, n_o$, $s \in \mathbb{Z}_+$.

Remark 5.2 Compared with sampled-data ILC in Chien (2000), Sun and Wang (2001), Abidi and Xu (2011), the advantages of our iterative controllers are

(1) Different from the synchronous iterative controllers in Chien (2000), Sun and Wang (2001), Abidi and Xu (2011), iterative controllers in this chapter are designed to be asynchronous in (5.9) and (5.19). That is, each controller component updates the feedback data it receives according to its iteration channel. Since it is nearly impossible for the controller to update their feedback information concurrently due to the unreliability of information channels and the capacity of the bandwidth of networks, it is obvious that our asynchronous protocol is more logical.

(2) The system states are measurable in those models of Chien (2000), Sun and Wang (2001), Abidi and Xu (2011), while the state $x(t,k)$ in model (5.16) is not

measurable due to external and/or internal disturbances. The unmeasurable state assumption considered in this chapter is more suitable for practical implementation.

In this chapter, the following assumptions are necessary.

Assumption 5.1 The boundary conditions for system (5.1) are given as $x_i(0, k) = x_{i0}$, where x_{i0} is a constant, for $i = 1, \ldots, n$, $\forall\, k \in \mathbb{Z}_+$.

Assumption 5.2 The boundary conditions for system (5.16) are given as $\widehat{x}_i(0, k) = \widehat{x}_{i0}$, where $\widehat{x}_{i0}$ is a constant, for $i = 1, 2, \ldots, n$, $\forall\, k \in \mathbb{Z}_+$.

Assumption 5.3 $k_0^i = 0,\ i = 1, 2, \ldots, n_u$.

5.4 Convergence Analysis

In this section, we aim to achieve the desired tracking of the fuzzy system (5.1). The two cases, i.e., (1) the measurability of the state and (2) the immeasurability of the state, are considered in Sects. 5.4.1 and 5.4.2, respectively.

5.4.1 Tracking with Measurability of the States

In this case, we will first design the synchronous iterative controllers for the fuzzy system (5.1) in the following.

Part I. Convergence under the synchronous iterative control scheme
In this part, the fuzzy synchronous controllers (5.14) and (5.8) will first be considered in the following theorem.

Theorem 5.1 *Consider the fuzzy system (5.1) with the fuzzy iterative learning controllers (5.14) and (5.8). Under Assumptions 5.1 and 5.3, the tracking error $e(t, k)$ has the following property:*

$$\lim_{k\to\infty} e(t, k) = \mathbf{0},\ \forall\, t \in \{1, \ldots, T\}, \tag{5.21}$$

if the learning gains Γ_j and Q_j $(j = 1, 2, \ldots, r)$ are chosen to satisfy

$$\rho\big(I_{n_o} + C_i B_p \Gamma_j Q_j\big) < 1,\ i, p = 1, \ldots, r, \tag{5.22}$$

where I_{n_o} is the n_o-dimensional identity matrix.

Proof We shall show the proof in two cases as follows:
Case 1 Here, we consider that there exists $\forall\, k \in \mathbb{Z}_+$ such that $k_s \le k,\ k + 1 < k_{s+1}$, $s \in \mathbb{Z}_+$. One has that from (5.2), (5.4), (5.7), and Assumption 5.1

$$
\begin{aligned}
&e(t+1,k+1)-e(t+1,k)\\
&=\sum_{i=1}^{r}h_i(x(t+1))C_i(x(t+1,k+1)-x(t+1,k))\\
&=\sum_{i=1}^{r}\sum_{j=1}^{r}\sum_{p=1}^{r}h_i(x(t+1))h_j(x(t))h_p(x(t))C_i\\
&\quad\cdot\big[A_j(x(t,k+1)-x(t,k))+B_p(u(t,k+1)-u(t,k))\big]\\
&=\prod_{s=0}^{t}\Bigg(\sum_{i=1}^{r}\sum_{j=1}^{r}h_i(x(s))h_j(x(s))C_iA_j\Bigg)\big(x(0,k+1)-x(0,k)\big)\\
&=\mathbf{0},
\end{aligned}
\tag{5.23}
$$

where $\prod_{s=0}^{t}h_i(x(s))=h_i(x(0))h_i(x(1))\ldots h_i(x(t))$. Equation (5.23) means that $e(t,k+1)=e(t,k)$, that is,

$$
e(t,k)=e(t,k_s),\quad k\in[k_s,k_{s+1}), \tag{5.24}
$$

where $t\in\{1,\ldots,T\}$, $s\in\mathbb{Z}_+$.

Case 2 In this case, we shall analyze how $e(t,k)$ changes at the sampled-data sequence $\{k_s\}$, $s\in Z_+$. From (5.2), one has

$$
\begin{aligned}
&e(t+1,k_{s+1})-e(t+1,k_s)\\
&=\sum_{i=1}^{r}h_i(x(t+1))C_i\big(x(t+1,k_{s+1})-x(t+1,k_s)\big)\\
&=\sum_{i=1}^{r}\sum_{j=1}^{r}\sum_{p=1}^{r}h_i(x(t+1))h_j(x(t))h_p(x(t))C_i\\
&\quad\cdot\big[A_j\beta(t,k_s)+B_p\big(u(t,k_{s+1})-u(t,k_s)\big)\big],
\end{aligned}
\tag{5.25}
$$

where $\beta(t,k_s)\triangleq x(t,k_{s+1})-x(t,k_s)$. According to (5.2), (5.6), and (5.8), one can obtain

$$
\beta(t,k_s)=\sum_{i=1}^{r}\sum_{p=1}^{r}h_i(x(t))h_p(x(t))\big[A_i\beta(t-1,k_s)+B_p\Gamma_jQ_je(t,k_s)\big]. \tag{5.26}
$$

Note that $\beta(0,k)=\mathbf{0},\ \forall\, k\in\mathbb{Z}_+$, one has from (5.26)

$$
\beta(t,k_s)=\sum_{\pi=0}^{t-1}\Bigg[\prod_{s=\pi+1}^{t-1}\Bigg(\sum_{i=1}^{r}h_i(x(s))A_i\Bigg)\cdot\sum_{p=1}^{r}h_p(x(\pi))B_p\Gamma_jQ_je(1+\pi,k_s)\Bigg].
\tag{5.27}
$$

Let $E(k) = (e^{\mathrm{T}}(1,k), \ldots, e^{\mathrm{T}}(T,k))^{\mathrm{T}}$, one has that from (5.6), (5.25), and (5.27)

$$E(k_{s+1}) = \Delta E(k_s),\ s \in \mathbb{Z}_+, \tag{5.28}$$

where the matrix

$$\Delta' = \begin{pmatrix} \Delta_{11} & O & O & \cdots & O & O \\ \Delta_{21} & \Delta_{11} & O & \cdots & O & O \\ \Delta_{31} & \Delta_{32} & \Delta_{11} & \cdots & O & O \\ \vdots & \vdots & \vdots & \cdots & \vdots & \vdots \\ \Delta_{T1} & \Delta_{T2} & \Delta_{T3} & \cdots & \Delta_{T(T-1)} & \Delta_{11} \end{pmatrix}$$

with

$$\Delta_{11} = \sum_{i=1}^{r}\sum_{j=1}^{r}\sum_{p=1}^{r} h_i(x(t+1))h_j(x(t))h_p(x(t))\big(I_{n_o} + C_i B_p \Gamma_j Q_j\big),$$

and Δ_{pq} $(p > q,\ p \in \{2, 3, \ldots, T\},\ q \in \{1, 2, \ldots, T-1\})$ can be obtained from (5.27) and are not listed here because none of them is needed in the following discussion. According to (5.28), one has

$$E(k_s) = \Delta^s E(0),\ s \in \mathbb{Z}_+. \tag{5.29}$$

Note that Δ is lower-triangular, one has $E(k_s) \to 0$, as $s \to \infty$, if condition (5.22) is satisfied. Combining the results in Case 1 and Case 2, one can achieve (5.21) if condition (5.22) is satisfied. □

Remark 5.3 Compared with the existing ILC literatures for T–S fuzzy systems (Liu and Kong 2013; Li and Li 2014), the contribution of our work is discussed as follows:

(1) Different from Eq. (16) in Liu and Kong (2013), the sampled iterative learning protocols (5.6) and (5.7) are designed in this chapter. In the sampling protocols, the iterative controllers do not need to be updated in each iteration step k, which is a costly operation. In practice, the approach in this chapter has the advantage that it is not necessary for the controller to have the redundant update in some successive iterations.

(2) In Li and Li (2014), a fuzzy logical system is used to approximate the nonlinearity of each follower agent. Compared with Li and Li (2014), T–S fuzzy rules are introduced to describe dynamic systems with variable structure. Moreover, the obtained criteria in Theorem 5.1 are independent of those T–S fuzzy rules.

Part II. Convergence under the asynchronous iterative control scheme

In this part, the asynchronous controllers (5.15) with (5.11) will be considered in the following.

For the asynchronous protocol in Fig. 5.2, all asynchronous iterations are denoted as a sequence $\{k_s\},\ s \in \mathbb{Z}_+$. Here, the protocol components may occur $k_q^i = k_s^j$, for

$\forall\, i,\ j \in \{1, \ldots, n_u\},\ i \neq j,\ q,\ s \in \mathbb{Z}_+$. In this case, we regard k_q^i and k_s^j as two different terms k_s and k_{s+1} for some $s \in \mathbb{Z}_+$. Then, in the interval $\{k_s, \ldots, k_{s+1} - 1\}$, the T–S fuzzy system (5.2) can be rewritten as

$$\begin{cases} x(t+1,k) = \sum\limits_{i=1}^{r}\sum\limits_{j=1}^{r} h_i(x(t))h_j(x(t))\big[A_i x(t,k) + B_j G_{k_s} u(t,k)\big], \\ y(t,k) = \sum\limits_{i=1}^{r} h_i(x(t))C_i x(t,k), \end{cases} \tag{5.30}$$

where $i = 1, \ldots, r,\ G_{k_s} = \mathrm{diag}(g_1^{k_s}, \ldots, g_{n_u}^{k_s}),\ \ \forall\ s \in \mathbb{Z}_+$. If there exists an $i \in \{1, \ldots, n\}$ such that $k_\omega^i = k_s$ ($\forall\, \omega, s \in \mathbb{Z}_+$), then $g_i^{k_s} = 1$. And $g_j^{k_s} = 0,\ j \neq i,\ \forall\, j \in \{1, \ldots, n_u\}$.

Remark 5.4 According to Fig. 5.2 and the above discussion, it is easy to know that G_{k_s} is a diagonal matrix with one nonzero element (here, the nonzero diagonal element is 1), and the other elements are all zero. Define $\Xi = \{G_{k_s} : \forall\, s \in \mathbb{Z}_+\}$, one can obtain that Ξ is a finite set and the number of elements in Ξ is n.

Theorem 5.2 *Consider the fuzzy system (5.1) with the fuzzy iterative learning controllers (5.15) with (5.11). Under Assumptions 5.1 and 5.3, the tracking error $e(t,k)$ such that*

$$\lim_{k\to\infty} e(t,k) = \mathbf{0},\ \forall\, t \in \{1, 2, \ldots, T\}, \tag{5.31}$$

if the learning gains Γ_j and Q_j $(j = 1, 2, \ldots, r)$ are satisfied to

$$\rho\left(\widetilde{\Theta}_{k_{s_d}}^{ip}\right) < 1,\ i, p = 1, 2, \ldots, r, \tag{5.32}$$

where the matrix

$$\widetilde{\Theta}_{k_{s_d}}^{ip} = \Theta_{k_{s_{d+1}-1}}^{ip}\Theta_{k_{s_{d+1}-2}}^{ip}\cdots\Theta_{k_{s_d+1}}^{ip}\Theta_{k_{s_d}}^{ip},\ \ \Theta_{k_{s_d}}^{ip} = I_{n_o} + C_i B_p G_{k_{s_d}}\Gamma_j Q_j,\ d \in \mathbb{Z}_+.$$

Proof The proof is also shown in two different cases:
Case 1 We consider in this case that there exists $\forall\, k \in \mathbb{Z}_+$ such that $k_s \leq k,\ k+1 < k_{s+1}, s \in \mathbb{Z}_+$. Similar to the proof in Theorem 5.1, one also has for $\forall\, t \in \{1, 2, \ldots, T\}$

$$\widehat{e}(t,k+1) - \widehat{e}(t,k) = \mathbf{0}. \tag{5.33}$$

Then, one can obtain

$$e(t,k) = e(t,k_s),\ \forall\, k_s \leq k < k_{s+1}, \tag{5.34}$$

where $\forall\, t \in \{1, \ldots, T\}, s \in \mathbb{Z}_+$.

Case 2 In this case, we also analyze how $e(t,k)$ changes at the sampled-data sequence $\{k_s\}, s \in \mathbb{Z}_+$. Let $E(k) = (e^{\mathrm{T}}(1,k), \ldots, e^{\mathrm{T}}(T,k))^{\mathrm{T}}$, similar to Theorem 5.1, one has

$$E(k_{s+1}) = \Omega_{k_s} E(k_s), \ s \in \mathbb{Z}_+, \tag{5.35}$$

where the matrix

$$\Omega_{k_s} = \begin{pmatrix} \Omega_{11}^{k_s} & O & O & \cdots & O & O \\ \Omega_{21}^{k_s} & \Omega_{11}^{k_s} & O & \cdots & O & O \\ \Omega_{31}^{k_s} & \Omega_{32}^{k_s} & \Omega_{11}^{k_s} & \cdots & O & O \\ \vdots & \vdots & \vdots & \cdots & \vdots & \vdots \\ \Omega_{T1}^{k_s} & \Omega_{T2}^{k_s} & \Omega_{T3}^{k_s} & \cdots & \Omega_{T(T-1)}^{k_s} & \Omega_{11}^{k_s} \end{pmatrix}$$

with

$$\Omega_{11}^{k_s} = \sum_{i=1}^{r}\sum_{j=1}^{r}\sum_{p=1}^{r} h_i(x(t+1))h_j(x(t))h_p(x(t))\big(I_{n_o} + C_i B_p G_{k_s}\Gamma_j Q_j\big),$$

and $\Omega_{\eta q}^{k_s}$ $(\eta > q, \ \eta \in \{2,3,\ldots,T\}, \ q \in \{1,2,\ldots,T-1\})$ are not listed here because none of them is needed in the following discussion.

In (5.35), it is difficult to obtain that $\rho(\Omega_{k_s}) < 1$ since matrix B_{k_s} is a matrix with one nonzero diagonal element. Hence, one obtains from (5.35)

$$E(l_{s_{d+1}}) = \widetilde{\Omega}_{k_{s_d}} E(k_{s_d}), \ \ d \in \mathbb{Z}_+, \tag{5.36}$$

where $\widetilde{\Omega}_{k_{s_d}} = \Omega_{k_{s_{d+1}-1}}\Omega_{k_{s_{d+1}-2}}\cdots\Omega_{k_{s_d+1}}\Omega_{k_{s_d}}$. Note that $\widetilde{\Omega}_{k_{s_d}}$ has a lower-triangular structure, and hence one has $E(k_{s_d}) \to 0$ as $d \to \infty$, if condition (5.32) is satisfied. Combining the results in Case 1 and Case 2, one can obtain (5.31) if the condition (5.32) holds. □

Remark 5.5 Since G_{k_s} is a matrix with one nonzero diagonal element, condition (5.32) can be obtained if one chooses $G_{k_{s_d}} + G_{k_{s_d+1}} + G_{k_{s_d+2}} + \cdots + G_{k_{s_{d+1}-1}} \geq I_{n_o}$, i.e., $k_{s_{d+1}} - k_{s_d} \geq n_o$, $\forall\, d \in \mathbb{Z}_+$, and designs the appropriate learning gain matrices $\Gamma_j, Q_j, j = 1, \ldots, r$.

5.4.2 Tracking with Immeasurability of the States

In this case, for the fuzzy system (5.16), the purpose is to design the synchronous and asynchronous iterative controllers to achieve that

$$\lim_{k\to\infty} e(t,k) = \mathbf{0}, \tag{5.37}$$

and

$$\lim_{k\to\infty}(\widehat{y}(t,k)-y(t,k))=\mathbf{0}, \tag{5.38}$$

$\forall\, t\in\{1,2,\ldots,T\}$. Define $\zeta(t,k)=(\widehat{e}^{\mathrm{T}}(t,k),\ e^{\mathrm{T}}(t,k))^{\mathrm{T}}$, one has that (5.37) and (5.38) can be satisfied if

$$\lim_{k\to\infty}\zeta(t,k)=\mathbf{0}, \tag{5.39}$$

$\forall\, t\in\{1,2,\ldots,T\}$.

Part I. Convergence under the synchronous iterative control scheme
In this part, the fuzzy synchronous controllers (5.17) with (5.18) will be considered in the following theorem.

Theorem 5.3 *Consider the fuzzy system (5.16) with the fuzzy iterative learning controllers (5.17) with (5.18). Under Assumptions 5.1–5.3, one can obtain*

$$\lim_{k\to\infty}\zeta(t,k)=\mathbf{0},\ \forall\, t\in\{1,\ldots,T\}, \tag{5.40}$$

if the learning gains Γ_j, Q_j, Υ_j, *and* Λ_j $(j=1,\ldots,r)$ *are chosen to satisfy that*

$$\rho\big(I_{n_o}+C_iB_p\Gamma_jQ_j\big)<1,\ i,p=1,2,\ldots,r, \tag{5.41}$$

and

$$\rho\big(I_{n_o}+C_iB_p\Upsilon_j\Lambda_j\big)<1,\ i,p=1,\ldots,r. \tag{5.42}$$

Proof We complete the proof into two cases as follows:
Case 1 Here, we consider in this case that there exists $\forall\, k\in\mathbb{Z}_+$ such that $k_s\le k,\ k+1<k_{s+1}$, $s\in\mathbb{Z}_+$. Similar to (5.23), one has for $\forall\, t\in\{1,\ldots,T\}$

$$\widehat{e}(t,k+1)-\widehat{e}(t,k)=\mathbf{0}. \tag{5.43}$$

Hence, one gets from (5.23) and (5.43)

$$\zeta(t,k)=\zeta(t,k_s),\ \forall\, k_s\le k<k_{s+1}, \tag{5.44}$$

where $\forall\, t\in\{1,\ldots,T\}$, $s\in\mathbb{Z}_+$.

Case 2 In this case, we shall analyze how $\zeta(t,k)$ changes at the sampled-data sequence $\{k_s\}$, $s\in\mathbb{Z}_+$. Similar to (5.25) and (5.27), one has from (5.2) and (5.16)

$$
\begin{aligned}
&\zeta(t+1,k_{s+1})-\zeta(t+1,k_s)\\
&=\begin{pmatrix}\widehat{e}(t+1,k_{s+1})-\widehat{e}(t+1,k_s)\\ e(t+1,k_{s+1})-e(t+1,k_s)\end{pmatrix}\\
&=\begin{pmatrix}\sum\limits_{i=1}^{r}\sum\limits_{j=1}^{r}\sum\limits_{p=1}^{r}h_i(\widehat{x}(t+1))h_j(\widehat{x}(t))h_p(\widehat{x}(t))\\ \times C_i\big[A_j\widehat{\beta}(t,k_s)+B_p(u(t,k_{s+1})-u(t,k_s))\big]\\ \sum\limits_{i=1}^{r}\sum\limits_{j=1}^{r}\sum\limits_{p=1}^{r}h_i(x(t+1))h_j(x(t))h_p(x(t))\\ \times C_i\big[A_j\beta(t,k_s)+B_p(u(t,k_{s+1})-u(t,k_s))\big]\end{pmatrix},
\end{aligned}
\tag{5.45}
$$

where $\widehat{\beta}(t,k_s)\triangleq\widehat{x}(t,k_{s+1})-\widehat{x}(t,k_s)$, and

$$
\widehat{\beta}(t,k_s)=\sum_{\pi=0}^{t-1}\Bigg[\prod_{s=\pi+1}^{t-1}\Bigg(\sum_{i=1}^{r}h_i(\widehat{x}(s))A_i\Bigg)\cdot\sum_{p=1}^{r}h_p(\widehat{x}(\pi))B_p\Upsilon_j\Lambda_j e(1+\pi,k_s)\Bigg].
\tag{5.46}
$$

Let $\widehat{\zeta}(k)=(\zeta^{\mathrm{T}}(1,k),\ldots,\zeta^{\mathrm{T}}(T,k))^{\mathrm{T}}$, one has from (5.28), (5.45), and (5.46)

$$
\widehat{\zeta}(k_{s+1})=\Delta_1\widehat{\zeta}(k_s),\ s\in\mathbb{Z}_+,
\tag{5.47}
$$

where $\Delta_1=\begin{pmatrix}\widehat{\Delta} & O\\ O & \Delta\end{pmatrix}$, Δ has been given in (5.28), and

$$
\widehat{\Delta}=\begin{pmatrix}
\widehat{\Delta}_{11} & O & O & \cdots & O & O\\
\widehat{\Delta}_{21} & \widehat{\Delta}_{11} & O & \cdots & O & O\\
\widehat{\Delta}_{31} & \widehat{\Delta}_{32} & \widehat{\Delta}_{11} & \cdots & O & O\\
\vdots & \vdots & \vdots & \cdots & \vdots & \vdots\\
\widehat{\Delta}_{T1} & \widehat{\Delta}_{T2} & \widehat{\Delta}_{T3} & \cdots & \widehat{\Delta}_{T(T-1)} & \widehat{\Delta}_{11}
\end{pmatrix}
$$

with

$$
\widehat{\Delta}_{11}=\sum_{i=1}^{r}\sum_{j=1}^{r}\sum_{p=1}^{r}h_i(\widehat{x}(t+1))h_j(\widehat{x}(t))h_p(\widehat{x}(t))\big(I_{n_o}+C_iB_p\Upsilon_j\Lambda_j\big),
$$

and $\widehat{\Delta}_{\eta\theta}$ ($\eta>\theta,\ \eta\in\{2,3,\ldots,T\},\ \theta\in\{1,\ldots,T-1\}$) can be obtained from (5.46) and are not listed here because none of them is needed in the following discussion. According to (5.47), one has

$$
E(k_s)=(\Delta_1)^s E(0),\ s\in\mathbb{Z}_+.
\tag{5.48}
$$

Note that Δ_1 has a lower-triangular structure, and hence one has $E(k_s) \to 0$ as $s \to \infty$, if conditions (5.41) and (5.42) are satisfied. Combining the results in Case 1 and Case 2, one can achieve (5.40) if conditions (5.41) and (5.42) are satisfied.

Part II. Convergence under the asynchronous iterative control scheme
In this part, for the fuzzy system (5.16), the asynchronous controllers (5.19) with (5.20) will be considered in the following.

Similar to (5.30), in the interval $\{k_s, \ldots, k_{s+1} - 1\}$, the T S fuzzy system (5.16) can be rewritten as

$$\begin{cases} \widehat{x}(t+1,k) = \sum\limits_{i=1}^{r}\sum\limits_{j=1}^{r} h_i(\widehat{x}(t))h_j(\widehat{x}(t))\big[A_i\widehat{x}(t,k) + B_jG_{k_s}u(t,k)\big], \\ \widehat{y}(t,k) = \sum\limits_{i=1}^{r} h_i(\widehat{x}(t))C_i\widehat{x}(t,k), \end{cases} \tag{5.49}$$

where $i = 1, \ldots, r$, G_{k_s} has been given in (5.30), and the asynchronous controllers have been defined in (5.19) and (5.20).

Theorem 5.4 *Consider the fuzzy system (5.49) with the fuzzy iterative learning controllers (5.19) and (5.20). Under Assumptions 5.1–5.3, one can obtain*

$$\lim_{k\to\infty} \zeta(t,k) = \mathbf{0},\ \forall\, t \in \{1, 2, \ldots, T\}, \tag{5.50}$$

if the learning gains Γ_j, Q_j, Υ_j, and Λ_j $(j = 1, 2, \ldots, r)$ are chosen to satisfy that

$$\rho\left(\widetilde{\Theta}^{ip}_{k_{sd}}\right) < 1,\ i, p = 1, 2, \ldots, r, \tag{5.51}$$

and

$$\rho\left(\widetilde{\Phi}^{ip}_{k_{sd}}\right) < 1,\ i, p = 1, 2, \ldots, r, \tag{5.52}$$

where the matrices

$$\begin{aligned} \widetilde{\Theta}^{ip}_{k_{sd}} &= \Theta^{ip}_{k_{sd+1}-1}\Theta^{ip}_{k_{sd+1}-2}\cdots\Theta^{ip}_{k_{sd}+1}\Theta^{ip}_{k_{sd}},\ \ \Theta^{ip}_{k_{sd}} = I_{n_o} + C_iB_pG_{k_{sd}}\Gamma_jQ_j, \\ \widetilde{\Phi}^{ip}_{k_{sd}} &= \Phi^{ip}_{k_{sd+1}-1}\Phi^{ip}_{k_{sd+1}-2}\cdots\Phi^{ip}_{k_{sd}+1}\Phi^{ip}_{k_{sd}},\ \ \Phi^{ip}_{k_{sd}} = I_{n_o} + C_iB_pG_{k_{sd}}\Upsilon_j\Lambda_j,\ \ d \in \mathbb{Z}_+. \end{aligned}$$

Proof Similar to Theorem 5.3, we prove it in two cases.

Case 1 Here, we consider in this case that there exists $\forall\, k \in \mathbb{Z}_+$ such that $k_s \le k,\ k+1 < k_{s+1}$, $s \in \mathbb{Z}_+$. Similar to the proof in Theorem 5.3, one has

$$\widehat{e}(t,k+1) - \widehat{e}(t,k) = \mathbf{0}. \tag{5.53}$$

Hence, one gets from (5.23) and (5.53)

$$\zeta(t,k)=\zeta(t,k_s),\ \forall\, k_s\le k<k_{s+1}, \tag{5.54}$$

where $\forall\, t\in\{1,2,\ldots,T\}$, $s\in\mathbb{Z}_+$.

Case 2 In this case, we also analyze how $\zeta(t,k)$ changes at the sampled-data sequence $\{k_s\}$, $s\in\mathbb{Z}_+$. Let $\widehat{\zeta}(k)=(\zeta^{\mathrm{T}}(1,k),\ldots,\zeta^{\mathrm{T}}(T,k))^{\mathrm{T}}$, similar to Theorem 5.3, one has

$$\widehat{\zeta}(k_{s+1})=\Omega^1_{k_s}\widehat{\zeta}(k_s),\ s\in\mathbb{Z}_+, \tag{5.55}$$

where $\Omega^1_{k_s}=\begin{pmatrix}\widehat{\Omega}_{k_s} & O\\ O & \Omega_{k_s}\end{pmatrix}$, Ω_{k_s} is given in (5.35), and the matrix

$$\widehat{\Omega}_{k_s}=\begin{pmatrix}\widehat{\Omega}^{k_s}_{11} & O & O & \cdots & O & O\\ \widehat{\Omega}^{k_s}_{21} & \widehat{\Omega}^{k_s}_{11} & O & \cdots & O & O\\ \widehat{\Omega}^{k_s}_{31} & \widehat{\Omega}^{k_s}_{32} & \widehat{\Omega}^{k_s}_{11} & \cdots & O & O\\ \vdots & \vdots & \vdots & \cdots & \vdots & \vdots\\ \widehat{\Omega}^{k_s}_{T1} & \widehat{\Omega}^{k_s}_{T2} & \widehat{\Omega}^{k_s}_{T3} & \cdots & \widehat{\Omega}^{k_s}_{T(T-1)} & \widehat{\Omega}^{k_s}_{11}\end{pmatrix}$$

with

$$\widehat{\Omega}^{k_s}_{11}=\sum_{p=1}^{r}h_i(\widehat{x}(t+1))h_j(\widehat{x}(t))h_p(\widehat{x}(t))\big(I_{n_o}+C_iB_pG_{k_s}\Upsilon_j\Lambda_j\big),$$

and $\widehat{\Omega}^{k_s}_{\eta\theta}$ $(\eta>\theta,\ \eta\in\{2,3,\ldots,T\},\ \theta\in\{1,2,\ldots,T-1\})$ are not listed here because none of them is needed in the following discussion.

In (5.55), it is difficult to obtain that $\rho(\Omega^1_{k_s})<1$ since matrix G_{k_s} is a matrix with one nonzero diagonal element. Hence, one obtains from (5.55)

$$\widehat{\zeta}(k_{s_{d+1}})=\widetilde{\Omega}^1_{k_{s_d}}\widehat{\zeta}(k_{s_d}),\ \ d\in\mathbb{Z}_+, \tag{5.56}$$

where $\widetilde{\Omega}^1_{k_{s_d}}=\Omega^1_{k_{s_{d+1}-1}}\Omega^1_{k_{s_{d+1}-2}}\cdots\Omega^1_{k_{s_d+1}}\Omega^1_{k_{s_d}}$. Note that $\widetilde{\Omega}^1_{k_{s_d}}$ is lower-triangular, one has $\widehat{\zeta}(k_{s_d})\to 0$, as $d\to\infty$, if conditions (5.51) and (5.52) are satisfied. Combining the results in Case 1 and Case 2, one can obtain (5.50) if conditions (5.51) and (5.52) are satisfied.

Remark 5.6 In Huang et al. (2015), the synchronous and asynchronous sampling protocols have also been considered for a kind of T–S fuzzy model. Compared with Huang et al. (2015), the advantages of our work are as follows:

(1) The immeasurability of the state is introduced in the sampling protocols (see (5.16), (5.17), and (5.19)). Since the immeasurability phenomena typically needs to be taken into account in actual applications due to external and/or internal disturbances, such as wind gusts and electronic noises, the consideration is more realistic.

(2) Compared with the given LMI conditions (10), (11), (20), and (21) in Huang et al. (2015), our obtained conditions in Theorems 5.1–5.4 are easier to be verified even for high-dimensional system.

Remark 5.7 The measurability and immeasurability of the states have also been discussed for a class of sampled-data fuzzy systems in Zhu et al. (2016). Compared with Zhu et al. (2016), the innovations of our work are as follows:

(1) The fuzzy controllers (5.9) and (5.19) are designed to be asynchronous. That is, based on its own iteration channel, each controller component may transmit the feedback data. Clearly, the unreliability of information channels makes it feasible for the controller to communicate feedback data asynchronously.

(2) Compared with LMI conditions (28), (29), (50), and (51) in Zhu et al. (2016), our obtained criteria in Theorems 5.1–5.4 are easier to verify when the dimension of the system is high.

5.5 Simulation Example

In this section, a numerical example is presented to demonstrate the usefulness of the result in Theorems 5.1–5.4.

Example 5.1 Fuzzy systems have really been used in a variety of domains, including risk analysis, industrial process control, medical diagnosis, and spacecraft altitude control. The following example discusses a fuzzy iterative learning system for an air conditioner, see Fig. 5.3.

For the purposes of this example, we assume there is just one air conditioner. For the sake of simplicity, we assume that the room temperature is the only factor affecting the air conditioner's fuzzy logic system. That is, the fuzzy set has two distinct rules. Take five sensor points in the room and the temperature of the ith $(1 \leq i \leq 5)$ points is defined as $x_i(t,k)$. Choose three measurable points and define their output temperatures to be $y_i(t,k)$ $(i = 1, 2, 3)$. Their relationship between the points and the fuzzy rules is shown as follows:

Plant Rule 1:
IF $x_1(t,k)$ is $\theta_1(t,k)$,
THEN

$$\begin{cases} x(t+1,k) = A_1 x(t,k) + B_1 u(t,k), \\ y(t,k) = C_1 x(t,k), \end{cases} \tag{5.57}$$

Plant Rule 2:
IF $x_2(t,k)$ is $\theta_2(t,k)$,
THEN

$$\begin{cases} x(t+1,k) = A_2 x(t,k) + B_2 u(t,k), \\ y(t,k) = C_2 x(t,k), \end{cases} \tag{5.58}$$

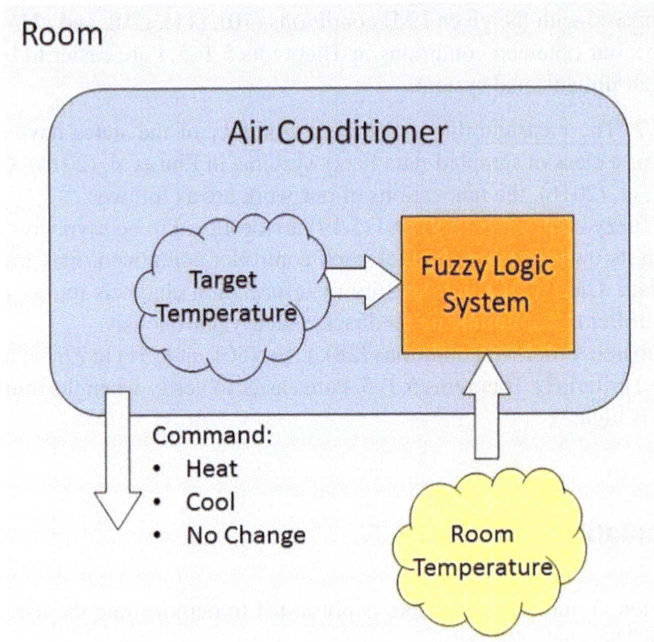

Fig. 5.3 The fuzzy logic control for automatic air-condition system

where $t \in \{0, 1, \ldots, T\}$ $(T = 10)$, $k \in \mathbb{Z}_+$, $x(t,k) = (x_1(t,k), \ldots, x_5(t,k))^{\mathrm{T}}$ and $x_i(t,k)$ is the temperature of point i. The membership function is assumed to be

$$\begin{aligned} h_1(x_1(t,k)) &= \frac{1 - \sin^2(x_1(t,k))}{2}, \\ h_2(x_2(t,k)) &= \frac{1 + \sin^2(x_1(t,k))}{2}. \end{aligned} \tag{5.59}$$

Matrices A_1, A_2, B_1, B_2, C_1, and C_2 are as follows:

$$A_1 = \begin{pmatrix} -1 & 0 & 0 & 0 & 0 \\ 1 & -2 & 0 & 0 & 0 \\ 0 & 1 & -2 & 0 & 0 \\ 1 & -1 & 0 & -3 & 0 \\ 0 & 1 & 0 & 1 & -2 \end{pmatrix}, \quad A_2 = \begin{pmatrix} -2 & 0 & 0 & 0 & 0 \\ -1 & -2 & 0 & 0 & 0 \\ 2 & 1 & -4 & 0 & 0 \\ 0 & 0 & 1 & -2 & 0 \\ -1 & 1 & 0 & 0 & -3 \end{pmatrix},$$

$$B_1 = B_2 = \begin{pmatrix} 0.5 & 0 & 0 \\ 0.5 & 0 & 0 \\ 1 & 1 & 0 \\ 0 & 1 & 0.5 \\ 0 & 0 & 0.5 \end{pmatrix}, \quad C_1 = C_2 = \begin{pmatrix} 1 & 1 & 0 & 0 & 0 \\ 1 & 0 & 1 & 0 & 0 \\ 0 & 0 & 0 & 1 & 1 \end{pmatrix}.$$

The synchronous iterative sampling sequence $\{k_s\}$ is denoted as $k_s = 2s,\ s \in \mathbb{Z}_+$. The asynchronous iterative sampling sequence $\{k_s^i\}$ is denoted as $k_s^i = is,\ i = 1, \ldots, 5,\ s \in \mathbb{Z}_+$. $G_{k_s} \in \{G_1, G_2, G_3\}$ with $G_1 = \text{diag}(1, 0, 0)$, $G_2 = \text{diag}(0, 1, 0)$, $G_3 = \text{diag}(0, 0, 1)$, $\forall\ s \in \mathbb{Z}_+$. We define the desired reference trajectory $x^*(t) = \sin(t)$ and $y^*(t) = C_1 \sin(t),\ t \in \{0, 1, \ldots, 10\},\ x_i(0, k) = x_{i0} = i/5,\ \widehat{x}_i(0, k) = \widehat{x}_{i0} = -i/5,\ i = 1, 2, \ldots, 5, \forall\, k \in \mathbb{Z}_+$.

For the synchronous and asynchronous controllers in (5.12) and (5.13), we design the learning gain matrices $\Gamma_i = \frac{1}{2} I_3$, and

$$Q_i = \begin{pmatrix} -1 & 0 & 0 \\ 2 & -1 & 0 \\ 1 & 3 & -1 \end{pmatrix}, \quad i = 1, 2.$$

For the synchronous and asynchronous controllers in (5.17) and (5.19), we design the learning gain matrices $\Upsilon_i = -\frac{1}{3} I_3$, and

$$\Lambda_i = \begin{pmatrix} 2 & 0 & 0 \\ 1.5 & 2 & 0 \\ 3 & 1 & 2 \end{pmatrix}, \quad i = 1, 2.$$

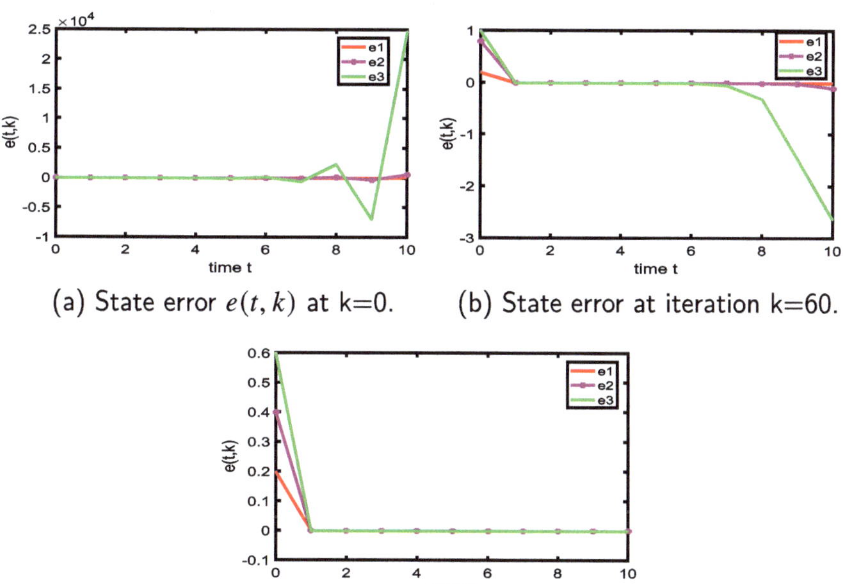

(a) State error $e(t, k)$ at k=0.

(b) State error at iteration k=60.

(c) State error at iteration k=200.

Fig. 5.4 The trajectories of state output error $e(t)$ with synchronous sampled-data iterative learning controller

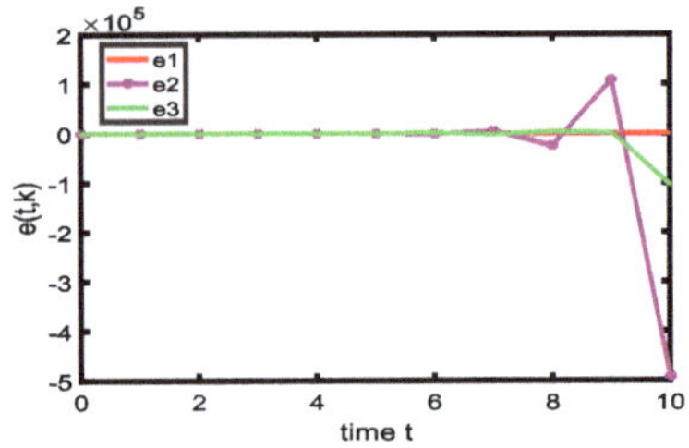

(a) State error $e(t,k)$ at all iterations $k_i = 0$, $i = 1, 2, 3, 4, 5$.

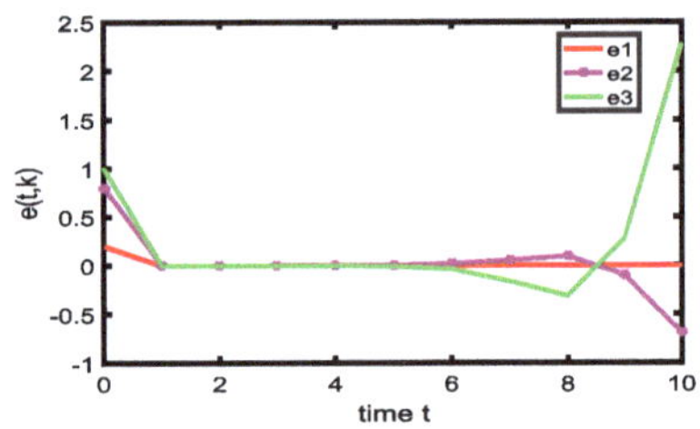

(b) State error at iterations $k_1 = 200$, $k_2 = 100$, $k_3 = 65$, $k_4 = 50$, $k_5 = 40$.

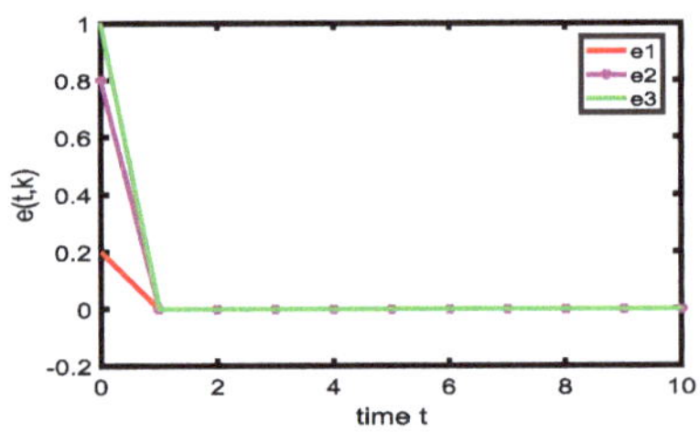

(c) State error at iterations $k_1 = 300$, $k_2 = 150$, $k_3 = 100$, $k_4 = 75$, $k_5 = 60$.

Fig. 5.5 The trajectories of state output error $e(t)$ with asynchronous sampled-data iterative learning controller

As a result, one has that $\rho\big(I_3 + C_i B_p \Gamma_j Q_j\big) = 0.5 < 1$, and $\rho\big(I_3 + C_i B_p \Upsilon_j \Lambda_j\big) = \frac{1}{3} < 1, i, p, j = 1, 2$. Hence, the conditions in Theorems 5.1 and 5.3 are satisfied. In addition, one has $\rho\big(\Theta_1^{ip}\Theta_2^{ip}\cdots\Theta_5^{ip}\big) = 0.5 < 1$ with $\Theta_\eta^{ip} = I_3 + C_i B_p G_\eta \Gamma_j Q_j$, $\eta = 1, 2, 3,\ i, p = 1, 2$. Also, one has $\rho\big(\Phi_1^{ip}\Phi_2^{ip}\cdots\Phi_5^{ip}\big) = 0.3333 < 1$ with $\Phi_\eta^{ip} = I_3 + C_i B_p G_\eta \Upsilon_j \Lambda_j$, $\eta = 1, 2, 3,\ i, p = 1, 2$. As mentioned in Remark 5.5, conditions (5.32), (5.51), and (5.52) in Theorems 5.2 and 5.4 could be satisfied if we choose $G_{k_{s_d}} + G_{k_{s_d+1}} + \cdots + G_{k_{s_{d+1}-1}} \geq I_3$, i.e., $k_{s_{d+1}} - k_{s_d} \geq 3$. Hence, one can achieve (5.31) and (5.50) from Theorems 5.1–5.4.

The corresponding error trajectories can be illustrated in Figs. 5.4, 5.5, 5.6, and 5.7. Figure 5.4 shows the state error trajectories $e(t)$ with synchronous sampled-data iterative learning controller. The trajectories of state output error $e(t)$ with asynchronous sampled-data iterative learning controller are shown in Fig. 5.5. Further, Fig. 5.6 shows the trajectories of observer output error $\widehat{e}(t)$ with synchronous sampled-data iterative learning controller. The trajectories of observer output error $\widehat{e}(t)$ with asynchronous sampled-data iterative learning controller are shown in Fig. 5.7.

Compared with the results between Figs. 5.4, 5.5, 5.6, and 5.7, one can obtain that the convergence rate of errors is not affected by the measurability or immeasurability of the states. However, compared with the results between Figs. 5.4 and 5.5

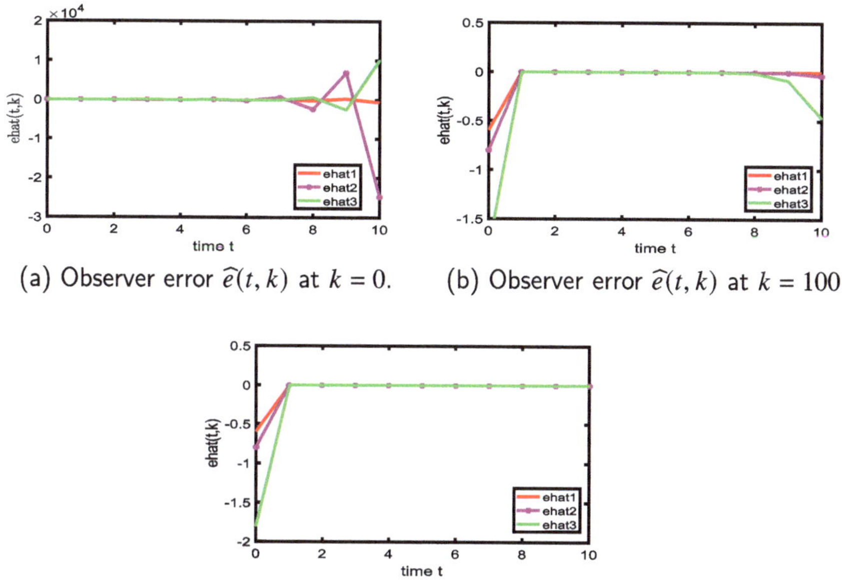

(a) Observer error $\widehat{e}(t,k)$ at $k = 0$. (b) Observer error $\widehat{e}(t,k)$ at $k = 100$.

(c) Observer error $\widehat{e}(t,k)$ at $k = 200$.

Fig. 5.6 The trajectories of observer output error $\widehat{e}(t)$ with synchronous sampled-data iterative learning controller

(or Figs. 5.6 and 5.7), one can get that the convergence rate of the state error (or the observer output error) with the synchronous controller is faster than the convergence rate of the error with the asynchronous controller. In the existing ILC literatures, see Chien (2000), Sun and Wang (2001), Abidi and Xu (2011), the iterative controllers are always designed to be synchronous. More general situations, such as the asynchronous controller and immeasurability of the states, have been considered in this example. In addition, the results in Figs. 5.5, 5.6, and 5.7 demonstrate the effectiveness of the designed controllers in such situations.

Remark 5.8 In conditions (5.32), (5.51), and (5.52), the learning gains can be easily chosen to satisfy the three conditions. In this example, one can simply choose the learning gains Γ_j, Q_j, Υ_j, and Λ_j to satisfy $\rho(\widetilde{\Theta}^{ip}_{k_{s_d}}) < 1$, and $\rho(\widetilde{\Phi}^{ip}_{k_{s_d}}) < 1$, $p = 1, 2, \ldots, r$, if one selects $G_{k_{s_d}} + G_{k_{s_d+1}} + G_{k_{s_d+2}} + \cdots + G_{k_{s_{d+1}-1}} \geq I_{n_o}$, i.e., $k_{s_{d+1}} - k_{s_d} \geq n_o$, $\forall\, d \in \mathbb{Z}_+$.

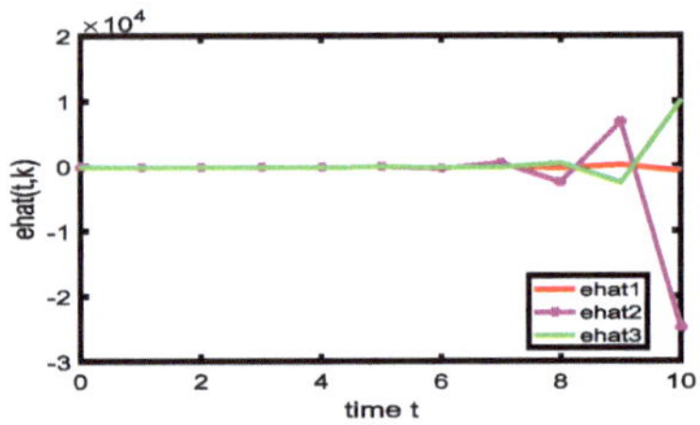

(a) Observer error $\widehat{e}(t, k)$ at all iterations $k_i = 0$, $i = 1, 2, 3, 4, 5$.

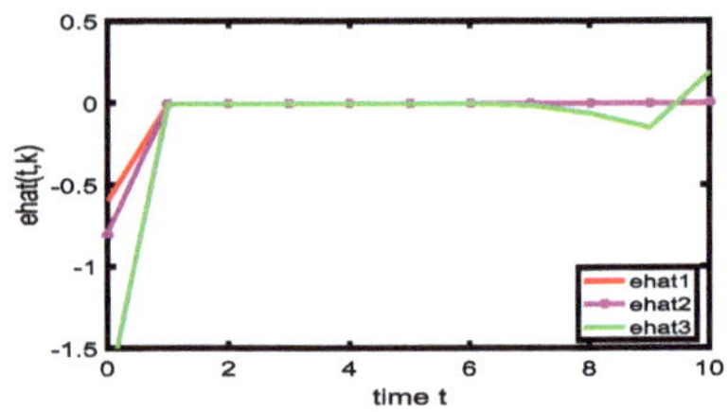

(b) Observer error $\widehat{e}(t, k)$ at iterations $k_1 = 200$, $k_2 = 100$, $k_3 = 65$, $k_4 = 50$, $k_5 = 40$.

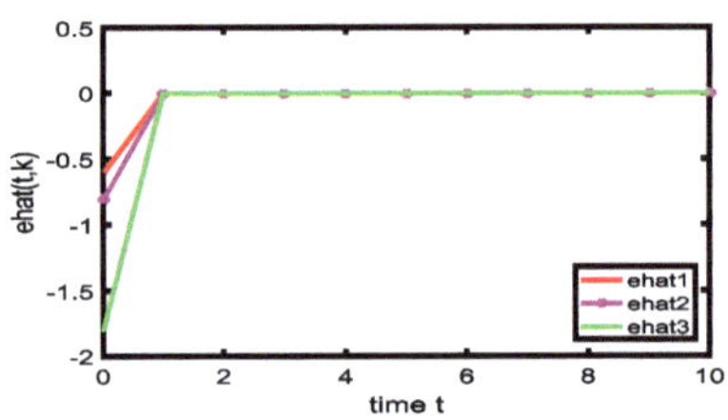

(c) Observer error $\widehat{e}(t, k)$ at iterations $k_1 = 300$, $k_2 = 150$, $k_3 = 100$, $k_4 = 75$, $k_5 = 60$.

Fig. 5.7 The trajectories of observer output error $\widehat{e}(t)$ with asynchronous sampled-data iterative learning controller

5.6 Summary

This chapter has discussed the tracking problem of T–S fuzzy systems with measurable and unmeasurable state information. First, we have proposed the T–S fuzzy model to describe the dynamic systems with variable structure. Then, some criteria have been proposed to achieve the tracking of T–S fuzzy systems with measurable and unmeasurable state information by designing synchronous and asynchronous iterative learning control schemes. Finally, numerical examples with the simulations have been provided to illustrate the effectiveness of the main results.

References

Abidi K, Xu JX (2011) Iterative learning control for sampled-data systems: From theory to practice. IEEE Trans Ind Electron 57(8):3002–3015

Chi RH, Liu Y, Hou ZS, Jin ST (2015) Data-driven terminal iterative learning control with high-order learning law for a class of nonlinear discrete-time multiple-input-multiple output systems. IET Control Theory Appl 9(7):1075–1082

Chi RH, Huang B, Wang DW, Zhang RK, Feng YJ (2016) Data-driven optimal terminal iterative learning control with initial value dynamic compensation. IET Control Theory Appl 10(12):1357–1364
Chien CJ (2000) A sampled-data iterative learning control using fuzzy network design. Int J Control 73(10):902–913
Chien CJ, Yao CY (2004) Iterative learning of model reference adaptive controller for uncertain nonlinear systems with only output measurement. Automatica 40(5):855–864
Choi HH (2008) Robust stabilization of uncertain fuzzy systems using variable structure system approach. IEEE Trans Fuzzy Syst 16(3):715–724
Choi DJ, Park P (2004) Guaranteed cost controller design for discrete-time switching fuzzy systems. IEEE Trans Syst Man Cybern 34(1):110–119
Huang C, Ho DWC, Lu JQ, Kurths J (2015) Pinning synchronization in T-S fuzzy complex networks with partial and discrete-time couplings. IEEE Trans Fuzzy Syst 23(4):1274–1285
Li JM, Li JS (2014) Adaptive fuzzy iterative learning control with initial-state learning for coordination control of leader-following multi-agent systems. Fuzzy Sets Syst 248:122–137
Li JS, Li JM (2014) Adaptive iterative learning control for coordination of second-order multi-agent systems. Intern J Robust Nonlin Control 24(18):3282–3299
Li H, Wu CW, Yin S, Lam HK (2009) Observer-based fuzzy control for nonlinear networked systems under unmeasurable premise variables. IEEE Trans Fuzzy Syst 24(5):1233–1245
Liu XJ, Kong XB (2013) Nonlinear fuzzy model predictive iterative learning control for drum-type boiler-turbine system. J Process Contr. 23(8):1023–1040
Liu M, Cao XB, Shi P (2013) Fault estimation and tolerant control for fuzzy stochastic systems. IEEE Trans Fuzzy Syst 21(2):221–229
Meng DY, Moore KL (2016) Learning to cooperate: networks of formation agents with switching topologies. Automatica 64:278–293
Meng DY, Jia YM, Du JP, Zhang J (2014) On iterative learning algorithms for the formation control of nonlinear multi-agent systems. Automatica 50(1):291–295
Meng DY, Jia YM, Du JP (2015) Robust consensus tracking control for multiagent systems with initial state shifts, disturbances, and switching topologies. IEEE Trans Neural Netw Learn Syst. 26(4):809–824
Niu B, Li L (2017) Adaptive backstepping-based neural tracking control for MIMO nonlinear switched systems subject to input delays. IEEE Trans Neural Netw Learn Syst. 29(6):2638–2644
Niu B, Karimi HR, Wang HQ, Liu YL (2016) Adaptive output-feedback controller design for switched nonlinear stochastic systems with a modified average dwell-time method. IEEE Trans Syst Man Cybern 47(7):1371–1382
Qiu JB, Feng G, Yang J (2009) A new design of delay-dependent robust filtering for discrete-time T-S fuzzy systems with time-varying delay. IEEE Trans Fuzzy Syst 17(5):1044–1058
Su XJ, Wu LG, Shi P, Chen CP (2015) Model approximation for fuzzy switched systems with stochastic perturbation. IEEE Trans Fuzzy Syst 23(5):1458–1473
Sun MX, Wang DW (2001) Sampled-data iterative learning control for nonlinear systems with arbitrary relative degree. Automatica 37(2):283–289
Sun M, Ge S, Sam Mareels I (2006) Adaptive repetitive learning control of robotic manipulators without the requirement for initial repositioning. IEEE Trans Robot 22(3):563–568
Takagi T, Sugeno M (1985) Fuzzy identification of systems and its applications to modeling and control. IEEE Trans Syst Man Cybern 1:116–132
Tanaka K, Iwasaki M, Wang HO (2000) Stability and smoothness conditions for switching fuzzy systems. In: Proceedings of the 2000 American Control Conference, vol 4, pp 2474–2478
Tayebi A (2004) Adaptive iterative learning control for robot manipulators. Automatica 40(7):1195–1203
Wu SC, Tomizuka M (2010) An iterative learning control design for self-servowriting in hard disk drives. Mechatronics 20(1):53–58
Xiong WJ, Yu WW, Lu JH, Yu XH (2014) Fuzzy modelling and consensus of nonlinear multiagent systems with variable structure. IEEE Trans Circuits Syst. 61(4):1183–1191

Xiong WJ, Yu XH, Patel R, Yu WW (2016) Iterative learning control for discrete-time systems with event-triggered transmission strategy and quantization. Automatica 72:84–91
Yang DR, Lee KS, Ahn HJ, Lee JH (2003) Experimental application of a quadratic optimal iterative learning control method for control of wafer temperature uniformity in rapid thermal processing. IEEE Trans Semicond Manuf 16(1):36–44
Yoneyama J (2009) H_∞ filtering for fuzzy systems with immeasurable premise variables: an uncertain system approach. Fuzzy Sets Syst 160(12):1738–1748
Zheng F, Wang QG, Lee TH (2002) Output tracking control of MIMO fuzzy nonlinear systems using variable structure control approach. IEEE Trans Fuzzy Syst 510(6):686–697
Zhu XL, Lin H, Xie XP (2016) Sampled-data fuzzy stabilization of nonlinear systems under nonuniform sampling. IEEE Trans Fuzzy Syst 24(6):1654–1667

Chapter 6
Tracking Under Saturated Finite Interval and HNN-Structural Output

6.1 Introduction

ILC has received attention up until this point due to its many industrial uses, including robotic manipulators, hard disk drives, quick thermal processing, and chemical polymerization/crystallization, see Kurek and Zaremba (1993), Wang et al. (2004), Liu and Jia (2012), Meng et al. (2012), Li and Li (2012). The goal of ILC designs is to use repetition to enhance tracking performance in a finite-time interval despite having incomplete knowledge of the dynamics structure and parameter values. A direct adaptive iterative learning control is provided in Wang et al. (2004) for a class of repeated nonlinear systems with unknown nonlinearities and variable initial resetting errors. This control is based on a new output-recurrent fuzzy neural network. The formation problem is equivalently transformed into a stability control problem over finite-time intervals by using the iterative learning approach, as demonstrated in Liu and Jia (2012). According to Li and Li (2012), all follower agents maintain the necessary distance from the leader and reach velocity consensus uniformly on the finite interval $[0, T]$ for the formation problem, whereas all follower agents track the leader uniformly on the finite interval $[0, T]$ for the consensus problem.

Because of the strict constraints, many interface devices, actuators, and sensors are susceptible to saturation in real-world applications. Input saturation may reduce control precision, produce oscillations, or potentially induce system instability. Due to its integral nature along the iteration axis and the presence of non-repeatable components in an iterative process, input saturation is particularly significant in the field of ILC. As a result, the study of ILC with saturation has received a great deal of attention and is both useful and essential in practice (Park and Bien 2002; Driessen and Sadegh 2002; Tan et al. 2011; Xu et al. 2004). In order to guarantee good tracking performance when the reference trajectory cannot be realized due to input saturation, two reference governors are integrated with the ILC controllers in Tan et al. (2011). In Xu et al. (2004) show that the nonlinearity generated by input saturation does not influence the perfect tracking performance with the iterative

W. Xiong et al., *Iterative Learning Control for Network Systems Under Constrained Information Communication*, Intelligent Control and Learning Systems 12,
https://doi.org/10.1007/978-981-97-0926-7_6

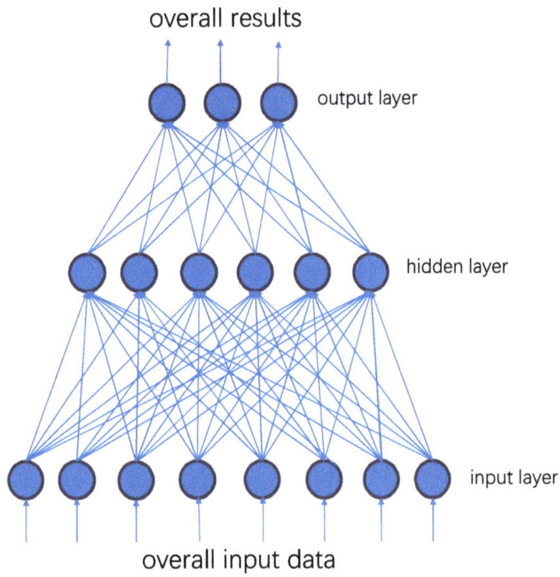

Fig. 6.1 A HNN with three layers

learning control scheme at the steady state. They do this by examining properties connected with the saturation function. However, it may be necessary (Xu et al. 2004) that the dynamic system is BIBO stable in [0,T], i.e., the system's state is bounded if the control input is bounded. It is obvious that many nonlinear functions, such as $F(x(t,k)) = \frac{1}{sint}x(t,k)$, do not satisfy the BIBO stable condition. We, therefore, aim to offer less conservative requirements to achieve the desired tracking in a finite interval with saturated ILCs, motivated by Meng et al. (2013), Hu et al. (2002), Zheng and Wu (2008).

In several fields, including signal processing, fault diagnosis, pattern recognition, image processing, and some optimization issues, many neural networks have been extensively studied and successfully implemented (Wang et al. 2006; Lu et al. 2009; Zeng and Zheng 2013; Cao and Wan 2014; Liu et al. 2014). Neural networks are employed for closed-loop output feedback control in recent years in order to do away with the necessity for numerical differentiation while maintaining high control precision (Li et al. 2011; Calise et al. 2001). Another goal of this chapter is to incorporate the output with a neural network structure to achieve the necessary tracking in a finite interval, which was inspired by the works of Li et al. (2011), Calise et al. (2001). The neural network topology is also hierarchical in order to deal with the nonlinear function in the output. In reality, neural networks frequently display hierarchical organization (Mavrovouniotis and Chang 1992; Zhang et al. 2005; Wu and Er 2000). The input layer, the hidden layer, and the output layer are the three levels of nodes that are often present (see Fig. 6.1). Hierarchical neural networks (HNNs) are neural networks having a hierarchical structure.

6.2 Problem Formulation

Consider the networked systems consisting of n nodes. Each node must manage two separate dynamic processes: the first process shows the dynamics of the system about time t; the second process describes the dynamics of the system about the iterative learning k. Hence, the ith node performing repetitive tasks over a fixed time interval $t \in [0, T]$ has the following dynamics over $k \in \mathbb{Z}_+$ ($\mathbb{Z}_+$ is the set of nonnegative integers).

$$x_i(t+1,k) = f_i(x_i(t,k)) + u_i(t,k), \tag{6.1}$$

where $x_i(t+1) \in \mathbb{R}^{m_i}$ ($i = 1, 2, \ldots, n,\ \sum_{i=1}^{n} m_i = m$) is the state vector of node i, and $f_i(x_i(t,k)) \in \mathbb{R}^{m_i}$ is a continuously nonlinear function. $u_i(t,k)$ are the iterative learning controllers. The output $y_i(t,k)$ of node i is given based on HNN structure, which is

$$\begin{cases} y_1(t,k) = A_1 x_1(t,k) + C_1 g_1(x_1(t,k)), \\ y_2(t,k) = B_1 x_1(t,k) + A_2 x_2(t,k) + C_2 g_2(x_2(t,k)), \\ \cdots\cdots\cdots\cdots\cdots\cdots\cdots, \\ y_{n-1}(t,k) = B_{n-2} x_{n-2}(t,k) + A_{n-1} x_{n-1}(t,k) + C_{n-1} g_{n-1}(x_{n-1}(t,k)), \\ y_n(t,k) = A_n x_n(t,k) + B_{n-1} x_{n-1}(t,k) + C_n g_n(x_n(t,k)), \end{cases} \tag{6.2}$$

where $A_i = \mathrm{diag}(a_1^i, \ldots, a_{m_i}^i)$ is a diagonal matrix with positive entries $a_j^i > 0$. $C_i \in \mathbb{R}^{m_i \times m_i}$ is the connection weight matrix and is supposed to be lower-triangular. The connection weight matrices B_i $(i = 1, 2, \ldots, n)$ show the connections between the states $x_i(\cdot)$ and $x_{i+1}(\cdot)$. Moreover, $g_i(x_i(t)) \in \mathbb{R}^{m_i}$ denotes the neuron activation function vector.

Remark 6.1 In model (6.2), the output is given based on the HNN structure. The states of the output vector $y_i(t)$ in the ith layer $(i = 2, \ldots, n)$ are related to the states of the output vector $y_i(t)$ and $y_{i-1}(t)$. However, the states of the output vector $y_i(t)$ $(i = 2, 3, \ldots, n)$ are independent to the states of the output vector $y_{i+1}(t)$. Which means each output neuron in the ith layer $(i = 2, 3, \ldots, n-1)$ receives information from the ith and $(i-1)$th layers, and sends information to the $(i+1)$th layer. The benefit of using the HNN model structure (6.2) is that it allows for the achievement of a straightforward necessary and sufficient condition to ensure interval tracking across a finite interval.

6.2.1 ILC Design

To study the tracking problem, the ILC scheme in this chapter is designed as follows:

$$u_i(t,k+1) = u_i(t,k) + \Delta u_i(t,k),\ i = 1, 2, \ldots, n, \tag{6.3}$$

where $\Delta u_i(t,k)$ will be given later. From systems (6.1) and (6.2), one has

$$\begin{cases} x(t+1,k) = F(x(t,k)) + u(t,k), \\ y(t,k) = Ax(t,k) + CG(x(t,k)), \end{cases} \tag{6.4}$$

with

$$u(t,k+1) = u(t,k) + \Delta u(t,k), \tag{6.5}$$

where $x(t,k) = (x_1^{\mathrm{T}}(t,k), \ldots, x_n^{\mathrm{T}}(t,k))^{\mathrm{T}} \in \mathbb{R}^m$, $u(t,k) = (u_1^{\mathrm{T}}(t,k), \ldots, u_n^{\mathrm{T}}(t,k))^{\mathrm{T}} \in \mathbb{R}^m$, $F(x(t,k)) = (f_1^{\mathrm{T}}(x_1(t,k)), \ldots, f_n^{\mathrm{T}}(x_n(t,k)))^{\mathrm{T}} \in \mathbb{R}^m$, $C = diag(C_1, \ldots, C_n) \in \mathbb{R}^m$, $\Delta u(t,k) = (\Delta u_1^{\mathrm{T}}(t,k), \ldots, \Delta u_n^{\mathrm{T}}(t,k))^{\mathrm{T}} \in \mathbb{R}^m$, $G(x(t,k)) = (g_1^{\mathrm{T}}(x_1(t,k)), \ldots, g_n^{\mathrm{T}}(x_n(t,k)))^{\mathrm{T}} \in \mathbb{R}^m$ and the matrix A is defined as

$$A = \begin{pmatrix} A_1 & O & O & \cdots & O \\ B_1 & A_2 & O & \cdots & O \\ \vdots & \vdots & \vdots & \cdots & \vdots \\ O & \cdots & B_{n-2} & A_{n-1} & O \\ O & O & \cdots & B_{n-1} & A_n \end{pmatrix} \in \mathbb{R}^{m \times m}.$$

Assumption 6.1 The functions $f_i(x_i(t,k))$ and $g_i(x_i(t,k))$ satisfy that $f_i(x_i(t,k)) = q_i(t)x_i(t,k)$ and $g_i(x_i(t,k)) = r_i(t)x_i(t,k)$ with $0 \le \underline{r}_i \le r_i(t) \le \overline{r}_i, \forall t \ge 0, i = 1, 2, \ldots, n$. Moreover, $f_i(\mathbf{0}) = \mathbf{0}$, $g_i(\mathbf{0}) = \mathbf{0}$, $\min_{1 \le i \le n} \underline{r}_i = \underline{r}$, $\max_{1 \le i \le n} \overline{r}_i = \overline{r}$.

According to Assumption 6.1, one has $F(x(t,k)) = q(t)x(t,k)$, $G(x(t,k)) = r(t)x(t,k)$ with $q(t) = \mathrm{diag}(q_1(t)I_{m_1}, \ldots, q_n(t)I_{m_n})$, $r(t) = \mathrm{diag}(r_1(t)I_{m_1}, \ldots, r_n(t)I_{m_n})$ and I_{m_i} $(i = 1, 2, \ldots, n)$ is the $m_i \times m_i$ identity matrix. Let $y^*(t) = Ax^*(t) + CG(x^*(t))$ be the desired control output status, and $x^*(t)$ be the desired trajectory of $x(t,k)$. For any desired trajectory $x^*(t)$ and appropriate boundary conditions (the boundary conditions will be provided later), there always exist control inputs $u^*(t)$ such that

$$x^*(t+1) = F(x^*(t)) + u^*(t). \tag{6.6}$$

Define $\theta(t,k) = x(t,k+1) - x(t,k)$ and the tracking error $e(t,k) = y^*(t) - y(t,k)$. In (6.5), design the ILC as

$$\Delta u(t,k) = Le(t+1,k), \tag{6.7}$$

where L is the learning gain and will be given in the following section. The boundary condition for system (6.4) is given as $x(0,k) = x_0$, for $k \in \mathbb{Z}_+$. $x^*(t) - x(t,0) = u_0(t)$ and $\|u_0(t)\| \le \varepsilon$ (ε is a constant), $t \in [0,T]$. Based on the given boundary conditions, the objective is to design the appropriate learning gain L to achieve that $e(t,k) \to \mathbf{0}$ over the finite interval $t \in [1,T]$, as $k \to 0$.

6.2.2 Definitions and Lemmas

In this chapter, we need the following definitions and lemma.

Definition 6.1 A scalar l is called an interval parameter if it lies between two boundaries according to $l \in [\underline{l}, \overline{l}]$, where the minimum value of l is $\underline{l}$ and the maximum value of l is $\overline{l}$.

Definition 6.2 A interval matrix S^I is defined as a matrix which is a member of the interval plant $\mathfrak{S}^I$ given by

$$\mathfrak{S}^I = \left\{ S^I : S^I = (s_{ij}^I),\ s_{ij}^I \in [\underline{s_{ij}}, \overline{s_{ij}}],\ i, j = 1, 2, \ldots, m \right\},$$

where $\overline{s_{ij}}$ is the maximum extreme value of the ith row and jth column elements of the uncertain plant, and $\underline{s_{ij}}$ is the minimum extreme value of the ith row and jth column elements of the uncertain plant.

Definition 6.3 The upper bound matrix $\overline{S}$ is a matrix whose elements are $\overline{s_{ij}}$. The lower bound matrix $\underline{S}$ is a matrix whose elements are $\underline{s_{ij}}$. The vertex matrices $\mathfrak{S}^v$ are defined by

$$\mathfrak{S}^v = \left\{ S^v : S^v = (s_{ij}^v),\ s_{ij}^v \in \{\underline{s_{ij}}, \overline{s_{ij}}\},\ i, j = 1, 2, \ldots, m \right\}.$$

Lemma 6.1 (Han and Lee 1994; Shih et al. 1998) *The spectral radius of a given interval matrix S^I is constrained by the highest value of the spectral radii of the vertex matrices S^v.*

6.3 Convergence Analysis

6.3.1 Convergence Under Interval ILC Scheme

Theorem 6.1 *The learning gain matrix L is designed to be lower-triangular. Then, one obtains that*

$$\rho\big(I - (A + r(t+1)C)L\big) < 1,\ \ \forall\, t = 0, 1, \ldots, T-1, \tag{6.8}$$

if and only if

$$\max \left\{ \rho\big(I - (A + \underline{r}C)L\big),\ \rho\big(I - (A + \overline{r}C)L\big) \right\} < 1. \tag{6.9}$$

Proof According to Assumption 6.1, one has that $\underline{r}I_m \le r(t+1) \le \overline{r}I_m,\ \forall\, t = 1, 2, \ldots, T-1$. Note that matrices A and C are both lower-triangular. Hence, from

Lemma 6.1, one has that the spectral radius of matrix $I-(A+r(t+1)C)L$ is bounded by the maximum value of the spectral radii of matrices $I-(A+\underline{r}C)L$ and $I-(A+\overline{r}C)L$. Hence, one obtains that spectral radius of matrix $I-(A+r(t+1)C)L$ is less than 1 if and only if condition (6.9) is satisfied. The proof is completed. □

Remark 6.2 Due to the HNN-structural output in system (6.4), it should be noted that matrices A and C are both lower-triangular. In Theorem 6.1 one has that the spectral radius of the time-varying matrix $I-\big(A+r(t+1)C\big)L$ is less than 1 if and only if the spectral radii of matrices $I-\big(A+\underline{r}C\big)L$ and $I-\big(A+\overline{r}C\big)L$ are both less than 1. Given this information, the HNN structure in the output plays a vital part in the design, making it possible and straightforward to apply the ILC to the tracking analysis. As a result, the interval tracking problem will be discussed in the discussion that follows.

Based on the result in Theorem 6.1, one has the following result.

Theorem 6.2 *Consider system (6.4). Under Assumption 6.1, in the finite interval* $[1,T]$*, the tracking error* $e(t,k)$ *such that*

$$e(t,k)\to\mathbf{0},\ \ when\ k\to+\infty, \tag{6.10}$$

if and only if the lower-triangular learning gain L is designed to satisfy (6.9).

Proof From Assumption 6.1, one has

$$y(t,k)=Ax(t,k)+CG(x(t,k))=\big(A+Cr(t)\big)x(t,k). \tag{6.11}$$

Then, one can get

$$\begin{aligned}
&e(t+1,k+1)-e(t+1,k)\\
&=y^*(t+1)-y(t+1,k+1)-\big(y^*(t+1)-y(t+1,k)\big)\\
&=-\big[y(t+1,k+1)-y(t+1,k)\big]\\
&=-\big(A+Cr(t+1)\big)\big[x(t+1,k+1)-x(t+1,k)\big]\\
&=-\big(A+Cr(t+1)\big)\big[F(x(t,k+1))+u(t,k+1)-F(x(t,k))-u(t,k)\big]\\
&=-\big(A+Cr(t+1)\big)q(t)\theta(t,k)-\big(A+Cr(t+1)\big)Le(t+1,k).
\end{aligned} \tag{6.12}$$

According to $\theta(t,k)=x(t,k+1)-x(t,k)$, one has

$$\begin{aligned}
\theta(t,k)&=F(x(t-1,k+1))+u(t-1,k+1)-F(x(t-1,k))-u(t-1,k)\\
&=q(t-1)\theta(t-1,k)+Le(t,k),\ \ t\geq 1.
\end{aligned} \tag{6.13}$$

From (6.13) and $\theta(0,k)=\mathbf{0}$, $\forall\, k\in Z_+$, one can get

$$\theta(t,k) = \sum_{l=0}^{t-1} \prod_{j=l+1}^{t-1} q(j)Le(l+1,k),\ t \geq 1. \tag{6.14}$$

From (6.12) and (6.14), one has

$$\begin{aligned} e(t+1,k+1) = &-\big(A + Cr(t+1)\big)q(t)\theta(t,k) \\ &+\big(I_m - (A + Cr(t+1))L\big)e(t+1,k), \end{aligned} \tag{6.15}$$

with $\theta(t,k) = \sum_{l=0}^{t-1} \prod_{j=l+1}^{t-1} q(j)Le(l+1,k)$.

Let $U(k) = (e(1,k)^{\mathrm{T}}, \ldots, e(T,k)^{\mathrm{T}})^{\mathrm{T}}$. Then, one has

$$U(k+1) = \Omega \cdot U(k),\ k \in \mathbb{Z}_+, \tag{6.16}$$

where the matrix

$$\Omega = \begin{pmatrix} \pi_1 & O & O & \cdots & O & O \\ *_{21} & \pi_2 & O & \cdots & O & O \\ *_{31} & *_{32} & \pi_3 & \cdots & O & O \\ \vdots & \vdots & \vdots & \ldots & \vdots & \vdots \\ *_{T1} & *_{T2} & *_{T3} & \cdots & *_{T,T-1} & \pi_T \end{pmatrix}$$

with $\pi_i = I_m - (A + Cr(i))L,\ i = 1, \ldots, T$. Moreover, $*_{ij}\ (i = 2, \ldots, T,\ j = 1, \ldots, T-1)$ can be given from (6.14) and are not listed here since none of them are required for the following analysis. From (6.16), one gets $U(k) \leq \Omega^k U(0)$. Note that Ω is lower-triangular and $U(0)$ is bounded, one has $U(k) \to \mathbf{0}$ if and only if $\Omega^k \to O$, $k \to +\infty$. Moreover, $\Omega^k \to O$ if and only if $\rho(\pi_i) < 1,\ i = 1, 2, \ldots, T$. According to Theorem 6.1, in the finite interval $[1, T]$, one has $e(t,k) \to \mathbf{0},\ as\ k \to +\infty$ if and only if condition (6.9) is satisfied. □

Remark 6.3 In Theorem 6.2, one has that $e(t,k) \to \mathbf{0}$, when $k \to +\infty$. From (6.5) and (6.7), note that the ILC $u(t,k)$ is independent of the iteration k when $e(t,k) \to 0$. As a result, the ILC $u(t,k)$ is independent of the iteration k when $k \to +\infty$. However, the ILC $u(t,k)$ is still related to time t when $k \to +\infty$. The corresponding simulation results can be seen in Examples of Sect. 6.4.

Remark 6.4 The ILC method has two benefits over the pinning control (Lu et al. 2009) and the adaptive control (Cao and Lu 2006): (1) The application of a repeating type of ILC will enable high-accuracy tracking performance within a limited time period. The ILC scheme can therefore be designed for various real-world applications. (2) It can be seen from condition (6.9) that it is independent of the nonlinear function $F(x(t,k))$. If the nonlinear function $F(x(t,k))$ is considered a disturbance, it can be shown that the system (6.4) can always accomplish the desired tracking with any disturbance by designing the ILC strategy (6.7).

6.3.2 Convergence Under Saturated Interval ILC Scheme

In this part, we consider the following model with saturated ILC scheme

$$\begin{cases} x(t+1,k) = F(x(t,k)) + D\text{sat}(u(t,k)), \\ y(t,k) = Ax(t,k) + CG(x(t,k)), \end{cases} \tag{6.17}$$

Here, the coefficients of system (6.17) except for the term $D\text{sat}(u(t,k))$ have the same definitions as system (6.4). $D \in \mathbb{R}^{m\times m}$ does not need to be lower-triangular, and the saturation function used here is the standard symmetric one defined as follows for $w = (w_1, \ldots, w_m)^{\mathrm{T}}$

$$\text{sat}(w) = (\text{sat}(w_i)) = \begin{cases} 1, & if\ \ w_i > 1; \\ w_i, & if\ \ -1 \le w_i \le 1; \\ -1, & if\ \ w_i < -1. \end{cases} \tag{6.18}$$

For matrix $\Lambda \in \mathbb{R}^{m\times 2m}$, we define that

$$\begin{aligned} \mathcal{L}(\Lambda) &= \mathcal{L}(\Lambda^1, \Lambda^2) \\ &= \Big\{(\varsigma_1(t), \varsigma_2(t)) : \big|\Lambda_i^1 \varsigma_1(t) + \Lambda_i^2 \varsigma_2(t)\big| \le 1,\ i = 1, \ldots, m\Big\}, \end{aligned} \tag{6.19}$$

where $\varsigma_j(t) \in \mathbb{R}^m,\ j = 1, 2,\ \Lambda = (\Lambda^1, \Lambda^2) \in \mathbb{R}^{m\times 2m},\ \Lambda^1, \Lambda^2 \in \mathbb{R}^{m\times m}$ are lower-triangular, and Λ_i^j $(i = 1, 2, \ldots, m,\ j = 1, 2)$ represents the ith row of matrix Λ^j. The following technical lemma is required.

Lemma 6.2 (Hu et al. 2002) *Let $u \in \mathbb{R}^m$ and $v \in \mathbb{R}^m$, suppose that $|v_i| \le 1,\ i = 1, 2, \ldots, m$, then*

$$\text{sat}(u) \in co\Big\{E_s u + E_s^- v\Big\},\ \ s \in [1, N],$$

where E_s are all the different diagonal matrices with elements either 1 or 0, $E_s^- = I_m - E_s$, $N = 2^m$ and "co" stands for the convex hull. In such a case, there exist $\rho_1 \ge 0,\ \ldots, \rho_N \ge 0$ with $\sum_{s=1}^N \rho_s = 1$ to satisfy

$$\text{sat}(u) = \sum_{s=1}^{N} \rho_s \big(E_s u + E_s^- v\big). \tag{6.20}$$

Define that

$$\Xi = \Big\{\rho \in \mathbb{R}^N : \rho = (\rho_1, \ldots, \rho_N)^{\mathrm{T}}, \sum_{s=1}^{N} \rho_s = 1,\ 0 \le \rho_s \le 1,\ s = 1, \ldots, N\Big\}. \tag{6.21}$$

Lemma 6.2 rewrites the saturated control problem by using an auxiliary control v that fulfills $|v_i| < 1$. Assume $(x, x) \in \mathcal{L}(\Lambda)$, one has $\text{sat}(u) = \sum_{s=1}^{N} \rho_s(E_s u + E_s^-(\Lambda^1 + \Lambda^2)x)$ according to (6.19). In general, there are many choices of $\rho_s's$ satisfying (6.20). One choice of such $\rho_s's$ of particular interest has been shown in Wu et al. (2007), which is

$$\rho_s(\varsigma_1(t), \varsigma_2(t)) = \prod_{i=1}^{m} \Big[\pi_i(1 - \lambda_i(\varsigma_1(t), \varsigma_2(t))) + (1 - \pi_i)\lambda_i(\varsigma_1(t), \varsigma_2(t))\Big], \tag{6.22}$$

where $s \in \{1, \ldots, N\}$.

Here, for each $s \in N$, let $\pi_i \in \{0, 1\}$ satisfy $s = \pi_1 2^{m-1} + \pi_2 2^{m-2} + \cdots + \pi_m \lambda_i's$ are defined as

$$\lambda_i(\varsigma_1(t), \varsigma_2(t)) = \begin{cases} 1, & if u = \Lambda^1\varsigma_1(t) + \Lambda^2\varsigma_2(t), \\ \frac{\text{sat}(u) - \Lambda^1\varsigma_1(t) - \Lambda^2\varsigma_2(t)}{u - \Lambda^1\varsigma_1(t) - \Lambda^2\varsigma_2(t)}, & otherwise, \end{cases} \tag{6.23}$$

for $i \in \{1, 2, \ldots, m\}$. As shown in Wu et al. (2007), $\rho_s(\varsigma_1(t), \varsigma_2(t))'s$ are Lipschitz functions about $\varsigma_1(t)$ and $\varsigma_2(t)$ with $\rho = (\rho_1, \ldots, \rho_N)^{\mathrm{T}} \in \Xi$. It should be noted that the values of $\rho_s's$ are available for use in gain-scheduling control in real time and in some ways represent the severity of control saturation.

Assume $(x, x) \in \mathcal{L}(\Lambda)$, the result of Lemma 6.2 can be extended to apply in system (6.17). That is, system (6.17) can be rewritten as

$$\begin{cases} y(t,k) = Ax(t,k) + CG(x(t,k)), \\ x(t+1,k) = F(x(t,k)) + D\sum\limits_{s=1}^{N} \rho_s\big[E_s u(t,k) + E_s^-(\Lambda^1 + \Lambda^2)x(t,k)\big]. \end{cases} \tag{6.24}$$

Similar to the above part, let $y^*(t)$ be the expected control output status, $e(t, k) = y^*(t) - y(t, k)$ and $\theta(t, k) = x(t, k+1) - x(t, k)$. Applying the similar control rule (6.5) with $\Delta u(t, k) = L_1 e(t + 1, k)$ (here, L_1 does not need to be lower-triangular), one has the following result.

Theorem 6.3 *Consider system (6.24). Under Assumption 6.1, in the finite interval* $[1, T]$, *the tracking error* $e(t, k)$ *such that*

$$\big\|e(t,k)\big\| \to 0, \ when\ k \to +\infty, \tag{6.25}$$

if the learning gain L_1 *and matrix* D *are designed to satisfy*

$$\sup_{i\in\{0,1,\ldots,T\}} \big\|I - (A + Cr(i))DE_sL_1\big\| < 1, \ s \in [1, N], \ \ E_s \neq \mathbf{0}. \tag{6.26}$$

Proof Similar to Theorem 6.2, from systems (6.12) and (6.24), one has

$$\begin{aligned}
&e(t+1,k+1)-e(t+1,k)\\
&=-\big(A+Cr(t+1)\big)\Big[F(x(t,k+1))+D\sum_{s=1}^{N}\rho_s\Big(E_s u(t,k+1)\\
&\quad+E_s^{-}(\Lambda^1+\Lambda^2)x(t,k+1)\Big)-F(x(t,k))\\
&\quad-D\sum_{s=1}^{N}\rho_s\Big(E_s u(t,k)+E_s^{-}(\Lambda^1+\Lambda^2)x(t,k)\Big)\Big]\\
&=-\big(A+Cr(t+1)\big)\Big[q(t)+\sum_{s=1}^{N}\rho_s E_s^{-}(\Lambda^1+\Lambda^2)\Big]\theta(t,k)\\
&\quad-\big(A+Cr(t+1)\big)D\sum_{s=1}^{N}\rho_s E_s L_1 e(t+1,k).
\end{aligned}\tag{6.27}$$

According to $\theta(t,k)=x(t,k+1)-x(t,k)$, one has

$$\begin{aligned}
\theta(t,k)&=F(x(t-1,k+1))+D\sum_{s=1}^{N}\rho_s\Big[E_s u(t-1,k+1)\\
&\quad+E_s^{-}(\Lambda^1+\Lambda^2)x(t-1,k+1)\Big]\\
&\quad-F(x(t-1,k))-D\sum_{s=1}^{N}\rho_s\Big[E_s u(t-1,k)+E_s^{-}(\Lambda^1+\Lambda^2)x(t-1,k)\Big]\\
&=\Big[q(t-1)+D\sum_{s=1}^{N}\rho_s E_s^{-}(\Lambda^1+\Lambda^2)\Big]\theta(t-1,k)\\
&\quad+D\sum_{s=1}^{N}\rho_s E_s L_1 e(t,k),\ t\geq 1.
\end{aligned}\tag{6.28}$$

From (6.28) and $\theta(0,k)=\mathbf{0}$, $\forall\, k\in Z_+$, one can get

$$\theta(t,k)=\sum_{l=0}^{t-1}\prod_{j=l+1}^{t-1}\Big[q(j)+D\sum_{s=1}^{N}\rho_s E_s^{-}(\Lambda^1+\Lambda^2)\Big]D\sum_{s=1}^{N}\rho_s E_s L_1 e(l+1,k),\ t\geq 1. \tag{6.29}$$

From (6.27) and (6.29), one has

$$
\begin{aligned}
e(t+1,k+1) = &-\big(A+Cr(t+1)\big)\Bigg[q(t)+\sum_{s=1}^{N}\rho_s E_s^-(\Lambda^1+\Lambda^2)\Bigg]\theta(t,k)\\
&+\Bigg[I-\big(A+Cr(t+1)\big)D\sum_{s=1}^{N}\rho_s E_s L_1\Bigg]e(t+1,k). \quad (6.30)
\end{aligned}
$$

Let $\Delta U(k) = \big(\|e(1,k)\|,\ldots,\|e(T,k)\|\big)^{\mathrm{T}}$, one obtains

$$
\Delta U(k+1) \le \Omega_1 \Delta U(k),\ k \in \mathbb{Z}_+, \quad (6.31)
$$

where $\Psi_i = \big\| I - (A+Cr(i))D\sum_{s=1}^{N}\rho_s E_s L_1 \big\|$, and the matrix

$$
\Omega_1 = \begin{pmatrix}
\Psi_1 & 0 & 0 & \cdots & 0 & 0\\
\Upsilon_{21} & \Psi_2 & 0 & \cdots & 0 & 0\\
\Upsilon_{31} & \Upsilon_{32} & \Psi_3 & \cdots & 0 & 0\\
\vdots & \vdots & \vdots & \cdots & \vdots & \vdots\\
\Upsilon_{T1} & \Upsilon_{T2} & \Upsilon_{T3} & \cdots & \Upsilon_{T,T-1} & \Psi_T
\end{pmatrix},\quad i = 1,\cdots,T.
$$

Υ_{ij} $(i = 2,\ldots,T,\ j = 1,\ldots,T-1)$ can be given from (6.29) and are not listed here since none of them needs to be used in the following analysis.

From (6.31), one gets $\Delta U(k) \le \Omega_1^k \Delta U(0)$. Note that Ω_1 is lower-triangular and $\Delta U(0)$ is bounded, one has $\Delta U(k) \to 0$ if and only if $\Psi_i < 1,\ i = 1,2,\ldots,T$, as $k \to +\infty$. From Lemma 6.2 and $N = 2^m$, there exists a $j \in \{1,\ldots,N\}$ such that $E_j = 0$. According to (6.21), one has

$$
\begin{aligned}
\Psi_i &= \Bigg\| I - (A+Cr(i))D\sum_{s=1}^{N}\rho_s E_s L_1 \Bigg\|\\
&= \Bigg\| \sum_{s=1,s\ne j}^{N} \rho_s\big[I-(A+Cr(i))DE_sL_1\big] \Bigg\|\\
&\le \Bigg| \sum_{s=1,s\ne j}^{N}\rho_s \Bigg| \cdot \Big\| I-(A+Cr(i))DE_sL_1 \Big\|\\
&\le \big\| I-(A+Cr(i))DE_sL_1 \big\|,\ s \ne j,\ i = 1,2,\ldots,T. \quad (6.32)
\end{aligned}
$$

Combining (6.26) and (6.32), one achieves that $\Psi_i < 1,\ i = 1,2,\ldots,T$. Then, $\Delta U(k) \to \mathbf{0}$ as $k \to +\infty$. That is, $\|e(t,k)\| \to 0,\ k \to +\infty$. □

Remark 6.5 Xu et al. (2004) also takes the input saturation into account. With the input saturation, Theorem 2 in Xu et al. (2004) requires that the dynamic system (17) is BIBO stable in [0, T], i.e., the state of the system is bounded if the control input is bounded. Clearly, the BIBO stable condition is not satisfied if $F(x(t,k)) = \frac{1}{\sin t}x(t,k)$

in (6.17). In order to achieve the desired tracking in a finite interval, a less conservative condition (6.26) is given in this chapter.

6.4 Simulation Examples

In this section, two numerical examples are presented to demonstrate the effectiveness of the results in systems (6.4) and (6.17), respectively.

Example 6.1 Consider system (6.4), define $T = 30$, $L = I_3$, $f_i(x_i(t,k)) = \cos t \cdot x_i(t,k)$, $x(t,k) = (x_1(t,k), x_2(t,k), x_3(t,k))^{\mathrm{T}} \in \mathbb{R}^3$, $g_i(x_i(t,k)) = (0.5 \cdot \sin t + 1) \cdot x_i(t,k)$, $i = 1, 2, 3$. In addition, the matrices

$$A = \begin{pmatrix} 0.25 & 0 & 0 \\ 0.5 & 0.3 & 0 \\ 0 & 1 & 0.1 \end{pmatrix}, \quad C = \begin{pmatrix} 0.5 & 0 & 0 \\ 0.4 & 0.4 & 0 \\ 1 & 0.75 & 0.3 \end{pmatrix}.$$

Then, one has $0.5 \le r(t) \le 1.5$, $\forall t \ge 1$,

$$I - (A + \underline{r}C)L = \begin{pmatrix} 0.5 & 0 & 0 \\ -0.7 & 0.5 & 0 \\ -0.5 & -1.375 & 0.75 \end{pmatrix}, \quad I - (A + \overline{r}C)L = \begin{pmatrix} 0 & 0 & 0 \\ -1.1 & 0.1 & 0 \\ -1.5 & -2.125 & 0.45 \end{pmatrix}.$$

It is clear to know that $\max\{\rho(I - (A + \underline{r}C)L), \rho(I - (A + \overline{r}C)L)\} = 0.75 < 1$. According to Theorem 6.2, in the finite interval $[1, T]$, one has $e(t,k) \to 0$, as $k \to +\infty$.

The corresponding simulation results can be seen in Figs. 6.2 and 6.3. According to Figs. 6.2a, 6.3a and b, it is clear that the tracking error asymptotically reaches 0 over time interval $[1, T]$ as the iteration index k increases. From Remark 6.3, one has that the ILC $u(t,k)$ is independent of k when $k \to \infty$. The corresponding simulation results can be seen in Figs. 6.2b and 6.3c. Moreover, from Fig. 6.3d, one can see that the variation of $u(t,k)$ is not large when it changes with t (at the iteration $k = 60$).

Example 6.2 Consider system (6.17), define $T = 30$, $x(t,k) = (x_1(t,k), x_2(t,k))^{\mathrm{T}} \in \mathbb{R}^3$, $f_i(x_i(t,k)) = \cos t \cdot x_i(t,k)$, $g_i(x_i(t,k)) = (0.5 \cdot \sin t + 1) \cdot x_i(t,k)$, $i = 1, 2$, and the matrices

$$A = \begin{pmatrix} 0.13 & 0 \\ 0.5 & 0.8 \end{pmatrix}, \quad C = \begin{pmatrix} 0.4 & 0 \\ 0.3 & 0.7 \end{pmatrix}, \quad D = \begin{pmatrix} 0.5 & 0.5 \\ 0.5 & 0.5 \end{pmatrix}, \quad L_1 = \begin{pmatrix} 0.5 & 0.5 \\ 0.5 & 0.5 \end{pmatrix}.$$

One has $0.5 \le r(t) \le 1.5$, $\forall t \ge 1$. Moreover, E_s $(s = 1, 2, 3, 4)$ are all the 2×2 different diagonal matrices with elements either 1 or 0, and $E_s^- = I_2 - E_s$. Then, using a straightforward computation, the condition (6.26) in Theorem 6.3 is satisfied.

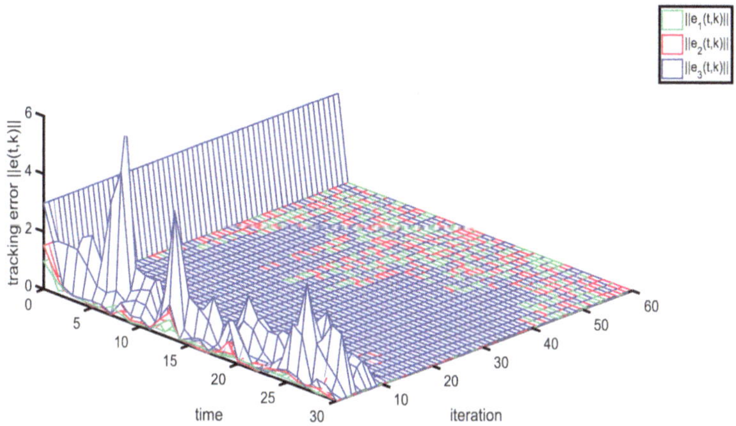

(a) The cubic diagram of error $\|e(t,k)\|$.

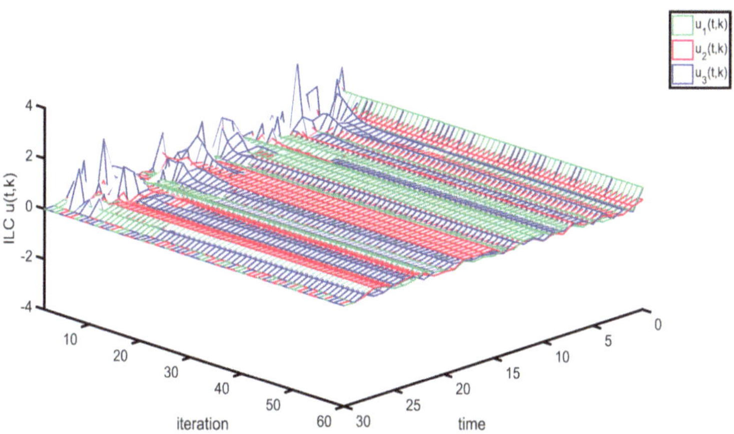

(b) ILC u(t,k) with respect to iteration k and time t.

Fig. 6.2 The error $\|e(t,k)\|$ and ILC u(t, k) in system (6.4)

Which means in the finite interval $[1, T]$, one has $e(t,k) \to 0$, as $k \to +\infty$. The corresponding simulation results can be seen in Figs. 6.4 and 6.5.

Figure 6.4 shows the tracking error and ILC controller with respect to iteration k and time t. According to Fig. 6.4, one knows that our saturated ILC scheme can effectively achieve the tracking. In addition, from Fig. 6.5a and b, it is evident that as the iteration index k rises, the tracking error asymptotically decreases to 0 over the time interval $[1, T]$. From Remark 6.3, one has that the saturated ILC $u(t,k)$ is also independent of the iteration k when $k \to \infty$. The corresponding simulation results

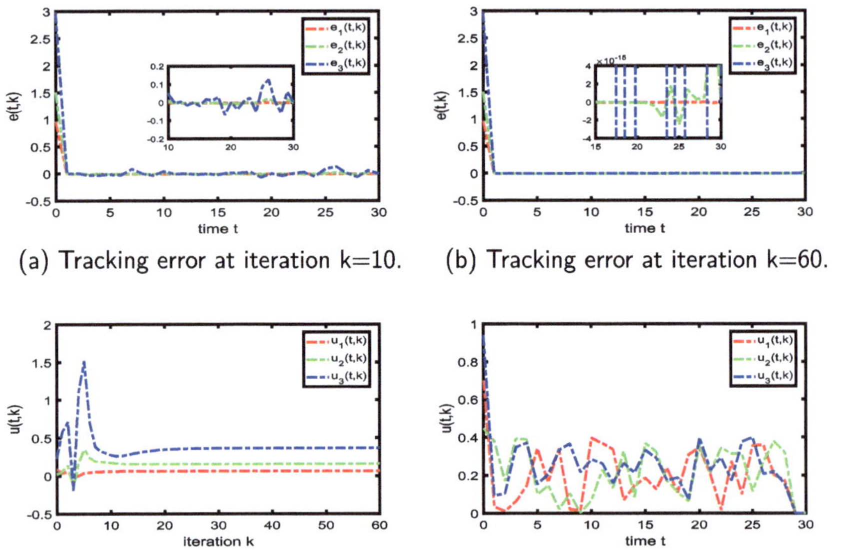

(a) Tracking error at iteration k=10. (b) Tracking error at iteration k=60.

(c) u(t,k) with respect to k (t=10). (d) u(t,k) with respect to time t (k=60).

Fig. 6.3 The trajectories of tracking error $e(t, k)$ and u(t, k) in system (6.4)

can be seen in Fig. 6.5c (at the time t = 20) or Fig. 6.4c. Furthermore, Fig. 6.4c and Fig. 6.5d (at the iteration k = 60) show that the saturated ILC $u(t, k)$ lies between [0, 1] as t varies.

6.5 Summary

This chapter has studied the problem of interval iterative learning for dynamic systems with HNN-structural output. A necessary and sufficient condition has been devised to achieve interval tracking in a finite interval using the output of a dynamic system with an HNN structure. Then, utilizing ILCs with saturation, a less conservative criterion has been presented to achieve tracking in a finite interval. Finally, numerical examples that go along with the simulations have been given to show how useful the results are.

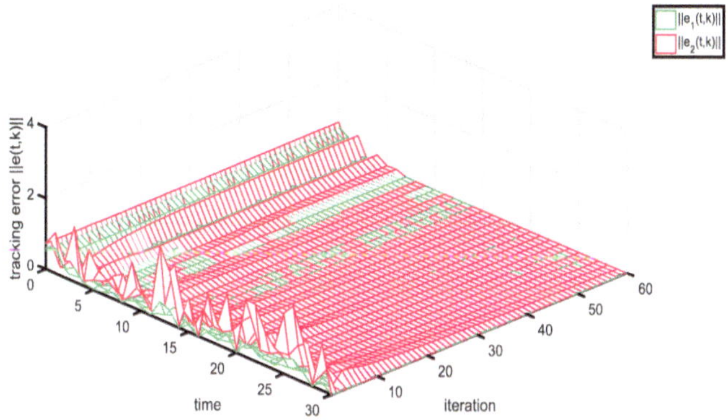

(a) The cubic diagram of error $\|e(t,k)\|$.

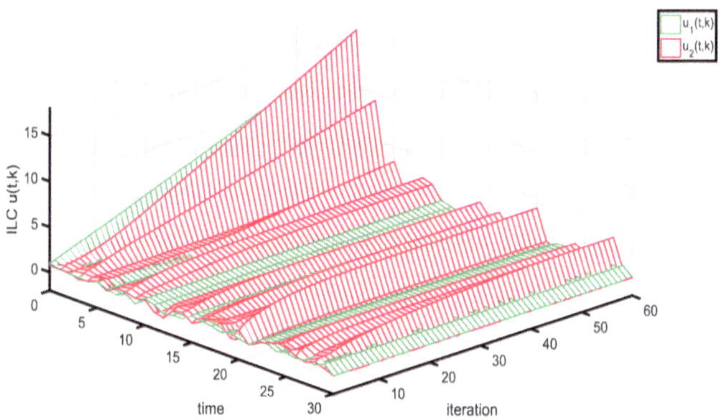

(b) ILC u(t,k) with respect to iteration k and time t.

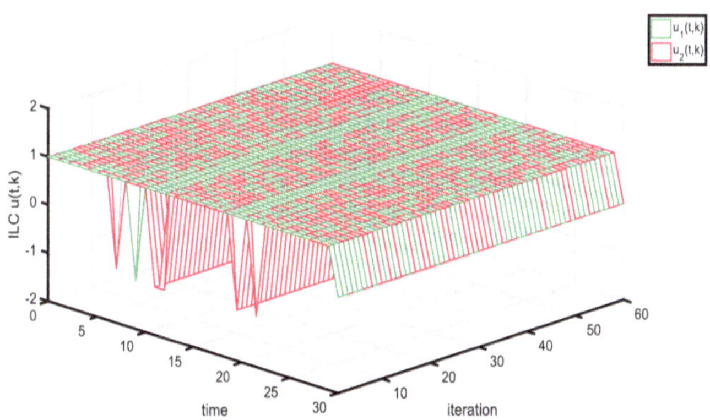

(c) Saturated ILC u(t,k) with respect to iteration k and time t.

Fig. 6.4 The error $\|e(t,k)\|$ and ILC u(t, k) in system (6.17)

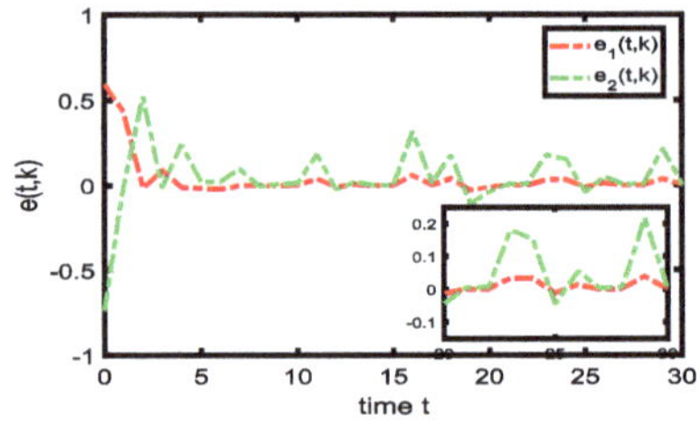

(a) Tracking error at iteration k=10.

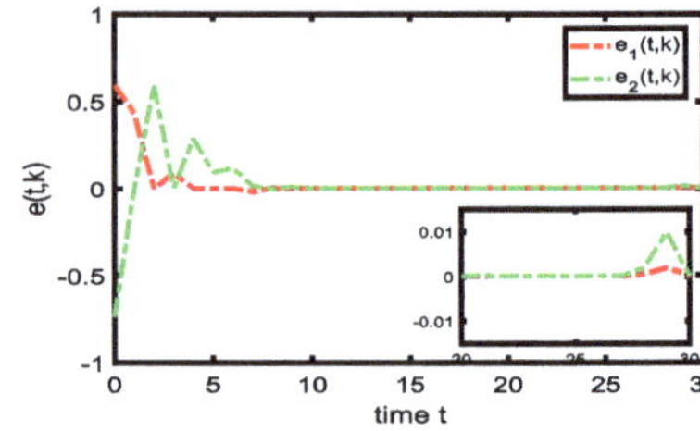

(b) Tracking error at iteration k=60.

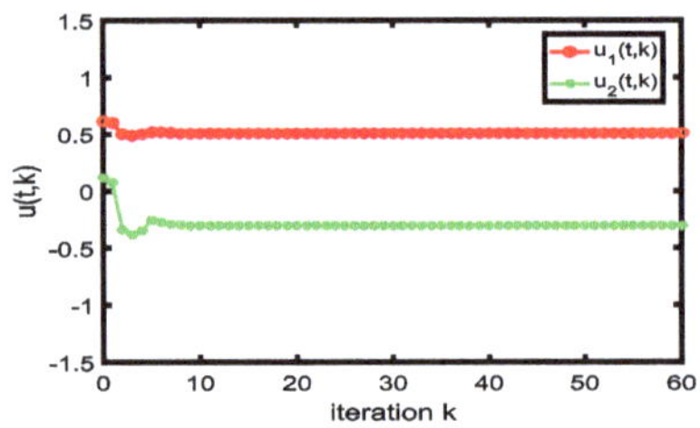

(c) Saturated u(t,k) with respect to iteration k (t=20).

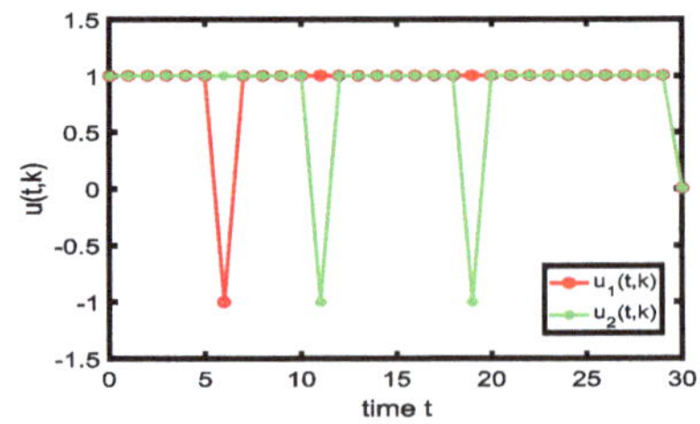

(d) Saturated u(t,k) with respect to time t (k=60).

Fig. 6.5 The trajectories of state error $e(t, k)$ and ILC u(t, k) in system (6.17)

References

Calise AJ, Hovakimyan N, Idan M (2001) Adaptive output feedback control of nonlinear systems using neural networks. Automatica 37(8):1201–1211

Cao JD, Lu JQ (2006) Adaptive synchronization of neural networks with or without time-varying delay. Chaos 16:013133

Cao JD, Wan Y (2014) Matrix measure strategies for stability and synchronization of inertial BAM neural network with time delays. Neural Netw 53:165–172

Driessen B, Sadegh N (2002) Multi-input square iterative learning control with input rate limits and bounds. IEEE Trans Syst Man Cybern B 32(4):545–550

Han H, Lee J (1994) Necessary and sufficient conditions for stability of time-varying discrete interval matrices. Int J Control 59(4):1021–1029

Hu TS, Lin ZL, Chen BM (2002) Analysis and design for discrete-time linear systems subject to actuator saturation. Syst Control Lett 45(2):97–112

Kurek J, Zaremba M (1993) Iterative learning control synthesis based on 2-D system theory. IEEE Trans Autom Control 38(1):121–125

Li JS, Li JM (2012) Adaptive iterative learning control for coordination of second-order multi-agent systems. Int J Robust Nonlin Syst. 24(18):3282–3299

Li J, Chen WS, Li JM (2011) Adaptive NN output-feedback decentralized stabilization for a class of large-scale stochastic nonlinear strict-feedback systems. Int J Robust Nonlin 21(4):452–472

Liu Y, Jia YM (2012) An iterative learning approach to formation control of multi-agent systems. Syst Control Lett 61(1):148–154

Liu XY, Park JH, Jiang N, Cao JD (2014) Nonsmooth finite-time stabilization of neural networks with discontinuous activations. Neural Netw 52:25–32

Lu JQ, Ho DWC, Wang ZD (2009) Pinning stabilization of linearly coupled stochastic neural networks via minimum number of controllers. IEEE Trans Neural Netw 20(10):1617–1629
Mavrovouniotis M, Chang S (1992) Hierarchical neural networks. Comput Chem Eng 16(4):347–369
Meng DY, Jia YM, Du JP, Yu FS (2012) Tracking control over a finite interval for multi-agent systems with a time-varying reference trajectory. Syst Control Lett 61(7):807–818
Meng ZY, Zhao ZY, Lin ZL (2013) On global leader-following consensus of identical linear dynamic systems subject to actuator saturation. Syst Control Lett 62(2):132–142
Meng DY, Jia YM, Du JM, Yu FS (2013) Tracking algorithms for multiagent systems. IEEE Trans Neural Netw Learn Syst. 24(10):1660–1676
Meng DY, Jia YM, Du JP, Zhang J (2014) On iterative learning algorithms for the formation control of nonlinear multi-agent systems. Automatica 50(1):291–295
Meng DY, Jia YM, Du JP (2014) Robust consensus tracking control for multiagent systems with initial state shifts, disturbances, and switching topologies. IEEE Trans Neural Netw Learn Syst. 26(4):809–824
Park K, Bien Z (2002) A study on robustness of iterative learning controller with input saturation against time-delay. In: The 4th Asian Control Conference. Singapore
Shih MH, Lur YY, Pang CY (1998) An inequality for the spectral radius of an interval matrix. Linear Algebra Appl 274(1):27–36
Tan Y, Xu JX, Norrlöf M, Freeman C (2011) On reference governor in iterative learning control for dynamic systems with input saturation. Automatica 47(11):2412–2419
Wang YC, Chien CJ, Teng CC (2004) Direct adaptive iterative learning control of nonlinear systems using an output-recurrent fuzzy neural network. IEEE Trans Syst Man Cybern B 34(3):1348–1359
Wang ZD, Liu YR, Li MZ, Liu XH (2006) Stability analysis for stochastic Cohen-Grossberg neural networks with mixed time delays. IEEE Trans Neural Netw 17(3):814–820
Wu S, Er M (2000) Dynamic fuzzy neural networks-a novel approach to function approximation. IEEE Trans Syst Man Cybern B 30(2):358–364
Wu F, Lin ZL, Zheng Q (2007) Output feedback stabilization of linear systems with actuator saturation. IEEE Trans Autom Control 52(1):122–128
Xu JX, Tan Y, Lee T (2004) Iterative learning control design based on composite energy function with input saturation. Automatica 40(8):1371–1377
Zeng ZG, Zheng WX (2013) Multistability of two kinds of recurrent neural networks with activation functions symmetrical about the origin on the phase plane. IEEE Trans Neural Netw Learn Syst. 24(11):1749–1762
Zhang CL, Jiang J, Kamel M (2005) Intrusion detection using hierarchical neural networks. Pattern Recogn Lett 26(6):779–791
Zheng Q, Wu F (2008) Output feedback control of saturated discrete-time linear systems using parameter-dependent Lyapunov functions. Syst Control Lett 57(11):896–903

Chapter 7
Tracking Based on Discontinuous Learning Control Strategy

7.1 Introduction

As an effective control strategy, ILC is famous for its ability to achieve perfect tracking of expected goals, which has a long research history, as seen from Xiong (2015), Meng et al. (2013), Meng and Jia (2014), Meng et al. (2014). ILC is a sophisticated technology used to improve the performance of controlled dynamic systems that are operated to repeatedly follow the same finite length expected trajectory over-and-over. It is shown in Meng and Moore (2016) that using the iterative learning update algorithm based on the nearest neighbor's relative distance, the multi-agent system can achieve the required cooperative formation control tasks under very loose conditions, which only requires the agent's interaction topology graph to jointly own a spanning tree on a finite length of iteration interval. In Zhang et al. (2016), an adaptive iterative learning control scheme is proposed for a class of nonlinear parametric systems with unknown time-varying parameters and input saturation. Using the adaptive iterative learning control method, in Li and Li (2014) discussed the consistency and formation control of distributed multi-agent system with second-order dynamics and unknown time-varying parameters.

The tracking problem of dynamical systems has attracted a lot of attention due to its potential applications in formation flying for spacecraft, sensor networks, formation control, and other areas. And there have been numerous research findings about the tracking issue (Chen et al. 2013; Yoo 2016; Wen et al. 2014; Xu et al. 2016; Yi and Zheng 2016). To ensure stability and demonstrate the quantitative bound of the tracking errors, in Cao et al. (2009) evaluate the condition on the communication graph, the sampling time and the control gain. Studying a distributed coordinated tracking problem for numerous networked Euler–Lagrange systems is the aim of Mei et al. (2011). Demigha et al. prefer collaborative target tracking over single node tracking in Demigha et al. (2013).

Notably, the above literature consistently discusses the tracking problem when time tends to infinity. However, one must accomplish the necessary tracking in a finite

W. Xiong et al., *Iterative Learning Control for Network Systems Under Constrained Information Communication*, Intelligent Control and Learning Systems 12,
https://doi.org/10.1007/978-981-97-0926-7_7

amount of time in many real-world applications, including batch reactor processes, wafer fabrication processes, IC welding procedures, and others. As a result, the tracking issues have been investigated during a limited time period in Forti et al. (2005), Cortés (2006), Xiao et al. (2009), Wang and Xiao (2010). The tracking techniques in these results, however, are connected to the sign function, which is challenging to use in practical contexts. Therefore, it makes sense to introduce ILC while talking about the tracking problem of continuous networks over a finite period of time. ILC has always been used in discrete-time systems in the most recent results (Meng and Jia 2014; Meng et al. 2014). Consequently, two difficult inquiries about ILC are raised: (1) How might suitable iterative learning controllers be designed for continuous systems? (2) Which types of ILC are more logical if realistic networks must take into account instantaneous disturbances and sudden change at ambiguous moments (i.e., impulsive effects Li et al. 2014)?

Two distinct ILCs are created in this chapter in order to address the aforementioned two questions. Consideration of impulsive phenomena into the ILC comes first. It is important to highlight that the ILC method with impulsive control has received relatively little attention from coupled networks. The second is to look at the ILC's sampled-data control technique. Due to the reliability of information channels, the capacity of transmission bandwidth, etc., it is quite challenging to collect continuous information about time t in actual applications. Therefore, implementing sampled-data control in the ILC is important and necessary. Sampled-data control has been researched and used in a number of real-world networks (Li et al. 2011; Wang and Wu 2016; Lu et al. 2016; Jiang 2015; Xu et al. 2014; Ge et al. 2015). A limited amount of work has been done on the ILC scheme with zero-order hold (ZOH) sampled-data control. Controllers have been running continuously for time t.

In light of the problems above, this chapter makes two contributions: (1) The coupled continuous networks are controlled by the impulsive and ZOH sampled-data ILCs. The ZOH sampled-data ILC, in this case, denotes that the most recent control signal will be utilized till the arrival of the newest signal. (2) To achieve the necessary tracking in a finite interval, two criteria are proposed based on the impulsive and ZOH sampled-data ILCs.

7.2 Problem Formulation

Consider an n-dimensional coupled continuous network, where $n > 0$ is an integer. The state vector of the ith node is $x_i(t) = (x_i^1(t), x_i^2(t), \ldots, x_i^N(t))^{\mathrm{T}} \in \mathbb{R}^N$, $i = 1, 2, \ldots, n$, and N is a positive integer, which is the dimension of $x_i(t)$. The network model is

$$\frac{dx_i(t)}{dt} = f_i(x_i(t)) + \sum_{j=1}^{n} a_{ij}x_j(t), \ i \in \{1, 2, \ldots, n\}, \tag{7.1}$$

where $t \in [0, T]$ is the fixed time interval with $T > 0$, $f_i(x_i(t,k)) = r_i(t)x_i(t,k) \in \mathbb{R}^N$ is a continuous function, and $A = (a_{ij})_{n\times n}$ is the coupling configuration of the network. If there is a directed connection from node j to node i (here, $i \neq j,\ i, j \in \{1, 2, \ldots, n\}$), then $a_{ij} > 0$; otherwise, $a_{ij} = 0$. Moreover,

$$a_{ii} = -\sum_{j=1, j\neq i}^{n} a_{ij},\quad i = 1, \ldots, n.$$

As was already said, the goal of this chapter is to design appropriate ILC strategies to investigate about the tracking of system (7.1) in a finite amount of time with repeatable control. Additionally, we will construct two different ILCs (the impulsive and ZOH sampled-data ILCs) for the system (7.1) in order to account for immediate disturbances and the challenges associated with measuring continuous information transfer.

Take into account two separate dynamic processes for iterative learning controllers: the first one explains the dynamics of time t; the second one displays the dynamics of iterative learning k. In light of this, the system with the impulsive and ZOH sampled-data ILCs can be represented as

$$\frac{dx_i(t,k)}{dt} = r_i(t)x_i(t,k) + \sum_{j=1}^{n} a_{ij}x_j(t,k) - \delta(t - t_l)u_i(t_l,k), \tag{7.2}$$

and

$$\frac{dx_i(t,k)}{dt} = r_i(t)x_i(t,k) + \sum_{j=1}^{n} a_{ij}x_j(t,k) - u_i(t,k), \tag{7.3}$$

where $\delta(s) = \begin{cases} +\infty, & s = 0 \\ 0, & s \neq 0 \end{cases}$ is the delta function, $i = 1, 2, \ldots, n,\ l,\ k \in \mathbb{Z}_+$ ($\mathbb{Z}_+ = \{0, 1, \ldots\}$). For model (7.2), every node receives the impulsive ILC information at the time sequence $t_0, \ldots, t_l, \ldots$, which is denoted by $\{t_l\}$. In system (7.3), the ZOH sampled-data ILC $u_i(t,k)$ only updates its information at the time sequence $\{t_l\}$. That is, $u_i(t,k) = u_i(t_l,k),\ \forall\, t \in [t_l, t_{l+1})$, which yields that $u_i(t,k)$ in system (7.3) is called the ZOH sampled-data ILC.

Remark 7.1 The difference between the impulsive and ZOH sampled-data ILCs is: the impulsive ILC is instantaneous, which only occurs at the time points $t_0, \ldots, t_l, \ldots$; however, the ZOH sampled-data ILC $u_i(t,k) = u_i(t_l,k),\ \forall\, t \in [t_l, t_{l+1})$, which is persistent and only updates its information at the time sequence $\{t_l\}$.

Define the desired trajectory of $x_i(t,k)$ as $x_i^r(t)$, and suppose that $x_i^r(t)$ satisfies

$$\frac{dx_i^r(t)}{dt} = r_i(t)x_i^r(t) + \sum_{j=1}^{n} a_{ij}x_i^r(t), \tag{7.4}$$

and

$$\frac{dx_i^r(t)}{dt} = f_i(x_i^r(t)) + \sum_{j=1}^{n} a_{ij}x_i^r(t) + u_i^r(t), \tag{7.5}$$

where $l,\ k \in \mathbb{Z}_+,\ i = 1, 2, \ldots, n$. The boundary conditions of models (7.2), (7.3), and (7.4) are $x_i(0,k) = x_i^r(0) = x_0, k \in \mathbb{Z}_+; u_i(t,0) = u_{i0}(t)$, and $\|u_{i0}(t)\| \le \varepsilon_i$ (ε_i is a positive scalar), $\forall\ i = 1, \ldots, n,\ \ \forall\ t \in [0, T]$. The tracking error is introduced as

$$e_i(t,k) = x_i^r(t) - x_i(t,k).$$

In the following, the objective is to design the impulsive and ZOH sampled-data ILCs $u_i(t_l,k)$ and $u_i(t,k)$ to obtain $e_i(t,k) \to \mathbf{0},\ \forall\ t \in [0,T],\ i = 1, 2, \ldots, n$, as $k \to +\infty$.

Remark 7.2 In comparison with the ILC schemes in Meng and Moore (2016), Li and Li (2014), one of the advantages is that the controllers $u_i(t)$ in systems (7.2) and (7.3) only receive or update the impulsive or sampled-data ILC message at the sequence $\{t_l\}$ ($l \in \mathbb{Z}_+$), rather than receiving the continuous information at any time t. Clearly, it is not economical and practical to transmit the controller $u_i(t)$'s feedback information persistently because of the capability of bandwidth, the unreliability of channels, etc.

In the interval $[0,T]$, the times that $u_i(t_l,k),\ \forall\ i \in \{1, 2, \ldots, n\}$, update its information must be limited. Hence, we also give the following assumptions.

Assumption 6.1 $t_0 = 0,\ \forall\ i = 1, 2, \ldots, n.$

Assumption 6.2 $t_l > t_{l-1}$, for $l \in \{1, 2, \ldots, \Pi\}$, and $t_\Pi = T$, $\Pi > 0$ is an integer, $\forall\ i = 1, 2, \ldots, n.$

Assumption 6.3 $\forall\ l \in \{0, 1, \ldots, \Pi - 1\}$, $0 < t_{l+1} - t_l \le \overline{\omega}$, where $\overline{\omega} \in \mathbb{R}$.

Lemma 7.1 (Gronwall (1919)) *Let $u(t)$ and $f(t)$ be nonnegative, continuous functions on $I = [0, +\infty)$ for which the inequality*

$$u(t) \le u_0 + \int_0^t f(s)u(s)ds,\ t \in I$$

holds, where u_0 is a nonnegative constant. Then

$$u(t) \leq u_0 \cdot \exp\left\{ \int_0^t f(s)ds \right\}, \ t \in I.$$

7.3 Finite-Time Tracking with Impulsive ILC

7.3.1 Control Strategy Design

For the sake of simplicity, we only take into account the scenario in which the dimension of $x_i(t)$ is 1. Keep in mind that if the model is described by the Kronecker product, our analysis works for any dimension as well. In this section, the tracking of the model (7.2) is discussed using the impulsive ILCs.

Let the states and parameters be as follows:

$$x(t,k) = \big(x_1(t,k), x_2(t,k), \ldots, x_n(t,k)\big)^{\mathrm{T}} \in \mathbb{R}^n,$$
$$r(t) = \operatorname{diag}\big(r_1(t), r_2(t), \ldots, r_n(t)\big) \in \mathbb{R}^n.$$

With impulsive ILCs, systems (7.2) and (7.4) are representable as

$$\frac{dx(t,k)}{dt} = r(t)x(t,k) + Ax(t,k) - \delta(t-t_l)u(t_l,k), \tag{7.6}$$

$$\frac{dx^r(t)}{dt} = r(t)x^r(t) + Ax^r(t), \tag{7.7}$$

where $x^r(t) = (x_1^r(t), \ldots, x_n^r(t))^{\mathrm{T}} \in \mathbb{R}^n$, $k \in \mathbb{Z}_+$, $l = 0, 1, \ldots, \Pi$.

Let $e(t,k) \triangleq x^r(t) - x(t,k) = (e_1(t,k), e_2(t,k), \ldots, e_n(t,k))^{\mathrm{T}} \in \mathbb{R}^n$. According to (7.6) and (7.7), one knows

$$\frac{de(t,k)}{dt} = r(t)e(t,k) + Ae(t,k) + \delta(t-t_l)u(t_l,k), \tag{7.8}$$

where $l = 0, 1, 2, \ldots, \Pi,\ k \in \mathbb{Z}_+$. The ILC vector $u(t_l,k)$ is designed as

$$u(t_l,k+1) - u(t_l,k) = L\big(e(\beta,k) - e(t_l,k)\big), \tag{7.9}$$

where β is a scalar, which can be arbitrarily chosen in (t_l, t_{l+1}), $l = 0, 1, 2, \ldots, \Pi$, $k \in \mathbb{Z}_+,\ e(t_l,k) = x^r(t_l) - x(t_l,k)$, and the learning gain L will be designed in the following theorem.

7.3.2 Convergence Analysis

Theorem 7.1 *With Assumptions 6.1–6.3, the tracking error* $e(t,k)$ *of model (7.8) satisfies the following inequality:*

$$\|e(t,k)\| \le e^{\mu\varpi}(e^{\mu T}+1)\|u(t_l,k)\| + \sum_{\varrho=0}^{l-1}\|u(t_\varrho,k)\|,\ \forall\, t\in[t_l,t_{l+1}), \tag{7.10}$$

where $t\in[0,T]$, $k\in\mathbb{Z}_+$, $l=0,1,\ldots,\Pi-1$, *and*

$$\mu \triangleq \bar{r} + \|A\|. \tag{7.11}$$

Proof $\forall\, t\in[t_l,t_{l+1})$, one has

$$\begin{aligned} e(t,k) &= x^r(t) - x(t,k) \\ &= x^r(t_l) - x(t_l,k) + \int_{t_l}^{t}\left(\frac{dx^r(\theta)}{d\theta} - \frac{dx(\theta,k)}{d\theta}\right)d\theta \\ &= e(t_l,k) + \int_{t_l}^{t}(r(\theta)+A)e(\theta,k)d\theta + u(t_l,k)\int_{t_l}^{t}\delta(s-t_l)ds \\ &= e(t_l,k) + \int_{t_l}^{t}(r(\theta)+A)e(\theta,k)d\theta + u(t_l,k), \end{aligned} \tag{7.12}$$

where I_n is the n-dimensional identity matrix. According to (7.11) and (7.12), one can get

$$\|e(t,k)\| \le \|e(t_l,k)\| + \|u(t_l,k)\| + \int_{t_l}^{t}\mu\|e(\theta,k)\|d\theta. \tag{7.13}$$

According to Lemma 7.1, one obtains

$$\|e(t,k)\| \le e^{\mu\varpi}\Big(\|e(t_l,k)\| + \|u(t_l,k)\|\Big),\quad \forall t\in[t_l,t_{l+1}). \tag{7.14}$$

In addition, $\forall\, t\in[0,T]$, if there exists υ ($\upsilon\in\{1,2,\ldots,\Pi\}$) such that $t_{\upsilon-1}<t\le t_\upsilon\le T$, one can get

$$\|e(t,k)\| \le \|e(0,k)\| + \left\|\int_0^t(r(\theta)+A)e(\theta,k)d\theta\right\| + \sum_{l=0}^{\upsilon}\|u(t_l,k)\|. \tag{7.15}$$

From $e(0,k)=x^r(0)-x(0,k)=0$ and Lemma 7.1, one has

$$\|e(t,k)\| \le e^{\mu T}\sum_{l=0}^{\upsilon}\|u(t_l,k)\|, \tag{7.16}$$

where $t_{\upsilon-1} < t \le t_\upsilon \le T, k, \upsilon \in \mathbb{Z}_+$. Let $t = t_l$ in (7.16), one obtains

$$\|e(t_l, k)\| \le e^{\mu T} \sum_{\varrho=0}^{l} \|u(t_\varrho, k)\|. \tag{7.17}$$

According to (7.11), (7.14), and (7.17), one can get

$$\|e(t, k)\| \le e^{\mu\varpi}(e^{\mu T} + 1)\|u(t_l, k)\| + \sum_{\varrho=0}^{l-1} \|u(t_\varrho, k)\|, \quad \forall\, t \in [t_l, t_{l+1}). \tag{7.18}$$

Hence, the proof is complete. □

According to Theorem 7.1 and the impulsive ILC (7.9), one has the following result.

Theorem 7.2 *With Assumptions 6.1–6.3, the impulsive controller (7.9) can ensure the error $e(t, k)$ of model (7.8) such that*

$$\|e(t, k)\| \to 0 \; as \; k \to +\infty, \tag{7.19}$$

if the learning gain L is designed to satisfy

$$\Upsilon = \|I_n - L\| + \varpi\mu e^{\mu\varpi}(e^{\mu T} + 1)\|L\| < 1, \tag{7.20}$$

where $t \in [0, T]$, μ is shown in (7.11).

Proof One has from $e(t, k) = x^r(t) - x(t, k)$,

$$\begin{aligned} e(\beta, k) &= x^r(\beta) - x(\beta, k) \\ &= x^r(t_l) - x(t_l, k) + \int_{t_l}^{\beta} \left(\frac{dx^r(\theta)}{d\theta} - \frac{dx(\theta, k)}{d\theta} \right) d\theta \\ &= e(t_l, k) + \int_{t_l}^{\beta} (r(\theta) + A)e(\theta, k)d\theta + u(t_l, k), \end{aligned} \tag{7.21}$$

where β is given in (7.9). It follows from (7.9) and (7.21) that

$$\begin{aligned} u(t_l, k+1) &= u(t_l, k) - L\left[\int_{t_l}^{\beta} (r(\theta) + A)e(\theta, k)d\theta + u(t_l, k) \right] \\ &= (I_n - L)u(t_l, k) - L \int_{t_l}^{\beta} (r(\theta) + A)e(\theta, k)d\theta. \end{aligned} \tag{7.22}$$

Then, by (7.22), one has

$$\left\|u(t_l, k+1)\right\| \leq \left\|I_n - L\right\| \cdot \left\|u(t_l, k)\right\| + \mu \left\|L\right\| \int_{t_l}^{\beta} \left\|e(\theta, k)\right\| d\theta. \tag{7.23}$$

Combining (7.23) and the result in Theorem 7.1 together gives

$$\left\|u(t_l, k+1)\right\| \leq \Upsilon \left\|u(t_l, k)\right\| + \Upsilon_1 \sum_{\varrho=0}^{l-1} \left\|u(t_\varrho, k)\right\|, \tag{7.24}$$

where Υ is given in (7.20), and $\Upsilon_1 = \overline{\omega}\mu \cdot \|L\|$. Define $\Delta U(k+1) = (\|u(t_0, k+1)\|, \|u(t_1, k+1)\|, \ldots, \|u(t_\Pi, k+1)\|)^{\mathrm{T}}$. One can get

$$\Delta U(k+1) \leq \Xi \cdot \Delta U(k),\ k \in \mathbb{Z}_+, \tag{7.25}$$

where the matrix

$$\Xi = \begin{pmatrix} \Upsilon & 0 & 0 & \cdots & 0 & 0 \\ \Upsilon_1 & \Upsilon & 0 & \cdots & 0 & 0 \\ \Upsilon_1 & \Upsilon_1 & \Upsilon & \cdots & 0 & 0 \\ \vdots & \vdots & \vdots & \cdots & \vdots & \vdots \\ \Upsilon_1 & \Upsilon_1 & \Upsilon_1 & \cdots & \Upsilon_1 & \Upsilon \end{pmatrix}.$$

Then, $\Delta U(k) \leq \Xi^k \Delta U(0), k \in \mathbb{Z}_+$. Since $\Delta U(0)$ is bounded from $\|u_{i0}(t)\| \leq \varepsilon_i$, one has from (7.20) that

$$\left\|u(t_l, k)\right\| \to 0 \ when\ \ k \to +\infty,\ l = 0, 1, \ldots, \Pi. \tag{7.26}$$

Combining (7.16) and (7.26) together, one can obtain

$$\left\|e(t, k)\right\| \to 0 \ when\ k \to +\infty,\ \forall\, t \in [0, T],$$

thereby completing the proof. □

7.4 Finite-Time Tracking with ZOH Sampled-Data ILC

7.4.1 Control Strategy Design

Similar to (7.6) and (7.7), systems (7.3) and (7.4) with ZOH sampled-data ILCs can be described as

$$\frac{dx(t, k)}{dt} = r(t)x(t, k) + Ax(t, k) - u(t, k), \tag{7.27}$$

$$\frac{dx^r(t)}{dt} = r(t)x^r(t) + Ax^r(t), \tag{7.28}$$

where $k \in \mathbb{Z}_+$, the ILC $u(t, k)$ updates its information at $\{t_l : l = 0, 1, 2, \ldots, \Pi\}$. And $\forall\, l = 1, 2, \ldots, \Pi$, the ILC $u(t_l, k)$ is

$$u(t_l, k+1) - u(t_l, k) = L_1\big(e(t_{l+1}, k) - e(t_l, k)\big), \tag{7.29}$$

where L_1 is also the learning gain and will be given in the following.

7.4.2 Convergence Analysis

From systems (7.27) and (7.28), one has

$$\frac{de(t,k)}{dt} = r(t)e(t,k) + Ae(t,k) + u(t,k),\ \ k \in \mathbb{Z}_+. \tag{7.30}$$

In analogy to Theorem 7.1, one can obtain the following theorem.

Theorem 7.3 *With Assumptions 6.1–6.3, in the interval* $[0, T]$, *the error* $e(t, k)$ *of (7.30) satisfies*

$$\big\|e(t,k)\big\| \leq \varpi e^{\mu\varpi}\Big(e^{\mu T}\sum_{\zeta=0}^{l-1}\big\|u(t_\zeta, k)\big\| + \big\|u(t_l, k)\big\|\Big), \tag{7.31}$$

where $\forall\, t \in [t_l, t_{l+1}),\ l = 1, 2, \ldots, \Pi - 1$, *and* μ *is shown in (7.11).*

Proof Similar to (7.14) in Theorem 7.1, one gets

$$\big\|e(t,k)\big\| \leq e^{\mu\varpi}\Big(\big\|e(t_l,k)\big\| + \big\|u(t_l,k)\big\|\varpi\Big),\ \ \forall t \in [t_l, t_{l+1}). \tag{7.32}$$

$\forall\, t \in [0, T]$, there exists $\sigma \in \{0, 1, 2, \ldots, \Pi - 1\}$ to satisfy that $t \in [t_\sigma, t_{\sigma+1})$. One can obtain

$$\begin{aligned}
e(t,k) &= x^r(t) - x(t,k)\\
&= x^r(0) - x(0,k) + \int_0^t \Big(\frac{dx^r(\theta)}{d\theta} - \frac{dx(\theta,k)}{d\theta}\Big)d\theta\\
&= e(0,k) + \int_0^t (r(\theta) + A)e(\theta,k)d\theta + \int_0^{t_1} u(t_\theta,k)d\theta + \int_{t_1}^{t_2} u(t_\theta,k)d\theta\\
&\quad + \cdots + \int_{t_\sigma}^{t} u(t_\theta,k)d\theta\\
&= e(0,k) + \int_0^t (r(\theta) + A)e(\theta,k)d\theta + \sum_{\zeta=0}^{\sigma} u(t_\zeta,k)\Delta t_\zeta,
\end{aligned} \tag{7.33}$$

where $\Delta t_\zeta = t_{\zeta+1} - t_\zeta,\ k \in \mathbb{Z}_+$. Since $e(0, k) = x^r(0) - x(0, k) = 0$, one has from Lemma 7.1 that

$$\|e(t,k)\| \le e^{\mu T} \sum_{\zeta=0}^{\sigma} \|u(t_\zeta, k)\| \overline{\omega},\ \forall\, t \in [0, T],\ k \in \mathbb{Z}_+. \tag{7.34}$$

Let $t = t_l$ in (7.34). One has

$$\|e(t_l,k)\| \le e^{\mu T} \sum_{\zeta=0}^{l-1} \|u(t_\zeta, k)\| \overline{\omega},\ k \in \mathbb{Z}_+. \tag{7.35}$$

According to (7.32) and (7.35), one has

$$\|e(t,k)\| \le \overline{\omega} e^{\mu \overline{\omega}} \left(e^{\mu T} \sum_{\zeta=0}^{l-1} \|u(t_\zeta, k)\| + \|u(t_l, k)\| \right), \tag{7.36}$$

where $\forall\, t \in [t_l, t_{l+1}),\ l = 0, 1, 2, \ldots, \Pi - 1$, and the proof is complete. □

Based on Theorem 7.3 and designing the ZOH sampled-data ILC (7.9), one has the following result.

Theorem 7.4 *With Assumptions 6.1–6.3 and the sampled-data ILC scheme (7.9), the error* $e(t, k)$ *of system (7.30) can achieve that in the interval* $[0, T]$

$$\|e(t,k)\| \to 0 \ when\ k \to +\infty, \tag{7.37}$$

if the following inequality holds

$$\pi_1 = \eta + \mu \|L_1\| \overline{\omega}^2 e^{\mu \overline{\omega}} < 1, \tag{7.38}$$

where $\eta = \sup_{l \in \{0,1,\ldots,\Pi\}} \|I_n - L_1(t_{l+1} - t_l)\|$, *and* μ *is given in (7.11).*

Proof Similar to (7.21), one has

$$e(t_{l+1}, k) = e(t_l, k) + \int_{t_l}^{t_{l+1}} (r(\theta) + A) e(\theta, k) d\theta + u(t_l, k)(t_{l+1} - t_l). \tag{7.39}$$

It follows from (7.9) and (7.39) that

$$\begin{aligned} u(t_l, k+1) &= u(t_l, k) - L_1 \left[\int_{t_l}^{t_{l+1}} (r(\theta) + A) e(\theta, k) d\theta + u(t_l, k) \Delta t_l \right] \\ &= (I_n - L_1 \Delta t_l) u(t_l, k) - L_1 \int_{t_l}^{t_{l+1}} \big(r(\theta) + A\big) e(\theta, k) d\theta, \end{aligned} \tag{7.40}$$

where $\Delta t_l = t_{l+1} - t_l$. Then one can achieve from (7.40) that

$$\left\|u(t_l, k+1)\right\| \le \eta\left\|u(t_l, k)\right\| + \mu\left\|L_1\right\| \int_{t_l}^{t_{l+1}} \left\|e(\theta, k)\right\| d\theta. \tag{7.41}$$

Combining (7.41) and the result in Theorem 7.3 together yields that $\forall\, t \in [t_l, t_{l+1})$,

$$\begin{aligned}\left\|u(t_l, k+1)\right\| &\le \eta\left\|u(t_l, k)\right\| + \mu\left\|L_1\right\|\overline{\varpi}^2 e^{\mu\overline{\varpi}}\left(e^{\mu T}\sum_{\zeta=0}^{l-1}\left\|u(t_\zeta, k)\right\| + \left\|u(t_l, k)\right\|\right)\\ &= \pi_1\left\|u(t_l, k)\right\| + \pi_2\sum_{\zeta=0}^{l-1}\left\|u(t_\zeta, k)\right\|,\end{aligned} \tag{7.42}$$

where $\pi_1 = \eta + \mu\|L_1\|\overline{\varpi}^2 e^{\mu\overline{\varpi}}$, and $\pi_2 = \mu\|L_1\|\overline{\varpi}^2 e^{\mu\overline{\varpi}} e^{\mu T}$.

Define $\Delta U(k+1) = \left(\|u(t_0, k+1)\|, \|u(t_1, k+1)\|, \ldots, \|u(t_\Pi, k+1)\|\right)^{\mathrm{T}}$. One has

$$\Delta U(k+1) \le \Omega \Delta U(k),\ k \in \mathbb{Z}_+, \tag{7.43}$$

where the matrix

$$\Omega = \begin{pmatrix} \pi_1 & 0 & 0 & \cdots & 0 & 0 \\ \pi_2 & \pi_1 & 0 & \cdots & 0 & 0 \\ \pi_2 & \pi_2 & \pi_1 & \cdots & 0 & 0 \\ \vdots & \vdots & \vdots & \cdots & \vdots & \vdots \\ \pi_2 & \pi_2 & \pi_2 & \cdots & \pi_2 & \pi_1 \end{pmatrix}.$$

Then, $\Delta U(k) \le \Omega^k \Delta U(0),\ k \in \mathbb{Z}_+$. Since $\Delta U(0)$ is bounded from $\|u_{i0}(t)\| \le \varepsilon_i$, one has from (7.38)

$$\left\|u(t_l, k)\right\| \to 0 \text{ when } k \to +\infty,\ l = 0, 1, \ldots, \Pi. \tag{7.44}$$

From (7.34), one also gets

$$\left\|e(t, k)\right\| \le e^{\mu T}\sum_{\zeta=0}^{\Pi}\left\|u(t_\zeta, k)\right\|\overline{\varpi},\ \ \forall\, t \in [0, T],\ k \in \mathbb{Z}_+. \tag{7.45}$$

Combining (7.44) with (7.45) produces

$$\left\|e(t, k)\right\| \to 0 \text{ when } k \to +\infty,\ \forall\, t \in [0, T],$$

thus completing the proof. □

Remark 7.3 The tracking problem can be solved in a finite amount of time under a repeatable control environment according to Theorems 7.2 and 7.4. The tracking problem is typically solved using the standard tracking control as $t \to +\infty$ (see Song et al. 2010). As a result, one goal of this chapter is to repeatedly use a method for achieving high tracking accuracy in a limited time frame while operating in a repeatable control environment. In fact, Cortés (2006), Xiao et al. (2009), Wang and Xiao (2010) contain a discussion of the tracking problem in a finite interval. However, those studies' tracking protocols relate to the sign function, which makes them challenging to use in actual applications.

Remark 7.4 A discrete ILC-based algorithm is described for a certain class of discrete models in Meng and Jia (2014). Our models (7.2) and (7.3), in contrast to Meng and Jia (2014), are hybrid systems. In other words, systems (7.2) and (7.3) consider impulsive and sampled-data iterative learning controllers for a particular class of coupled continuous networks. Designing suitable impulsive and sampled-data iterative learning controllers to achieve finite-time tracking is, therefore, quite challenging. Fortunately, checking the acquired conditions (7.20) and (7.38) in the application is not difficult.

Remark 7.5 From conditions (7.20) and (7.38) in Theorems 7.2 and 7.4, it can be seen that the two conditions are not difficult to obtain if one can choose appropriately $\overline{\omega}$ and the learning gains L and L_1. The detailed explanation can be seen in Example 1. In the example, when $\overline{\omega} = 0.1,\ L = 0.5I_3,\ L_1 = 5I_3$, one can easily obtain that $\Upsilon = 0.5038 < 1, \pi_1 = 0.5016 < 1$.

7.5 Simulation Example

In this section, an example is presented to illustrate the usefulness of our criteria.

Example 7.1 Consider systems (7.6), (7.7), (7.27), and (7.28). Here, define $T = 10$, $x(t,k) = (x_1(t,k), x_2(t,k), x_3(t,k))^{\mathrm{T}} \in \mathbb{R}^3$, $r_i(t)x_i(t,k) = 0.01 \cdot \sin t \cdot x_i(t,k)$, $t_{l+1} - t_l = 0.1$, $L = 0.5I_3$, $L_1 = 5I_3$, and the matrix

$$A = \begin{pmatrix} 0.015 & -0.01 & 0 \\ 0 & 0.01 & -0.015 \\ 0 & -0.01 & 0.012 \end{pmatrix}.$$

The boundary conditions are

$$x_i(0,k) = x_i^r(0) = x_0 = 0,\quad i = 1,2,3,\ k \in \mathbb{Z}_+,\quad \forall\, t \in [0,T],$$

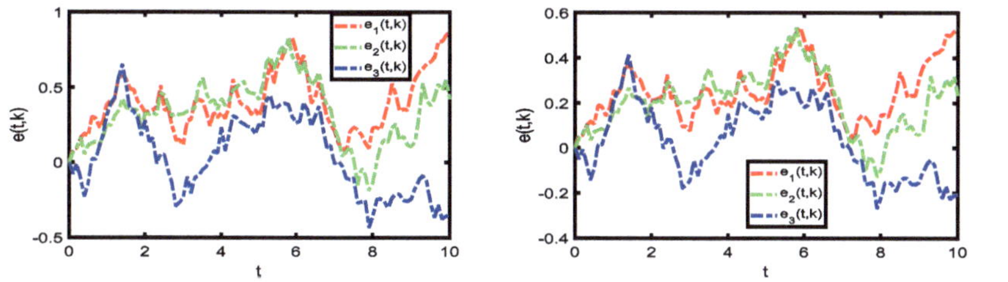

(a) The tracking error at iteration $k = 1$.(b) The tracking errorat iteration $k = 10$.

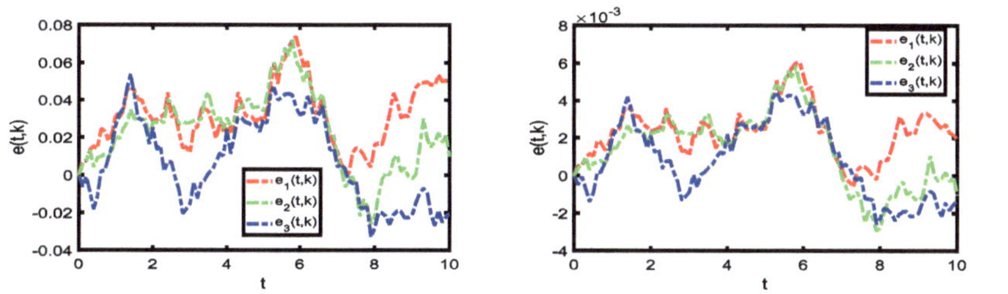

(c) The tracking errorat iteration $k = 50$.(d) The tracking errorat iteration $k = 100$.

Fig. 7.1 The trajectories of $e(t, k)$ of model (7.8) with impulsive ILCs

and $u_i(t, 0)$ are randomly chosen in [0, 3]. It follows that

$$
\begin{aligned}
&|u_i(t,0)| = |u_{i0}(t)| \leq 3, \quad \mu = \overline{r} + \left\|A\right\| = 0.01 + 0.0218 = 0.0318, \\
&\Pi = 100, \quad \overline{\omega} = 0.1, \quad \eta = \sup_{l \in \{0,1,\ldots,\Pi\}} \left\|I_n - L_1(t_{l+1} - t_l)\right\| = 0.5, \\
&\overline{\omega}\mu e^{\mu\overline{\omega}}(e^{\mu T} + 1) \cdot \left\|L\right\| = 0.0038, \quad \mu \cdot \left\|L_1\right\| \cdot \overline{\omega}^2 e^{\mu\overline{\omega}} = 0.0016.
\end{aligned}
$$

Then it is clear that

$$
\begin{aligned}
\Upsilon &= \left\|I_n - L\right\| + \overline{\omega}\mu e^{\mu\overline{\omega}}(e^{\mu T} + 1)\left\|L\right\| = 0.5038 < 1, \\
\pi_1 &= \eta + \mu\left\|L_1\right\| \cdot \overline{\omega}^2 e^{\mu\overline{\omega}} = 0.5016 < 1.
\end{aligned}
$$

From Theorems 7.2 and 7.4, by applying the ILC scheme (7.9), one has $\|e(t,k)\| \to 0$ as $k \to +\infty$. In Figs. 7.1 and 7.2, the state errors of models (7.8) and (7.30) are exhibited at different iteration k. According to Figs. 7.1 and 7.2, one can see that, over the finite interval [0, 10], the tracking errors tend to zero when the iteration index k increases.

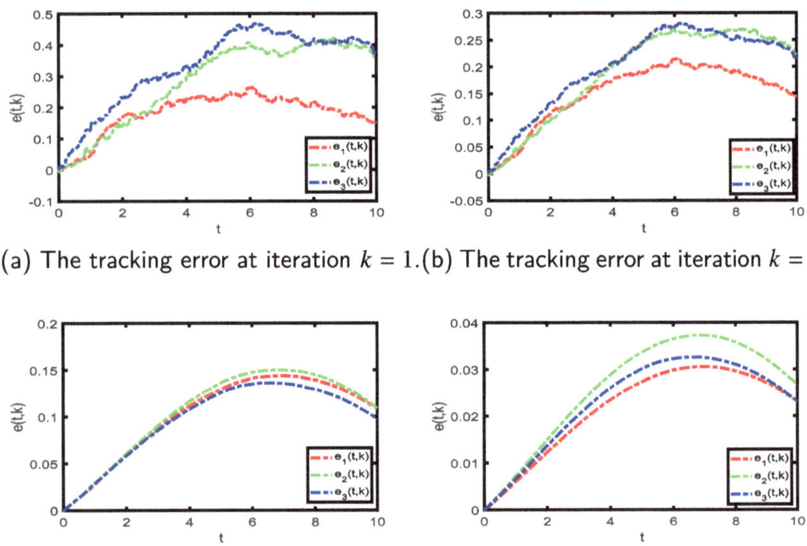

(a) The tracking error at iteration $k = 1$. (b) The tracking error at iteration $k = 10$.

(c) The tracking error at iteration $k = 50$. (d) The tracking error at iteration $k = 100$.

Fig. 7.2 The trajectories of $e(t, k)$ of model (7.30) with ZOH sampled-data ILCs

7.6 Summary

The tracking challenge for discrete-time systems with event-triggered mechanisms and quantization has been covered in this chapter. To accomplish the goal, iterative learning control techniques have also been developed. The number of iterative steps that need to be updated has been decreased thanks to the event-triggered mechanism in our iterative learning controllers. To overcome communication bandwidth restrictions, a quantization approach has also been included in the event-triggered controller. The tracking has been accomplished by employing the interval matrix properties under some relaxed conditions. The efficiency of the major outcomes has been demonstrated using a numerical example, which comes last.

References

Cao YC, Ren W, Li Y (2009) Distributed discrete-time coordinated tracking with a time-varying reference state and limited communication. Automatica 45(5):1299–1305

Chen Y, Lü JH, Lin ZL (2013) Consensus of discrete-time multi-agent systems with transmission nonlinearity. Automatica 49(6):1768–1775

Cortés J (2006) Finite-time convergent gradient flows with applications to network consensus. Automatica 42(11):1993–2000

Demigha O, Hidouci W, Ahmed T (2013) On energy efficiency in collaborative target tracking in wireless sensor network: a review. IEEE Commun Surv Tut 15(3):1210–1222

Forti M, Nistri P, Papini D (2005) Global exponential stability and global convergence in finite time of delayed neural networks with infinite gain. IEEE Trans Neur Netw 16(6):1449–1463

Ge C, Zhang WW, Li W, Sun XC (2015) Improved stability criteria for synchronization of chaotic Lur'e systems using sampled-data control. Neurocomputing 151:215–222

Gronwall TH (1919) Note on the derivatives with respect to a parameter of the solutions of a system of differential equations. Ann Math 20(4):292–296

Jiang XF (2015) On sampled-data fuzzy control design approach for T-S model-based fuzzy systems by using discretization approach. Inform Sci. 296:307–314

Li JS, Li JM (2014) Adaptive iterative learning control for coordination of second-order multi-agent systems. Inter J Robust Nonlin Control 24(18):3282–3299

Li N, Zhang YL, Hu JW, Nie ZY (2011) Synchronization for general complex dynamical networks with sampled-data. Neurocomputing 74:805–811

Li CJ, Yu WW, Huang TW (2014) Impulsive synchronization schemes of stochastic complex networks with switching topology: average time approach. Neur Netw 54:85–94

Li CJ, Yu XH, Huang TW, Chen G, He X (2016) A generalized Hopfield network for nonsmooth constrained convex optimization: Lie derivative approach. IEEE Trans Neur Netw Learn Syst 27(2):308–321

Lu L, Xing Z, He B (2016) Non-uniform sampled-data control for stochastic passivity and passification of Markov jump genetic regulatory networks with time-varying delays. Neurocomputing 171:434–443

Mei J, Ren W, Ma GF (2011) Distributed coordinated tracking with a dynamic leader for multiple Euler-Lagrange systems. IEEE Trans Automat Control 56(6):1415–1421

Meng DY, Jia YM (2014) Formation control for multi-agent systems through an iterative learning design approach. Intern J Robust Nonlin Control 24(2):340–361

Meng DY, Moore KL (2016) Learning to cooperate: networks of formation agents with switching topologies. Automatica 64:278–293

Meng DY, Jia YM, Du JP, Yu FS (2013) Tracking algorithms for multiagent systems. IEEE Trans Neural Netw Learn Syst. 24(10):1660–1676

Meng DY, Jia YM, Du JP, Zhang J (2014) On iterative learning algorithms for the formation control of nonlinear multi-agent systems. Automatica 50(1):291–295

Song Q, Cao JD, Yu WW (2010) Second-order leader-following consensus of nonlinear multi-agent systems via pinning control. Syst Control Lett 59(9):553–562

Wang ZP, Wu HN (2016) Finite dimensional guaranteed cost sampled-data fuzzy control for a class of nonlinear distributed parameter systems. Inform Sci 327(10):21–39

Wang L, Xiao F (2010) Finite-time consensus problems for networks of dynamic agents. IEEE Trans Automat Control 55(4):950–955

Wen GH, Duan ZS, Ren W, Chen GR (2014) Distributed consensus of multi-agent systems with general linear node dynamics and intermittent communications. Intern J Robust Nonlin Control 24(16):2438–2457

Xiao F, Wang L, Chen J, Gao YP (2009) Finite-time formation control for multi-agent systems. Automatica 45(11):2605–2611

Xiong WJ (2015) Zheng WX (2015) A new approach to finite-time tracking of coupled continuous networks. In: IEEE international symposium on circuits and systems 2015, pp 3076–3079

Xu GH, Guan ZH, He DX, Chi M, Wu YH (2014) Distributed tracking control of second-order multi-agent systems with sampled data. J Franklin Inst 351(10):4786–4801

Xu WY, Chen GR, Ho DWC (2016) A layered event-triggered consensus scheme. IEEE Trans Cybern. 47(8):2334–2340

Yi Y, Zheng WX (2016) Decentralized PI tracking control for non-Gaussian large-scale interconnected distribution systems. J Franklin Inst 353(1):235–255

Yoo SJ (2016) Adaptive tracking control for uncertain switched nonlinear systems in nonstrict-feedback form. J Franklin Inst 353(6):1409–1422

Yue D, Tian EG, Han QL (2013) A delay system method for designing event-triggered controllers of networked control systems. IEEE Trans Automat Control 58(2):475–481

Zhang RK, Hou ZS, Ji HH, Yin CK (2016) Adaptive iterative learning control for a class of nonlinearly parameterised systems with input saturations. Intern J Syst Sci. 47(5):1084–1094

Chapter 8
Finite-Iteration Learning Tracking with Packet Losses

8.1 Introduction

Since the ILC was proposed, it has developed into a powerful control tool for theoretical study and practical applications. Because of its simplicity and effectiveness, ILC has been applied in many fields, such as fault-tolerant control (Wang et al. 2016), Timoshenko Beam (He et al. 2017), linear systems (Amann et al. 1996; Bu and Hou 2016), nonlinear systems (Bu et al. 2017; Yu and Li 2016; Wei et al. 2016), hard disk drives (Graham and Callafon 2006; Wu and Tomizuka 2010), autonomous vehicles (Chen and Moore 2002), data-driven control (Janssens et al. 2013; Chi et al. 2015, 2016), and multi-agent systems (Xiong et al. 2016; Meng et al. 2015; Meng and Moore 2016; Li and Li 2014). As an illustration, data-driven iterative learning is demonstrated in Chi et al. (2015) for a class of nonlinear discrete-time systems with high-order learning law. For discrete-time models, iterative learning is suggested in Xiong et al. (2016) by using the event-triggered technique. In Meng and Moore (2016), the associated formation among nodes converges to the anticipated one exponentially as the iteration step grows if the union set of the interactive graphs includes a spanning tree.

Modeling the actual challenges of systems requires taking into account a number of constraints, as stated in Xiong et al. (2008). For instance, a traveler must have a valid ticket in order to board a flight. Single system models are offered to describe models with constraints. The control issues with singular models have been extensively studied recently (Zhang et al. 2017; Wu et al. 2010; Wang et al. 2013). For instance, in Lu and Ho (2006), the stability of a perturbed singular system is explored. In Tian et al. (2016), the iterative learning control for a particular class of singular systems is discussed. In Cao et al. (2016), an iterative learning control technique is suggested for solitary systems with repeating running characteristics. The iterative learning controllers in Tian et al. (2016), Cao et al. (2016) are unrelated to the singular matrix E. Matrix A_{22} must, therefore, be invertible, which is a

W. Xiong et al., *Iterative Learning Control for Network Systems Under Constrained Information Communication*, Intelligent Control and Learning Systems 12,
https://doi.org/10.1007/978-981-97-0926-7_8

conservative requirement. In order to reduce conservatism, the initial goal of this study is to construct suitable iterative learning controllers for solitary systems.

In recent years, the majority of ILC research achievements have concentrated on asymptotic tracking, which is when $k \to +\infty$, iterations are required to solve the tracking problem (see Xiong et al. 2016; Meng et al. 2015; Meng and Moore 2016). In many practical applications, tracking systems across a finite number of iterations is the major focus. Finite-iteration tracking research is actually important and useful. For instance, it is usually believed that the course of a spaceship will be completed in a finite number of repetitive steps. Additionally, as is well known, academics do not focus on creating suitable learning controllers for isolated systems that experience packet dropout. Thus, the following query is raised: Q1) For a given coupled system with restrictions, how might packet-dropping learning controllers be designed to achieve the necessary tracking in a finite-iteration? Finite-iteration tracking is used to describe tracking in this context. Thus, another goal of this study is to examine the finite-iteration tracking for singular coupled systems with packet-dropping learning controllers. This goal is motivated by the finite-time and packet dropout studies cited in Amato and Ariola (2005), Niu and Ho (2010), Song et al. (2015).

By developing packet-dropping learning controllers, it is extremely desirable to explore finite-iteration tracking for single coupled systems in light of the ideas above. The two primary goals of this chapter are as follows: (1) Discuss the tracking issue of the above systems using coupled models with some algebraic restrictions and create packet-dropping learning controllers; (2) Analyze the finite-iteration tracking of the aforementioned systems by designing packet-dropping learning controllers.

8.2 Problem Formulation

As stated in the introduction, discussing the finite-iteration tracking of coupled systems by creating packet-dropping iterative learning algorithms is a difficult and fascinating subject. Each node in the model contains two distinct dynamic processes: the first one displays time dynamics t and the second one displays iterative learning k dynamics. We introduce the following single coupled systems with iterative learning controllers in response to the citation (Xiong et al. 2008).

$$Ex_i(t+1,k) = Bx_i(t,k) + a\sum_{j=1}^{N} c_{ij}\Lambda x_j(t,k) + u_i(t,k), \tag{8.1}$$

where $t \in \{0, 1, \cdots, T\}$, $k \in \mathbb{Z}_+$, and $i = 1, 2, \cdots, N$. The state variables of node i are denoted as $x_i(t,k) = (x_{i1}(t,k), \cdots, x_{in}(t,k))^{\mathrm{T}} \in \mathbb{R}^n$. Matrix E is singular, and $0 < rank(E) = r < n$. $B \in \mathbb{R}^{n\times n}$ is a constant matrix, N is the number of nodes, and $u_i(t,k) \in \mathbb{R}^n$ is the iterative learning controller of node i. The constant $a > 0$ denotes the coupling strength, and $\Lambda = \mathrm{diag}(\Lambda_1, \cdots, \Lambda_n) \in \mathbb{R}^{n\times n}$ is a diagonal matrix with $\Lambda_i = 1$ or 0. $C = (c_{ij})_{N\times N}$ is the coupling matrix of the system, which has

$$\begin{cases} c_{ij} = c_{ji} = 1\ (i \neq j), \ \textit{there is a connection between nodes } i \textit{ and } j, \\ c_{ij} = c_{ji} = 0\ (i \neq j), \ \textit{otherwise}, \end{cases}$$

Moreover, we define matrix C to satisfy that

$$c_{ii} = -\sum_{j=1, j\neq i}^{N} c_{ij} = -\sum_{j=1, j\neq i}^{N} c_{ji}, \quad i = 1, 2, \cdots, N.$$

Without loss of generality, define matrix $E = \begin{pmatrix} I_r & 0 \\ 0 & 0 \end{pmatrix}$, where I_r is an identity matrix with r dimension, the system (8.1) can be rewritten as systems (8.2) and (8.3).

$$x_i^{(1')}(t+1, k) = B_1 x_i^{(1')}(t, k) + B_2 x_i^{(2')}(t, k) + a \sum_{j=1}^{N} c_{ij} \Lambda_1 x_j^{(1')}(t, k) + u_i^1(t, k), \tag{8.2}$$

and

$$0 = B_3 x_i^{(1')}(t, k) + B_4 x_i^{(2')}(t, k) + a \sum_{j=1}^{N} c_{ij} \Lambda_2 x_j^{(2')}(t, k) + u_i^2(t, k). \tag{8.3}$$

In (8.2) and (8.3), $i = 1, 2, \cdots, N$, matrix $B = \begin{pmatrix} B_1 & B_2 \\ B_3 & B_4 \end{pmatrix}$ with $B_1 \in \mathbb{R}^{r\times r}$ and $B_4 \in \mathbb{R}^{(n-r)\times(n-r)}$. $x_i(t, k) = \left((x_i^{(1')}(t, k))^{\mathrm{T}}, (x_i^{(2')}(t, k))^{\mathrm{T}}\right)^{\mathrm{T}}$ with $x_i^{(1')}(t, k) \in \mathbb{R}^r$, $x_i^{(2')}(t, k) \in \mathbb{R}^{n-r}$, $\Lambda = \mathrm{diag}(\Lambda_1, \Lambda_2)$ with $\Lambda_1 \in \mathbb{R}^{r\times r}$, $\Lambda_2 \in \mathbb{R}^{(n-r)\times(n-r)}$, and $u_i(t, k) = ((u_i^1(t, k))^{\mathrm{T}}, (u_i^2(t, k))^{\mathrm{T}})^{\mathrm{T}}$ with $u_i^1(t, k) \in \mathbb{R}^r$, $u_i^2(t, k) \in \mathbb{R}^{n-r}$.

The singular model (8.1) in this case is equivalent to the model (8.2) satisfying the condition (8.3). System (8.2) is still a coupled system, and Eq. (8.3) can be considered as a set of algebraic constraints of the model (8.2). Hence, the singular system (8.1) describes a coupled system satisfying some algebraic constraints.

As the preferred reference trajectory for node i, define $x_i^*(t) \in \mathbb{R}^n$. The intended tracking is always accomplished when the iteration $k \to +\infty$, according to the ILC references (Xiong et al. 2016; Meng and Moore 2016). To provide the suitable iterative learning controller $u_i(t, k)$ and achieve the desired finite-iteration system tracking, let $u(t, k) = (u_1(t, k), \cdots, u_N(t, k))^{\mathrm{T}}$, and $x(t, k) = (x_1(t, k), \cdots, x_N(t, k))$. Then, the coupled model (8.1) is rewritten as

$$Ex(t+1, k) = Bx(t, k) + a\Lambda x(t, k)C^{\mathrm{T}} + u(t, k). \tag{8.4}$$

Since C is a symmetric matrix, one can find a unitary matrix $w = (w_1, \cdots, w_N) \in \mathbb{R}^{N\times N}$ to satisfy $C = w\Gamma w^{\mathrm{T}}$ with $\Gamma = \mathrm{diag}(\lambda_1, \cdots, \lambda_N)$ and $ww^{\mathrm{T}} = I$. Define $y(t, k) = x(t, k)w$, one has along with (8.4)

$$Ey(t+1, k) = By(t, k) + a\Lambda y(t, k)\Gamma + \widehat{u}(t, k), \tag{8.5}$$

where $\widehat{u}(t, k) = (\widehat{u}_1(t, k), \cdots, \widehat{u}_N(t, k)) = u(t, k)w$, $\widehat{u}_i(t, k) = u(t, k)w_i \in \mathbb{R}^n$, $y(t, k) = (y_1(t, k), \cdots, y_N(t, k))$, $y_i(t, k) = x(t, k)w_i \in \mathbb{R}^n$, $i = 1, 2, \cdots, N$. Then, model (8.5) can be rewritten as

$$\begin{aligned} Ey_i(t+1, k) &= By_i(t, k) + a\lambda_i\Lambda y_i(t, k) + \widehat{u}_i(t, k) \\ &= \big(B + a\lambda_i\Lambda\big)y_i(t, k) + \widehat{u}_i(t, k). \end{aligned} \tag{8.6}$$

Some assumptions and lemmas are given in the following.

Assumption 8.1 There exist matrices P_i such that

$$E^T P_i = P_i^T E \geq 0, \tag{8.7}$$

and

$$\big(B + a\lambda_i\Lambda\big)^T P_i + P_i^{\mathrm{T}}\big(B + a\lambda_i\Lambda\big) < 0, \tag{8.8}$$

where $i = 1, 2, \cdots, N$.

Assumption 8.2 The initial state in each iteration is assumed as

$$x_i(0, k) = x_0^i \cdot (\underbrace{1, 1, \cdots, 1}_{n})^{\mathrm{T}}, \tag{8.9}$$

where $i = 1, 2, \cdots, N$, $\forall\, k \in \mathbb{Z}_+$, x_0^i $(i = 1, 2, \cdots, N)$ is a constant.

Lemma 8.1 (Horn and Johnson 2012; Xiong et al. 2016) *For a matrix $A \in \mathbb{R}^{N\times N}$, there must exist a matrix norm $\|\cdot\|$ such that*

$$\|A\| \leq \rho(A) + \zeta, \tag{8.10}$$

where $\forall\, \zeta > 0$.

Lemma 8.2 (Horn and Johnson (2012); Xiong et al. (2016)) *For a matrix $A \in \mathbb{R}^{N\times N}$, there must exist a vector norm $\|x\|_*$ such that*

$$\|Ax\|_* \leq \|A\| \cdot \|x\|_*, \tag{8.11}$$

where $\|A\|$ is an arbitrary matrix norm of matrix A, and $\forall\, x = (x_1, x_2, \cdots, x_N)^{\mathrm{T}} \in \mathbb{R}^N$ is a vector.

Remark 8.1 In this chapter, the matrix norm is defined as

$$\|A\| = \left\|D^{-1}P^{-1}APD\right\|_{\infty}, \tag{8.12}$$

where the definitions of matrices P and D are addressed in Lemma 2 in Xiong et al. (2016). This matrix norm can satisfy condition (8.10) in Lemma 8.1. Also, similar to Remark 3 of Xiong et al. (2016), for any vector $\chi \neq \mathbf{0} \in \mathbb{R}^N$, define the vector norm

$$\|x\|_* = \left\|x\chi^{\mathrm{T}}\right\|, \tag{8.13}$$

where $\|\cdot\|$ is the matrix norm in (8.12). Then, one has $\|Ax\|_* = \|Ax\chi^{\mathrm{T}}\| \leq \|A\| \cdot \|x\chi^{\mathrm{T}}\| = \|A\| \cdot \|x\|_*$. Other detailed descriptions are omitted here to save the space.

Based on Amato and Ariola (2005), we give the definition of finite-iteration tracking in the following.

Definition 8.1 (Finite-iteration tracking) System (8.1) is called to achieve the finite-iteration tracking if

$$\left\|\overline{E}_i(k)\right\|_* \leq e^{-Q}\left\|\overline{E}_i(0)\right\|_*, \ \forall k > K, \tag{8.14}$$

where $Q > 0$ is a given large number, the iteration $K > 0$ is a scalar which is related to Q, $\overline{e}_i(t,k) = x_i^*(t) - x_i(t,k)$, and $\overline{E}_i(k) = (\overline{e}_i^{\mathrm{T}}(1,k), \cdots, \overline{e}_i^{\mathrm{T}}(T,k))^{\mathrm{T}}$. Here, the scalar K is called the settling iteration.

Remark 8.2 One advantage of this article over the existing ILC results in Bu and Hou (2016), Bu et al. (2017), Meng et al. (2015), Meng and Moore (2016) is the definition of finite-iteration tracking that is provided (see Definition 8.1). Additionally, the settling iteration is provided in Definition 8.1 in order to calculate the ILC's rate of convergence. The general tracking issue that arises when the iteration $k \to +\infty$ was the only topic covered by the writers in the aforementioned literature. Therefore, it makes more sense to examine the system's finite-iteration tracking (8.1).

Remark 8.3 In Amann et al. (1996), the exponential convergence of the iterative learning error has been studied. The exponential convergence in Amann et al. (1996) is close to our definition of finite-iteration tracking. The difference between this chapter's finite-iteration tracking and the inequality (33) in Amann et al. (1996) is that this chapter's focus is on finding the settling iteration K to satisfy the inequality (8.14), whereas the exponential convergence (33) in Amann et al. (1996) is on the relationship between the iteration steps k and $k+1$.

8.3 ILC Scheme

The goal of this section is to construct suitable iterative learning controllers in order to track the model (8.1) through a finite number of iterations. When transmitting information, packet loss is taken into account. The following Bernoulli process is compliant with the packet loss $\phi(k)$ ($\phi(k) \in \{0, 1\}, k \in \mathbb{Z}_+$).

$$P\{\phi(k) = 1\} = \phi_1, \quad P\{\phi(k) = 0\} = 1 - \phi_1, \tag{8.15}$$

where ϕ_1 $(0 \leq \phi_1 \leq 1)$ is the probability that a data packet may be missing. Hence, for model (8.6), the iteration learning controllers $u_i(t, k)$ are designed in two situations.

(1) The controllers $u_i(t, k)$ are updated in each iteration step without the data packet loss

$$u_i(t, k+1) = u_i(t, k) + \Delta u_i(t, k), \tag{8.16}$$

and

$$\Delta u_i(t, k) = \alpha E\overline{e}_i(t+1, k) - \beta(B + a\lambda_i\Lambda)\overline{e}_i(t, k), \tag{8.17}$$

where $i = 1, 2, \cdots, N$. α and β are constants and will be designed later, where $k \in \mathbb{Z}_+, t \in \{0, 1, \cdots, T\}, i = 1, \cdots, N$.

(2) The controllers $u_i(t, k)$ are designed to consider the case of data packet losses

$$u_i(t, k+1) = u_i(t, k) + \Delta u_i(t, k), \tag{8.18}$$

(i) when the iteration $k = 0$,

$$\Delta u_i(t, k) = \alpha E\overline{e}_i(t+1, k) - \beta(B + a\lambda_i\Lambda)\overline{e}_i(t, k), \tag{8.19}$$

(ii) when the iteration $k \in \{1, 2, \cdots\}$,

$$\begin{cases} \Delta u_i(t, k) = \alpha E\overline{e}_i(t+1, k) - \beta(B + a\lambda_i\Lambda)\overline{e}_i(t, k), \\ \qquad \textit{if the data packet at iteration k is received}, \\ \Delta u_i(t, k) = \alpha E\overline{e}_i(t+1, k-1) - \beta(B + a\lambda_i\Lambda)\overline{e}_i(t, k-1), \\ \qquad \textit{if the data packet at iteration k is missing}, \end{cases} \tag{8.20}$$

where the same symbols have the same definitions with (8.17).

8.4 Main Results

8.4.1 The Finite-Iteration Tracking Without the Packet Dropout

Let $e_i(t, k) = y_i^*(t) - y_i(t, k)$ and $y_i^*(t) = x^*(t)w_i$, one has $e_i(t, k) = \overline{e}(t, k)w_i$. From (8.6) and (8.17), one can get

$$\begin{aligned}\Delta\widehat{u}_i(t,k) &\triangleq \widehat{u}_i(t,k+1)-\widehat{u}_i(t,k)\\&=[u(t,k+1)-u(t,k)]w_i\\&=\alpha E e_i(t+1,k)-\beta(B+a\lambda_i\Lambda)e_i(t,k),\end{aligned}\tag{8.21}$$

where $\overline{e}(t,k)=(\overline{e}_1(t,k),\cdots,\overline{e}_N(t,k))$. Based on (8.6) and (8.21), one has

$$\begin{aligned}&E\big[e_i(t+1,k+1)-e_i(t+1,k)\big]\\&=-E\big[y_i(t+1,k+1)-y_i(t+1,k)\big]\\&=\big(B+a\lambda_i\Lambda\big)\big[e_i(t,k+1)-e_i(t,k)\big]-\alpha E e_i(t+1,k)+\beta\big(B+a\lambda_i\Lambda\big)e_i(t,k).\end{aligned}\tag{8.22}$$

According to Assumption 8.1, it follows from Xiong et al. (2008), Xu et al. (2002), Lu and Ho (2006) that the pair $(E,\ B+a\lambda_i\Lambda)$ is regular and impulse free. As a result, one can find nonsingular matrices $W_i,\ H_i\in\mathbb{R}^{n\times n}$ $(i=1,2,\cdots,N)$ to satisfy that $W_iEH_i=\mathrm{diag}\{I_r,0\}$, $W_i(B+a\lambda_i\Lambda)H_i=\mathrm{diag}\{B_i,I_{n-r}\}$, where $B_i\in\mathbb{R}^{r\times r}$. Then, model (8.22) can be rewritten as

$$\begin{aligned}&\begin{pmatrix}I_r&0\\0&0\end{pmatrix}\begin{pmatrix}\widehat{e}_i^1(t+1,k+1)-\widehat{e}_i^1(t+1,k)\\\widehat{e}_i^2(t+1,k+1)-\widehat{e}_i^2(t+1,k)\end{pmatrix}\\&=\begin{pmatrix}B_i&0\\0&I_{n-r}\end{pmatrix}\begin{pmatrix}\widehat{e}_i^1(t,k+1)-\widehat{e}_i^1(t,k)\\\widehat{e}_i^2(t,k+1)-\widehat{e}_i^2(t,k)\end{pmatrix}-\alpha\begin{pmatrix}I_r&0\\0&0\end{pmatrix}\begin{pmatrix}\widehat{e}_i^1(t+1,k)\\\widehat{e}_i^2(t+1,k)\end{pmatrix}\\&\quad+\beta\begin{pmatrix}B_i&0\\0&I_{n-r}\end{pmatrix}\begin{pmatrix}\widehat{e}_i^1(t,k)\\\widehat{e}_i^2(t,k)\end{pmatrix},\end{aligned}\tag{8.23}$$

where $i=1,2,\cdots,N$, $\widehat{e}_i(t,k)=H_i^{-1}e_i(t,k)=\begin{pmatrix}\widehat{e}_i^1(t,k)\\\widehat{e}_i^2(t,k)\end{pmatrix}$, $\widehat{e}_i^1(t,k)\in\mathbb{R}^r$, $\widehat{e}_i^2(t,k)\in\mathbb{R}^{n-r}$.

Remark 8.4 In iterative learning controllers (8.17), the first term $\alpha E\overline{e}_i(t+1,k)$ is designed to achieve the finite-iteration tracking of $\widehat{e}_i^1(t,k)$ in (8.23). And the purpose of designing the second term $\beta(B+a\lambda_i\Lambda)\overline{e}_i(t,k)$ in (8.17) is to obtain the finite-iteration tracking of $\widehat{e}_i^2(t,k)$ in (8.23).

Theorem 8.1 *Consider the system (8.1). Under Assumption 8.1 and Assumption 8.2, the finite-iteration tracking of the system (8.1) can be achieved if one chooses scalars α $(0<\alpha<1)$ and β $(0<\beta<1)$ in the controllers (8.17) to satisfy that*

$$\Upsilon<1,\tag{8.24}$$

where $\Upsilon=\max\{1-\alpha,1-\beta\}+\zeta$ and $\zeta>0$ is appropriately small in Lemma 8.1. Moreover, one can choose the settling iteration

$$K\triangleq\log_{\Upsilon}^{\Psi}+1,\tag{8.25}$$

where

$$\Psi = e^{-Q} \cdot \left\|\mathrm{diag}(\underbrace{H_i, \cdots, H_i}_{T})\right\|^{-1} \cdot \left\|\mathrm{diag}(\underbrace{H_i^{-1}, \cdots, H_i^{-1}}_{T})\right\|^{-1},$$

and Q is the given large number in Definition 8.1.

Proof One can get from system (8.23)

$$\begin{aligned}&\widehat{e}_i^1(t+1,k+1) - \widehat{e}_i^1(t+1,k)\\ &= B_i(\widehat{e}_i^1(t,k+1) - \widehat{e}_i^1(t,k)) - \alpha\widehat{e}_i^1(t+1,k) + \beta B_i\widehat{e}_i^1(t,k),\end{aligned} \tag{8.26}$$

and

$$\widehat{e}_i^2(t,k+1) - \widehat{e}_i^2(t,k) + \beta\widehat{e}_i^2(t,k) = 0, \tag{8.27}$$

where $k \in \mathfrak{N}_+$, $t \in \{0, 1, 2, \cdots, T\}, i = 1, 2, \cdots, N$. Let $\vartheta_i(t,k) = \widehat{e}_i^1(t,k+1) - \widehat{e}_i^1(t,k)$, system (8.26) is

$$\vartheta_i(t+1,k) = B_i\vartheta_i(t,k) + \omega_i(t+1,k), \tag{8.28}$$

where $\omega_i(t+1,k) \triangleq -\alpha\widehat{e}_i^1(t+1,k) + \beta B_i\widehat{e}_i^1(t,k)$.

According to Assumption 8.2, it is easy to calculate that $\overline{e}_i(0,k) = \overline{e}_i(0,k+1)$, $i = 1, 2, \cdots, N, \forall\, k \in \mathbb{Z}_+$. One can get that $e_i(0,k) = e_i(0,k+1)$, and $\widehat{e}_i^1(0,k+1) = \widehat{e}_i^1(0,k)$, $i = 1, 2, \cdots, N, \forall\, k \in \mathbb{Z}_+$. That is, $\vartheta_i(0,k) = 0,\ \forall\, k \in \mathbb{Z}_+$. Hence, one can obtain

$$\vartheta_i(t,k) = \sum_{j=0}^{t-1}(B_i)^j\omega_i(t-j,k), \tag{8.29}$$

where $i = 1, 2, \cdots, N$, $t \in \{0, 1, \cdots, T\}$, $k \in \mathbb{Z}_+$. Combining (8.26) and (8.29), one can get

$$\begin{aligned}&\widehat{e}_i^1(t+1,k+1)\\ &= (1-\alpha)\widehat{e}_i^1(t+1,k) + \beta B_i\widehat{e}_i^1(t,k) + \sum_{j=0}^{t-1}(B_i)^{j+1}\omega_i(t-j,k),\end{aligned} \tag{8.30}$$

where $t \in \{0, 1, \cdots, T\}, k \in \mathbb{Z}_+$. Let $\widetilde{E}_i^1(k) = ((\widehat{e}_i^1(1,k))^{\mathrm{T}}, \cdots, (\widehat{e}_i^1(T,k))^{\mathrm{T}})^{\mathrm{T}}$, one has

$$\widetilde{E}_i^1(k+1) = \Omega\widetilde{E}_i^1(k), \tag{8.31}$$

where Ω is given in (8.32),

$$\Omega = \begin{pmatrix} (1-\alpha)I_r & O & O & \cdots & O & O \\ *_{21} & (1-\alpha)I_r & 0 & \cdots & O & O \\ *_{31} & *_{32} & (1-\alpha)I_r & \cdots & O & O \\ \vdots & \vdots & \vdots & \cdots & \vdots & \vdots \\ *_{T1} & *_{T2} & *_{T3} & \cdots & *_{T,T-1} & (1-\alpha)I_r \end{pmatrix}. \tag{8.32}$$

$*_{ij}$ $(i = 2, \cdots, T,\ j = 1, \cdots, T-1)$ can be given from (8.30) and are not detailed here because all of them will not be discussed in the following. In addition, one has from (8.27)

$$\widetilde{E}_i^2(k+1) = (1-\beta)\widetilde{E}_i^2(k),\ k \in \mathbb{Z}_+, \tag{8.33}$$

where $\widetilde{E}_i^2(k) = ((\widehat{e}_i^2(1,k))^{\mathrm{T}}, \cdots, (\widehat{e}_i^2(T,k))^{\mathrm{T}})^{\mathrm{T}}$. Let $\widehat{E}_i(k) = (\widehat{e}_i^{\mathrm{T}}(1,k), \cdots, \widehat{e}_i^{\mathrm{T}}(T,k))^{\mathrm{T}}$, $\widetilde{E}_i(k) = ((\widetilde{E}_i^1(k))^{\mathrm{T}}, (\widetilde{E}_i^2(k))^{\mathrm{T}})^{\mathrm{T}}$, one knows

$$\widehat{E}_i(k) = \Xi_i \widetilde{E}_i(k), \tag{8.34}$$

where matrix Ξ_i $(i = 1, 2, \cdots, N)$ can be obtained by implementing finite row exchanges on an identity matrix. As a result, one can have that $\Xi_i^{\mathrm{T}}\Xi_i = I_n$. Moreover, one also gets from (8.31) and (8.33)

$$\widetilde{E}_i(k+1) = \begin{pmatrix} \Omega & O \\ O & (1-\beta)I_{T(n-r)} \end{pmatrix} \widetilde{E}_i(k). \tag{8.35}$$

Combining Lemma 8.1 and (8.35), one obtains

$$\left\|\widetilde{E}_i(k)\right\|_* \le \Upsilon^k \left\|\widetilde{E}_i(0)\right\|_*, \tag{8.36}$$

where $\zeta > 0$ is arbitrarily small and $\Upsilon = \max\{1-\alpha, 1-\beta\} + \zeta$. Based on (8.34) and the characteristic of matrix Ξ_i, one has $\|\widetilde{E}_i(k)\| = \|\widehat{E}_i(k)\|$, for $\forall\, k \in \mathbb{Z}_+$. That is,

$$\left\|\widehat{E}_i(k)\right\|_* \le \Upsilon^k \left\|\widehat{E}_i(0)\right\|_*. \tag{8.37}$$

Let $E_i(k) = (e_i^{\mathrm{T}}(1,k), \cdots, e_i^{\mathrm{T}}(T,k))^{\mathrm{T}}$. Note that $\widehat{e}_i(t,k) = H_i^{-1}e_i(t,k)$, $\forall\, k \in \mathbb{Z}_+$, one has

$$E_i(k) = \mathrm{diag}(\underbrace{H_i, \cdots, H_i}_{T})\widehat{E}_i(k),\quad \widehat{E}_i(k) = \mathrm{diag}(\underbrace{H_i^{-1}, \cdots, H_i^{-1}}_{T})E_i(k).$$

Then, one obtains from (8.37)

$$\begin{aligned}\left\|E_i(k)\right\|_* &\le \|\mathrm{diag}(\underbrace{H_i, \cdots, H_i}_{T})\| \cdot \left\|\widehat{E}_i(k)\right\|_* \\ &\le \Upsilon^k \|\mathrm{diag}(\underbrace{H_i, \cdots, H_i}_{T})\| \cdot \left\|\widehat{E}_i(0)\right\|_* \\ &\le \Pi(k) \cdot \left\|E_i(0)\right\|_*, \end{aligned} \tag{8.38}$$

where $\Pi(k) = \Upsilon^k \|\mathrm{diag}(\underbrace{H_i, \cdots, H_i}_{T})\| \cdot \|\mathrm{diag}(\underbrace{H_i^{-1}, H_i^{-1}, \cdots, H_i^{-1}}_{T})\|$.

Note that $e_i(t, k) = \overline{e}_i(t, k)w$, one has $\overline{E}_i(k) = E_i(k)w^{-1}$. Since w is a unitary matrix such that $ww^{\mathrm{T}} = I$, it is easy to calculate that $\|w\| = \|w^{-1}\| = 1$. From (8.38), one has $\|\overline{E}_i(k)\|_* = \|E_i(k)\|_*$. That is,

$$\left\|\overline{E}_i(k)\right\|_* \le \Pi(k) \cdot \left\|\overline{E}_i(0)\right\|_*. \tag{8.39}$$

According to Definition 8.1 and (8.39), one can get an inequality $\Pi(k) \le e^{-Q}$. From the inequality, one has

$$\Upsilon^k \le e^{-Q} \cdot \left\|\mathrm{diag}(\underbrace{H_i, \cdots, H_i}_{T})\right\|^{-1} \cdot \left\|\mathrm{diag}(\underbrace{H_i^{-1}, \cdots, H_i^{-1}}_{T})\right\|^{-1}.$$

Then, from (8.24), one has $k \ge \log_{\Upsilon}^{\Psi}$ with

$$\Psi = e^{-Q} \cdot \left\|\mathrm{diag}(\underbrace{H_i, \cdots, H_i}_{T})\right\|^{-1} \cdot \left\|\mathrm{diag}(\underbrace{H_i^{-1}, \cdots, H_i^{-1}}_{T})\right\|^{-1}.$$

Defining the settling iteration $K = \log_{\Upsilon}^{\Psi} + 1$, one can obtain the finite-iteration tracking of the system (8.1) according to Definition 8.1. Clearly, question Q1 in the Introduction is answered. □

Remark 8.5 In this remark, we give an example to explain the relationship between $\widetilde{E}_i(k)$ and $\widehat{E}_i(k)$ in (8.34). Here, let $\widehat{e}_i(t, k) = \begin{pmatrix} \widehat{e}_i^1(t, k) \\ \widehat{e}_i^2(t, k) \end{pmatrix}$ with $\widehat{e}_i^1(t, k) = (\widehat{e}_i^{11}(t, k), \widehat{e}_i^{12}(t, k))^{\mathrm{T}} \in \mathbb{R}^2$ and $\widehat{e}_i^2(t, k) \in \mathbb{R}$. Also, we define $T = 2$, one has $\widetilde{E}_i(k) = (\widehat{e}_i^{11}(1, k), \widehat{e}_i^{12}(1, k), \widehat{e}_i^{11}(2, k), \widehat{e}_i^{12}(2, k), \widehat{e}_i^2(1, k), \widehat{e}_i^2(2, k))^{\mathrm{T}}$, and $\widehat{E}_i(k) = (\widehat{e}_i^{11}(1, k), \widehat{e}_i^{12}(1, k), \widehat{e}_i^2(1, k), \widehat{e}_i^{11}(2, k), \widehat{e}_i^{12}(2, k), \widehat{e}_i^2(2, k))^{\mathrm{T}}$. Then, one has $\widehat{E}_i(k) = \Xi_i \widetilde{E}_i(k)$ with the matrix

$$\Xi_i = \begin{pmatrix} 1 & 0 & 0 & 0 & 0 & 0 \\ 0 & 1 & 0 & 0 & 0 & 0 \\ 0 & 0 & 0 & 0 & 1 & 0 \\ 0 & 0 & 1 & 0 & 0 & 0 \\ 0 & 0 & 0 & 1 & 0 & 0 \\ 0 & 0 & 0 & 0 & 0 & 1 \end{pmatrix}.$$

Obviously, matrix Ξ_i can be obtained by implementing finite row exchanges on an identity matrix, and satisfies that $\Xi_i^T \Xi_i = I_6$.

Remark 8.6 Compared with singular systems (Tian et al. 2016; Cao et al. 2016), the advantages of this chapter are two twofold: (1) The iterative learning controllers (8.17), (8.19), and (8.20) are designed to be related to the singular matrix E. As a result, matrix B_4 does not need to be invertible. (2) This work presents the definition of the finite-iteration tracking for singular systems. Additionally, the settling iteration is provided to calculate the ILC's rate of convergence.

8.4.2 The Finite-Iteration Tracking with the Packet Dropout

From (8.19) and (8.20), one can get
(i) when the iteration $k = 0$,

$$\Delta u_i(t,k) = \alpha E\overline{e}_i(t+1,k) - \beta\big(B + a\lambda_i\Lambda\big)\overline{e}_i(t,k), \tag{8.40}$$

(ii) when the iteration $k \in \{1, 2, \cdots\}$,

$$\begin{aligned}\Delta\widehat{u}_i(t,k) = &\big(1-\phi(k)\big)\big[\alpha E e_i(t+1,k) - \beta\big(B + a\lambda_i\Lambda\big)e_i(t,k)\big]\\ &+\phi(k)\big[\alpha E e_i(t+1,k-1) - \beta\big(B + a\lambda_i\Lambda\big)e_i(t,k-1)\big],\end{aligned} \tag{8.41}$$

where $k \in \{1, 2, \cdots\}$. According to Assumption 8.1, similar to (8.23), one has the following equations from system (8.40) and (8.41).
(i) when the iteration $k = 0$,

$$\begin{aligned}&\begin{pmatrix} I_r & 0\\ 0 & 0\end{pmatrix}\begin{pmatrix}\widehat{e}_i^1(t+1,k+1) - \widehat{e}_i^1(t+1,k)\\ \widehat{e}_i^2(t+1,k+1) - \widehat{e}_i^2(t+1,k)\end{pmatrix}\\ =&\begin{pmatrix} B_i & 0\\ 0 & I_{n-r}\end{pmatrix}\begin{pmatrix}\widehat{e}_i^1(t,k+1) - \widehat{e}_i^1(t,k)\\ \widehat{e}_i^2(t,k+1) - \widehat{e}_i^2(t,k)\end{pmatrix} - \alpha\begin{pmatrix} I_r & 0\\ 0 & 0\end{pmatrix}\begin{pmatrix}\widehat{e}_i^1(t+1,k)\\ \widehat{e}_i^2(t+1,k)\end{pmatrix}\\ &+\beta\begin{pmatrix} B_i & 0\\ 0 & I_{n-r}\end{pmatrix}\begin{pmatrix}\widehat{e}_i^1(t,k)\\ \widehat{e}_i^2(t,k)\end{pmatrix},\end{aligned} \tag{8.42}$$

(ii) when the iteration $k \in \{1, 2, \cdots\}$,

$$\begin{aligned}&\begin{pmatrix} I_r & 0\\ 0 & 0\end{pmatrix}\begin{pmatrix}\widehat{e}_i^1(t+1,k+1) - \widehat{e}_i^1(t+1,k)\\ \widehat{e}_i^2(t+1,k+1) - \widehat{e}_i^2(t+1,k)\end{pmatrix}\\ =&\begin{pmatrix} B_i & 0\\ 0 & I_{n-r}\end{pmatrix}\begin{pmatrix}\widehat{e}_i^1(t,k+1) - \widehat{e}_i^1(t,k)\\ \widehat{e}_i^2(t,k+1) - \widehat{e}_i^2(t,k)\end{pmatrix} - (1-\phi(k))\alpha\begin{pmatrix} I_r & 0\\ 0 & 0\end{pmatrix}\begin{pmatrix}\widehat{e}_i^1(t+1,k)\\ \widehat{e}_i^2(t+1,k)\end{pmatrix}\\ &+(1-\phi(k))\beta\begin{pmatrix} B_i & 0\\ 0 & I_{n-r}\end{pmatrix}\begin{pmatrix}\widehat{e}_i^1(t,k)\\ \widehat{e}_i^2(t,k)\end{pmatrix} - \phi(k)\alpha\begin{pmatrix} I_r & 0\\ 0 & 0\end{pmatrix}\begin{pmatrix}\widehat{e}_i^1(t+1,k-1)\\ \widehat{e}_i^2(t+1,k-1)\end{pmatrix}\\ &+\phi(k)\beta\begin{pmatrix} B_i & 0\\ 0 & I_{n-r}\end{pmatrix}\begin{pmatrix}\widehat{e}_i^1(t,k-1)\\ \widehat{e}_i^2(t,k-1)\end{pmatrix},\end{aligned} \tag{8.43}$$

where $i = 1, 2, \cdots, N$, $\widehat{e}_i(t,k) = H_i^{-1} e_i(t,k) = \begin{pmatrix} \widehat{e}_i^1(t,k) \\ \widehat{e}_i^2(t,k) \end{pmatrix}$, $\widehat{e}_i^1(t,k) \in \mathbb{R}^r$, $\widehat{e}_i^2$ $(t,k) \in \mathbb{R}^{n-r}$. The definitions of other symbols are the same to Sect. 8.4.1. Then, one has the following theorem.

Theorem 8.2 *Consider of the system (8.1). Under Assumption 8.1 and Assumption 8.2, the finite-iteration tracking of the system (8.1) can be achieved if one chooses scalars $\alpha \in (0,1)$ and $\beta \in (0,1)$ in the controllers (8.18) to satisfy that*

$$\max\left\{1-\alpha, 1-\beta\right\} + \zeta = \Upsilon < 1, \tag{8.44}$$

where $\zeta > 0$ is appropriately small in Lemma 8.1. Moreover, the settling iteration can be chosen as

$$K \triangleq \log_{\Upsilon}^{\Psi} + 1, \tag{8.45}$$

where Ψ has the same definition as that in Theorem 8.1.

Proof One can get from system (8.42) and (8.43)
(i) when the iteration $k = 0$,

$$\begin{aligned} &\widehat{e}_i^1(t+1,k+1) - \widehat{e}_i^1(t+1,k) \\ &= B_i(\widehat{e}_i^1(t,k+1) - \widehat{e}_i^1(t,k)) - \alpha\widehat{e}_i^1(t+1,k) + \beta B_i\widehat{e}_i^1(t,k), \end{aligned} \tag{8.46}$$

and

$$\widehat{e}_i^2(t,k+1) - \widehat{e}_i^2(t,k) + \beta\widehat{e}_i^2(t,k) = 0, \tag{8.47}$$

where $k \in \mathfrak{N}_+$, $t \in \{0, 1, \cdots, T\}$, $i = 1, \cdots, N$.
(ii) when the iteration $k \in \{1, 2, \cdots\}$,

$$\begin{aligned} &\widehat{e}_i^1(t+1,k+1) - \widehat{e}_i^1(t+1,k) \\ &= B_i(\widehat{e}_i^1(t,k+1) - \widehat{e}_i^1(t,k)) - (1-\phi(k))\alpha\widehat{e}_i^1(t+1,k) \\ &\quad + (1-\phi(k))\beta B_i\widehat{e}_i^1(t,k) - \phi(k)\alpha\widehat{e}_i^1(t+1,k-1) + \phi(k)\beta B_i\widehat{e}_i^1(t,k-1), \end{aligned} \tag{8.48}$$

and

$$\widehat{e}_i^2(t,k+1) - \widehat{e}_i^2(t,k) + (1-\phi(k))\beta\widehat{e}_i^2(t,k) + \phi(k)\beta\widehat{e}_i^2(t,k-1) = 0, \tag{8.49}$$

where $k \in \mathbb{Z}_+$, $t \in \{0, 1, \cdots, T\}$, $i = 1, 2, \cdots, N$.

Let $\widetilde{E}_i^1(k) = ((\widehat{e}_i^1(1,k))^{\mathrm{T}}, \cdots, (\widehat{e}_i^1(T,k))^{\mathrm{T}})^{\mathrm{T}}$, similar with Theorem 8.1, one has

$$\widetilde{E}_i^1(1) = \Omega\widetilde{E}_i^1(0), \tag{8.50}$$

where the symbol Ω has the same definition as that in Theorem 8.1. When the iteration $k \in \{1, 2, \cdots\}$, let $\vartheta_i(t,k) = \widehat{e}_i^1(t,k+1) - \widehat{e}_i^1(t,k)$, one has from system (8.48)

$$\vartheta_i(t+1,k) = B_i\vartheta_i(t,k) + \widehat{\omega}_i(t+1,k), \tag{8.51}$$

where $\widehat{\omega}_i(t+1,k) \triangleq -(1-\phi(k))\alpha\widehat{e}_i^1(t+1,k) + (1-\phi(k))\beta B_i\widehat{e}_i^1(t,k) - \phi(k)\alpha\widehat{e}_i^1(t+1,k-1) + \phi(k)\beta B_i\widehat{e}_i^1(t,k-1)$.

Similar to (8.29), one obtains

$$\vartheta_i(t,k) = \sum_{j=0}^{t-1}(B_i)^j\widehat{\omega}_i(t-j,k), \tag{8.52}$$

where $i = 1, 2, \cdots, N$, $t \in \{0, 1, \cdots, T\}$, $k \in \{1, 2, \cdots\}$. Combining (8.48) and (8.52), one can obtain

$$\begin{aligned}
&\widehat{e}_i^1(t+1,k+1)\\
&= \left[1-(1-\phi(k))\alpha\right]\widehat{e}_i^1(t+1,k) - \phi(k)\alpha\widehat{e}_i^1(t+1,k-1)\\
&\quad +(1-\phi(k))\beta B_i\widehat{e}_i^1(t,k) + \phi(k)\beta B_i\widehat{e}_i^1(t,k-1) + \sum_{j=0}^{t-1}(B_i)^{j+1}\widehat{\omega}_i(t-j,k)
\end{aligned} \tag{8.53}$$

where $t \in \{0, 1, \cdots, T\}$, $k \in \{1, 2, \cdots\}$. Then, one has

$$\widetilde{E}_i^1(k+1) = \Omega_1(k)\widetilde{E}_i^1(k) + \Omega_2(k)\widetilde{E}_i^1(k-1), \tag{8.54}$$

where $\Omega_1(k)$ and $\Omega_2(k)$ are given in (8.55) and (8.56),

$$\Omega_1(k) = \begin{pmatrix} [1-(1-\phi(k))\alpha]I_r & O & \cdots & O\\ **_{21} & [1-(1-\phi(k))\alpha]I_r & \cdots & O\\ **_{31} & **_{32} & \cdots & O\\ \vdots & \vdots & \cdots & \vdots\\ **_{T1} & **_{T2} & \cdots & [1-(1-\phi(k))\alpha]I_r \end{pmatrix} \tag{8.55}$$

and

$$\Omega_2(k) = \begin{pmatrix} -\phi(k)\alpha I_r & O & O & \cdots & O\\ ***_{21} & -\phi(k)\alpha I_r & O & \cdots & O\\ ***_{31} & ***_{32} & -\phi(k)\alpha I_r & \cdots & O\\ \vdots & \vdots & \vdots & \cdots & \vdots\\ ***_{T1} & ***_{T2} & ***_{T3} & \cdots & *-\phi(k)\alpha I_r \end{pmatrix}, \tag{8.56}$$

$**_{ij}$ and $***_{ij}$ $(i = 2, \cdots, T,\ j = 1, \cdots, T-1)$ can be given from (8.53) and are not detailed here because all of them will not be discussed in the following. According to (8.50) and (8.54), one can obtain that

$$\widetilde{E}_i^1(k) = \widetilde{\Omega}(k)\widetilde{E}_i^1(0), \tag{8.57}$$

where $\widetilde{\Omega}(k)$ is a matrix which has the same dimension with matrices Ω_1 and Ω_2, $k \in \{1, 2, \cdots\}$. Since matrices Ω, $\Omega_1(k)$, and $\Omega_2(k)$ in (8.32) and (8.54) are all lower-triangular, one has that $\widetilde{\Omega}(k)$ is also a lower-triangular matrix. The diagonal element of matrix $\widetilde{\Omega}(k)$ is defined as $\varpi(k)$. In the following, we shall prove that

$$\varpi(k) \leq \varpi(k-1), \quad and \;\; 0 \leq \varpi(k) \leq (1-\alpha)^k, \tag{8.58}$$

where $\varpi(0) \triangleq 1$, $k \in \{1, 2, \cdots\}$. The proof will be completed according to the mathematical induction.
(i) When $k = 1$, the inequality (8.58) is satisfied according to (8.32) and (8.50).
(ii) When $k = 2$, one has from (8.50) and (8.54)

$$\widetilde{E}_i^1(2) = \Omega_1(1)\widetilde{E}_i^1(1) + \Omega_2(1)\widetilde{E}_i^1(0) = \big(\Omega_1(1)\Omega + \Omega_2(1)\big)\widetilde{E}_i^1(0). \tag{8.59}$$

Since $\varpi(k)$ is the diagonal element of matrix $\widetilde{\Omega}(k)$ in (8.57), according to (8.32), (8.55), and (8.56), one has

$$\varpi(2) = (1-\alpha)\varpi(1) + \phi(1)\alpha[\varpi(1) - \varpi(0)]. \tag{8.60}$$

Based on the result of $k = 1$, $\phi(1) \in \{0, 1\}$ and $0 < \alpha < 1$, the inequality (8.58) is satisfied when $k = 2$.
(iii) Assume the inequality (8.58) to be satisfied when $k = t$, $\forall\, t \in \mathfrak{N}_+$. We shall prove the inequality (8.58) is correct when $k = t + 1$. Similar to (8.59) and (8.60), one has from (8.50) and (8.54)

$$\varpi(k+1) = (1-\alpha)\varpi(k) + \phi(k)\alpha[\varpi(k) - \varpi(k-1)]. \tag{8.61}$$

According to the results of $k = t$, one has that $\varpi(k) \leq \varpi(k-1)$. With some simple calculation, one obtains that the inequality (8.58) is satisfied when $k = t + 1$. Hence, the inequality (8.58) is satisfied for $\forall\, k \in \{1, 2, \cdots\}$.

In addition, applying the above similar method, one has from (8.47) and (8.49)

$$\widetilde{E}_i^2(k) = \gamma(k)\widetilde{E}_i^2(0), \;\; k \in \mathbb{Z}_+, \tag{8.62}$$

where $\gamma(k) > 0$ is a scalar, which satisfies that $\gamma(k) \leq (1-\beta)^k$, $\widetilde{E}_i^2(k) = ((\widehat{e}_i^2(1,k))^{\mathrm{T}}, \cdots, (\widehat{e}_i^2(T,k))^{\mathrm{T}})^{\mathrm{T}}$. Let $\widehat{E}_i(k) = (\widehat{e}_i^{\mathrm{T}}(1,k), \;\; \cdots, \widehat{e}_i^{\mathrm{T}}(T,k))^{\mathrm{T}}$, $\widetilde{E}_i(k) = ((\widetilde{E}_i^1(k))^{\mathrm{T}}, (\widetilde{E}_i^2(k))^{\mathrm{T}})^{\mathrm{T}}$, one has

$$\widehat{E}_i(k) = \Xi_i \widetilde{E}_i(k), \tag{8.63}$$

where matrix Ξ_i $(i = 1, 2, \cdots, N)$ has the same definition with Theorem 8.1. Moreover, one also gets from (8.57) and (8.62)

$$\widetilde{E}_i(k+1) = \begin{pmatrix} \widetilde{\Omega}(k) & O \\ O & \gamma(k) I_{T(n-r)} \end{pmatrix} \widetilde{E}_i(0). \tag{8.64}$$

Combining Lemma 8.1 and (8.64), one obtains

$$\left\| \widetilde{E}_i(k) \right\|_* \leq \Upsilon^k \left\| \widetilde{E}_i(0) \right\|_*, \tag{8.65}$$

where $\zeta > 0$ is arbitrarily small and $\Upsilon = \max\{1-\alpha, 1-\beta\} + \zeta$. Similar to Theorem 8.1, one then has

$$\left\| \overline{E}_i(k) \right\|_* \leq \Pi(k) \cdot \left\| \overline{E}_i(0) \right\|_*, \tag{8.66}$$

where $\Pi(k) = \Upsilon^k \|\mathrm{diag}(\underbrace{H_i, H_i, \cdots, H_i}_{T})\| \cdot \|\mathrm{diag}(\underbrace{H_i^{-1}, H_i^{-1}, \cdots, H_i^{-1}}_{T})\|$.

According to Definition 8.1 and (8.66), one can get an inequality $\Pi(k) \leq e^{-Q}$. From the inequality, one has

$$\Upsilon^k \leq e^{-Q} \cdot \left\| \mathrm{diag}(\underbrace{H_i, \cdots, H_i}_{T}) \right\|^{-1} \cdot \left\| \mathrm{diag}(\underbrace{H_i^{-1}, \cdots, H_i^{-1}}_{T}) \right\|^{-1}.$$

Then, from (8.24), one has $k \geq \log_{\Upsilon}^{\Psi}$ with

$$\Psi = e^{-Q} \cdot \left\| \mathrm{diag}(\underbrace{H_i, \cdots, H_i}_{T}) \right\|^{-1} \cdot \left\| \mathrm{diag}(\underbrace{H_i^{-1}, \cdots, H_i^{-1}}_{T}) \right\|^{-1}.$$

Defining the settling iteration $K = \log_{\Upsilon}^{\Psi} + 1$, one can obtain the finite-iteration tracking of the system (8.1) according to Definition 8.1. □

Remark 8.7 Each iteration k of the control method (see Eq. (4) in Meng et al. 2015) makes the assumption that all of the information is received. However, packet loss frequently happens during data communication. As a result, another benefit of our ILC method (8.41) over cite Meng et al. (2015) is that it takes packet loss into account when determining the control strategy.

Remark 8.8 The benefit of Theorem 8.2 over Niu and Ho (2010), Song et al. (2015) is that the probability of the data packet in (8.15) is not required in the theorem. That is, even if every data packet is lost during the iteration k, the finite-iteration tracking of the system (8.1) can still be acquired. However, the conditions of the obtained results in Niu and Ho (2010), Song et al. (2015) contain the packet loss probability.

8.5 Simulation Example

In this part, an example with numerical simulations is addressed to illustrate the correctness of the results achieved in Theorems 8.1–8.2.

Example 8.1 According to Kirchhoff's voltage and current laws, a unique circuit system (5.3) has been discussed in Sun and Wang (2013). Environmental restrictions are crucial for urban transportation networks, according to Lin et al. (2016). In system design difficulties, several fixed requirements have been stated Friesz et al. (1992). It is obvious that discussing the finite-iteration tracking of systems with some algebraic constraints is both acceptable and meaningful. The equivalent iterative learning controller-equipped singular coupled systems are as follows:

$$Ex_i(t+1,k) = Bx_i(t,k) + a\sum_{j=1}^{N} c_{ij}\Lambda x_j(t,k) + u_i(t,k), \tag{8.67}$$

where $x_i(t,k) = (x_{i1}(t,k), x_{i2}(t,k), x_{i3}(t,k), x_{i4}(t,k))^{\mathrm{T}}$, the number of nodes $N = 4$, $a = 0.2$, $k \in \mathbb{Z}_+$,

$$E = \begin{pmatrix} 1 & 0 & 0 & 0 \\ 0 & 1 & 0 & 0 \\ 0 & 0 & 0 & 0 \\ 0 & 0 & 0 & 1 \end{pmatrix}, \quad B = (b_{ij})_{4\times 4} = \begin{pmatrix} -0.8 & 0.1 & 0 & 0 \\ 0.1 & -1 & 0.1 & 0 \\ 0 & 0.1 & -1 & 0 \\ 0 & 0 & 1 & -0.6 \end{pmatrix}$$

$$C = (c_{ij})_{4\times 4} = \begin{pmatrix} 1.3 & -0.5 & -0.8 & 0 \\ -0.5 & 0.8 & -0.2 & -0.1 \\ -0.8 & -0.2 & 1 & 0 \\ 0 & -0.1 & 0 & 0.1 \end{pmatrix},$$

$t \in \{0, 1, \cdots, 10\}$, $\Lambda = \mathrm{diag}(1, 1, 1, 0)$. Define $x_1(0,k) = x_0^1 = (-1, -2, 1, 3)^{\mathrm{T}}$, $x_2(0,k) = x_0^2 = (1, 2, -1, -2)^{\mathrm{T}}$, $x_3(0,k) = x_0^3 = (3, 2, 1, -2)^{\mathrm{T}}$, $x_4(0,k) = x_0^4 = (-2, 1, 4, -3)^{\mathrm{T}}$, $\forall\, k \in \mathbb{Z}_+$.

According to some simple calculations, one can obtain that Assumptions 8.1 and 8.2 are satisfied and $\lambda_1 = 0$, $\lambda_2 = 0.123$, $\lambda_3 = 1.0511$, $\lambda_4 = 2.0247$. Also, we define the desired reference trajectory $x_i^*(t) = 0.2 \cdot (\sin(t), \cos(t), t, t^{0.5})^{\mathrm{T}}$, $i = 1, 2, 3, 4$, $t \in \{0, 1, 2, \cdots, 10\}$.

For the iterative learning controllers (8.16), we design that

$$\Delta u_i(t,k) = \alpha E\overline{e}_i(t+1,k) - \beta(B + a\lambda_i\Lambda)\overline{e}_i(t,k), \tag{8.68}$$

where $\alpha = 0.5$, $\beta = 0.6$, $\overline{e}_i(t,k) = x_i^*(t) - x_i(t,k)$. For the iterative learning controller (8.20) with the packet loss, we define the probability of the packet loss as $\phi_1 = 0.25$. As a result, the conditions in Theorems 8.1 and 8.2 are satisfied. Hence, one can achieve the finite-iteration tracking of the system (8.67) from Theorems 8.1–8.2.

The corresponding error trajectories can be illustrated in Figs. 8.1, 8.2, 8.3, 8.4, 8.5, and 8.6. Figures 8.1, 8.2, and 8.3 show the state error trajectories $\overline{e}_i(t,k)$ ($i = 1, 2, 3, 4$, $\forall\, k \in \mathbb{Z}_+$) with iterative learning controller (8.68). Clearly, one can see

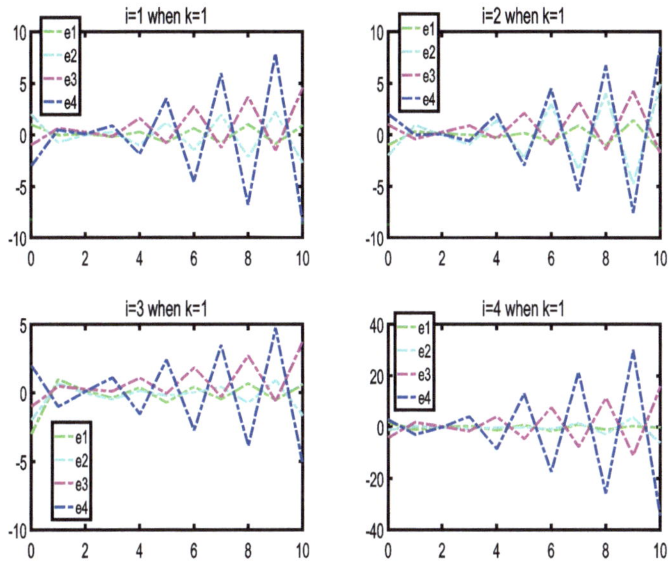

Fig. 8.1 The state error $\overline{e}_i(t, k)$ $(i = 1, 2, 3, 4)$ at iteration $k = 1$ with iterative learning controller (8.68)

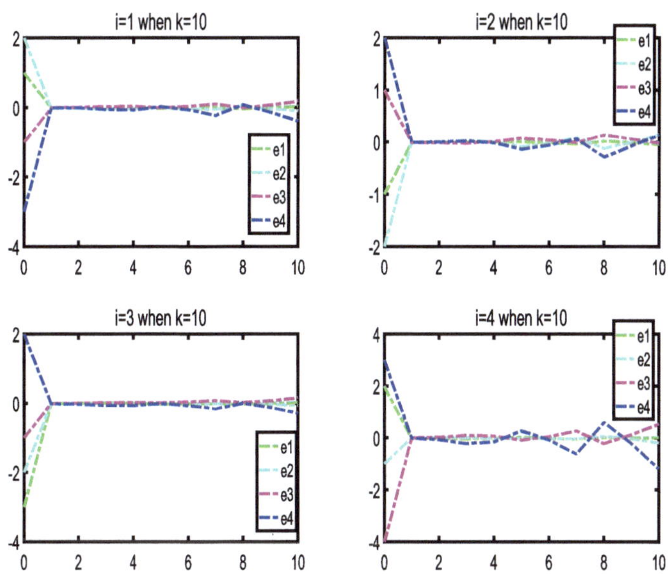

Fig. 8.2 The state error $\overline{e}_i(t, k)$ $(i = 1, 2, 3, 4)$ at iteration $k = 10$ with iterative learning controller (8.68)

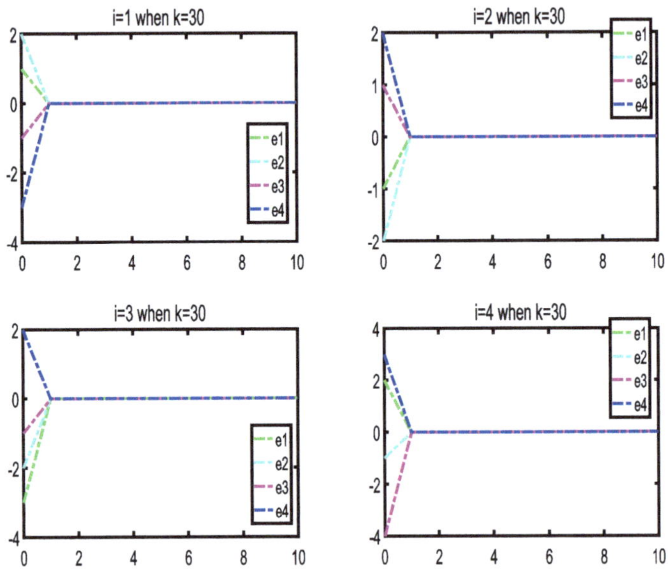

Fig. 8.3 The state error $\overline{e}_i(t, k)$ $(i = 1, 2, 3, 4)$ at iteration $k = 30$ with iterative learning controller (8.68)

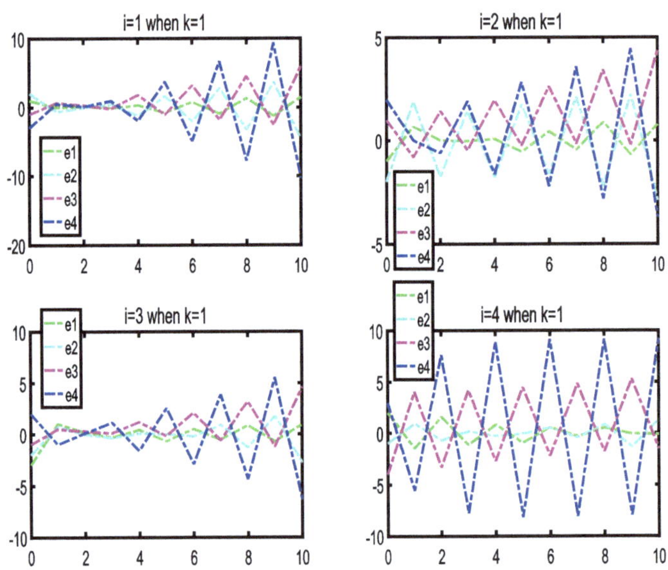

Fig. 8.4 The state error $\overline{e}_i(t, k)$ $(i = 1, 2, 3, 4)$ all iterations $k = 1$ with iterative learning controllers (8.19) and (8.20)

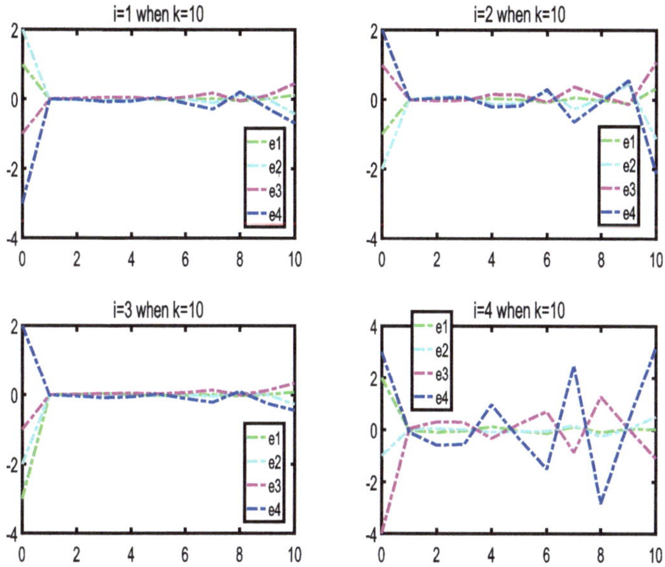

Fig. 8.5 The state error $\overline{e}_i(t, k)$ $(i = 1, 2, 3, 4)$ all iterations $k = 20$ with iterative learning controllers (8.19) and (8.20)

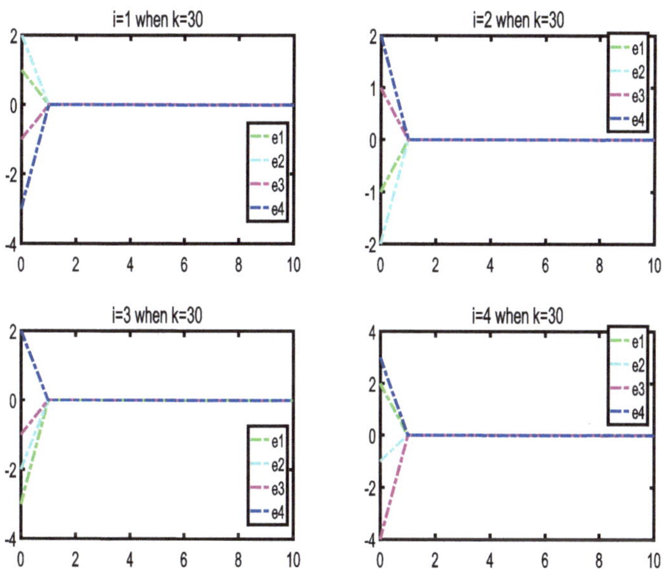

Fig. 8.6 The state error $\overline{e}_i(t, k)$ $(i = 1, 2, 3, 4)$ all iterations $k = 40$ with iterative learning controllers (8.19) and (8.20)

that the finite-iteration tracking of the system (8.67) is achieved from Fig. 8.3. The state output error $\overline{e}_i(t,k)$ ($i = 1, 2, 3, 4$, $\forall k \in \mathbb{Z}_+$) with iterative learning controllers (8.19) and (8.20) is shown in Fig. 8.4, 8.5 and 8.6. Also, one can see that the finite-iteration tracking of the system (8.67) is achieved from Fig. 8.6. From Figs. 8.3 and 8.6, one can see that, the rate of convergence for the system (8.67) with iterative learning controller (8.16) is faster than that with controller (8.18).

8.6 Summary

This chapter has studied the singular coupled systems' finite-iteration tracking problem. Systems having specific limitations have been proposed to be singularly linked systems. Then, by developing packet-dropping learning controllers, various criteria have been addressed to obtain the finite-iteration tracking of singular systems. Finally, a case study using numerical simulations has been provided to show that the suggested criteria are accurate. Future efforts may focus on providing a clearer definition of finite-iteration tacking and paying closer attention to examples of the tacking.

References

Amann N, Owens DH, Rogers E (1996) Iterative learning control for discrete-time systems with exponential rate of convergence. IEEE Proceed Contr Theor Appl. 143(2):217–224
Amato F, Ariola M (2005) Finite-time control of discrete-time linear systems. IEEE Trans Autom Control 55(5):724–729
Bu XH, Hou ZS (2016) Adaptive iterative learning control for linear systems with binary-valued observations. IEEE Trans Neur Netw Learn Syst. 29(1):232–237
Bu XH, Yu QX, Hou ZS, Qian W (2017) Model free adaptive iterative learning consensus tracking control for a class of nonlinear multiagent systems. IEEE Trans Syst Man Cybern Syst. 49(4):677–686
Cao W, Guo Y, Sun M (2016) Iterative learning control of a class of discrete-time singular system. Acta Phys Sin 65(12):120201
Chen YQ, Moore KL (2002) A practical iterative learning path-following control of an omni-directional vehicle. Asian J Contr 4(1):90–98
Chi RH, Liu Y, Hou ZS, Jin ST (2015) Data-driven terminal iterative learning control with high-order learning law for a class of nonlinear discrete-time multiple-input-multiple output systems. IET Control Theory Appl 9(7):1075–1082
Chi RH, Huang B, Wang DW, Zhang RK, Feng YJ (2016) Data-driven optimal terminal iterative learning control with initial value dynamic compensation. IET Control Theory Appl 10(12):1357–1364
Friesz TL, Cho HJ, Mehta NJ, Tobin RL, Anandalingam G (1992) A simulated annealing approach to the network design problem with variational inequality constraints. Transport Sci 26(1):18–26
Graham MR, Callafon RA (2006) An iterative learning design for repeatable runout cancellation in disk drives. IEEE Trans Control Syst Tech 14(3):474–482

He W, Meng TT, Zhang S, Liu JK, Li G, Sun CY (2017) Dual-loop adaptive iterative learning control for a Timoshenko beam with output constraint and input backlash. IEEE Trans Syst Man Cybern Syst. 49(5):1027–1038
Horn RA, Johnson CR (2012) Matrix Analysis. Cambridge University Press, Cambridge
Janssens P, Pipeleers G, Swevers J (2013) A data-driven constrained norm-optimal iterative learning control framework for lti systems. IEEE Trans Control Syst Tech 21(2):546–551
Li JS, Li JM (2014) Adaptive iterative learning control for coordination of second-order multi-agent systems. Int J Robust Nonlin Control 24(18):3282–3299
Lin X, Tampère CM, Viti F, Immers B (2016) The cost of environmental constraints in traffic networks: assessing the loss of optimality. Netw Spat Econ 16(1):349–369
Lu GP, Ho DWC (2006) Generalized quadratic stability for continuous-time singular systems with nonlinear perturbation. IEEE Trans Autom Control 51(5):818–823
Meng DY, Moore KL (2016) Learning to cooperate: Networks of formation agents with switching topologies. Automatica 64:278–293
Meng DY, Jia YM, Du J (2015) Robust consensus tracking control for multiagent systems with initial state shifts, disturbances, and switching topologies. IEEE Trans Neur Netw Learn Syst 26(4):809–824
Niu YG, Ho DWC (2010) Design of sliding mode control subject to packet losses. IEEE Trans Autom Control 55(11):2623–2628
Song J, Niu YG, Zou YY (2015) Robust finite-time bounded control for discrete-time stochastic systems with communication constraint. IET Control Theory Appl 9(13):2015–2021
Sun LY, Wang YZ (2013) An undecomposed approach to control design for a class of nonlinear descriptor systems. Int J Robust Nonlin Control 23(6):695–708
Tian SP, Liu Q, Dai XS, Zhang JX (2016) A PD-type iterative learning control algorithm for singular discrete systems. Adv Differ Equ 2016:1–9
Wang YY, Shi P, Wang QB, Duan DP (2013) Exponential H_∞ filtering for singular Markovian jump systems with mixed mode-dependent time-varying delay. IEEE Trans Circuits Syst I 60(9):2440–2452
Wang ZS, Liu L, Zhang HG, Xiao GY (2016) Fault-tolerant controller design for a class of nonlinear MIMO discrete-time systems via online reinforcement learning algorithm. IEEE Trans Syst Man Cybern Syst. 46(5):611–622
Wei QL, Liu DR, Lin HQ (2016) Value iteration adaptive dynamic programming for optimal control of discrete-time nonlinear systems. IEEE Trans Cybern 46(3):840–853
Wu SC, Tomizuka M (2010) An iterative learning control design for self-servowriting in hard disk drives. Mechatronics 20(1):53–58
Wu LG, Shi P, Gao HJ (2010) State estimation and sliding-mode control of Markovian jump singular systems. IEEE Trans Autom Control 55(5):1213–1219
Xiong WJ, Ho DWC, Cao JD (2008) Synchronization analysis of singular hybrid coupled networks. Phys Lett A 372(44):6633–6637
Xiong WJ, Yu XH, Patel R, Yu WW (2016) Iterative learning control for discrete-time systems with event-triggered transmission strategy and quantization. Automatica 72:84–91
Xiong WJ, Yu XH, Chen Y, Gao J (2016) Quantized iterative learning consensus tracking of digital networks with limited information communication. IEEE Trans Neur Netw Learn Syst 28(6):1473–1480
Xu S, Dooren PV, Stefan R, Lam J (2002) Robust stability and stabilization for singular systems with state delay and parameter uncertainty. IEEE Trans Autom Control 47(7):1122–1128
Yu M, Li CY (2016) Robust adaptive iterative learning control for discrete-time nonlinear systems with time-iteration-varying parameters. IEEE Trans Syst Man Cybern Syst 47(7):1737–1745
Zhang QL, Zhang JY, Wang YY (2017) Sliding-mode control for singular Markovian jump systems with Brownian motion based on stochastic sliding mode surface. IEEE Trans Syst Man Cybern Syst 49(3):494–505

Chapter 9
Finite-Iteration Learning Tracking with FlexRay Communication Protocol

9.1 Introduction

ILC is a useful method for enhancing system efficiency by taking into account prior execution. The main benefit of ILC is that it allows repeated operations to boost the system's efficiency. ILC has found extensive use in robotic manipulators He et al. (2015), Meng and He (2017), He et al. (2019), nonlinear systems (Bu et al. 2017; Shen et al. 2017; Meng and Moore 2020), multi-agent systems (Meng and Moore 2016; Xiong et al. 2016, 2020; Jin 2019), and data-driven control (Thomopoulos et al. 2013; Chi et al. 2018; Yu et al. 2019) owing to its high accuracy and efficacy in dealing with repeatable control. For instance, data missing and asynchronous control updating are explored in the ILC scheme to converge to the required tracking error for a certain class of large-scale systems in Shen and Chen (2012). A desired control work for multi-agent systems is accomplished in Meng and Moore (2016) by using an iterative technique with nearest neighbor distances.

The design of ILC has constraints according to the published literature. When a communication channel is established between two agents, as in the case of Bu et al. (2017); Meng and Moore (2016), the channel is constantly fixed, even if it is not utilized to communicate information or only transmits a tiny amount of information during a specific period of time. This circumstance can result in bandwidth being wasted. Additionally, as the iteration tends to infinity, the tracking is explored in Bu et al. (2017); Meng and Moore (2016).

In this chapter, a new type of ILC scheme will be created utilizing FlexRay to get over the aforementioned restrictions. In this case, FlexRay (Schmidt and Schmidt 2008; Murvay and Groza 2020; Park and Sunwoo 2011; Zeng et al. 2011; Tang et al. 2019) is a protocol used for car network connections. It benefits from supporting high-speed data transfer and offers two separate channels for fault-tolerant communication. In some cycles, the FlexRay technique is utilized. Each cycle is divided into four various segments: (1) *The static segment.* Some significant or crucial data, such as safety signals or vehicle control signals, can be accurately and promptly translated in

W. Xiong et al., *Iterative Learning Control for Network Systems Under Constrained Information Communication*, Intelligent Control and Learning Systems 12,
https://doi.org/10.1007/978-981-97-0926-7_9

this segment. (2) *dynamic segment*. Event-triggered data is able to be transferred in this segment while the priority is taken into consideration. (3) *The symbol window*. In this segment, unique symbols can get communicated. 4) *The network idle time*. In this time frame, a clock synchronization is typically carried out. The lengths of two segments, the symbol window and the network idle time, are significantly less than those in the static and dynamic segments, as noted in Tang et al. (2019). This practical fact makes it appropriate to disregard the symbol window and network idle time in the following explanation.

This chapter proposes designing a periodic ILC method based on the FlexRay communication protocol. In addition to meeting the goal of finite-iteration tracking, this periodic ILC architecture has the advantage of lowering the communication load. The iterative steps $\{0, 1, \cdots\}$ in the ILC scheme are split into multiples of periodic intervals. The designed iterative controllers' expressions are the same throughout all periodic intervals; the only variations are in the designed parameters. Moreover, the static segment and the dynamic segment are separated into each periodic interval. The dynamic segment will be reserved for additional signals and unique purposes, such as event-triggered signals. In contrast, the static segment will deal with signals (such as control signals and safety signals) to satisfy the general ILC design objectives. The suggested periodic method has two benefits: (1) This approach may offer more bandwidth and payload. If some agents have accomplished the desired tracking or have a rapid convergence rate, their idle bandwidths in the dynamic segment can be centralized and redistributed to other needed nodes with a slow convergence rate. Consequently, one can have more bandwidth and payload. (2) Since those necessary nodes with slow convergence rates have more bandwidth, the overall convergence rate of all agents will grow.

As a result, the finite-iteration tracking of discrete networks will be covered in this chapter to show our control schemes' benefits clearly. The majority of the ILC criteria currently in use (see Meng and Moore 2016; Tang et al. 2019) actually place a strong emphasis on asymptotic tracking, for instance, the tracking that is gained when the iteration goes to infinity. However, many real-life applications only allow a finite number of iterations in practice. For instance, the necessary tracking is accomplished as the iterations approach a finite number. A significant need for the design is to avoid endless iterations, and the number of iterations should be kept to a minimum. The clarification of the finite-iteration tracking is therefore important and vital. Additionally, in several practical applications, like transportation and formation control, a safe distance between two agents is typically considered to prevent collision. Therefore, it is essential to establish a safe distance between any two agents. Consequently, the following three goals are crucial for designing a periodic ILC scheme: (1) Each network agent succeeds in finishing the finite-iteration tracking. (2) Any two agents are separated from one another by a distance that exceeds a safe threshold. (3) Finite-iteration tracking has the smallest possible settling iterations.

9.2 Problem Formulation

A discrete network with n agents is considered

$$x(t+1,k) = Ax(t,k) + u(t,k), \tag{9.1}$$

where $x(t,k) = (x_1(t,k), \cdots, x_n(t,k))^{\mathrm{T}} \in \mathbb{R}^n$ and the position of the ith agent is $x_i(t,k)$, $i = 1, 2, \cdots, n$. Here, the position $x_i(t,k)$ depends on two independent variables: the first one is the time step t and $t \in \{0, 1, \cdots, T\}$ with T being a given positive integer and the second is the iterative step k and $k \in \{0, 1, 2, \cdots\}$. Matrix $A = (a_{ij})_{n\times n}$ is the connection matrix, and $u(t,k) = (u_1(t,k), \cdots, u_n(t,k))^{\mathrm{T}} \in \mathbb{R}^n$ with $u_i(t,k)$ being the controller of the ith agent, $i = 1, 2, \cdots, n$. Let $x_i^*(t)$ be the reference trajectory of $x_i(t,k)$, and the tracking error $\Delta x_i(t,k) = x_i(t,k) - x_i^*(t)$, $i = 1, 2, \cdots, n,\ t \in \{0, 1, \cdots, T\}$.

The following provides a definition.

Definition 9.1 The finite-iteration tracking is stated to be reached by model (9.1) if

$$\|\Delta X(k)\| \le e^{-\zeta} \|\Delta X(0)\|,\ \forall\, k \ge W, \tag{9.2}$$

where $\Delta x(t,k) = (\Delta x_1(t,k), \cdots, \Delta x_n(t,k))^{\mathrm{T}}$, $\Delta X(k) = (\Delta x^{\mathrm{T}}(1,k), \cdots, \Delta x^{\mathrm{T}}(T,k))^{\mathrm{T}}$, and $\zeta > 0$ is a given scalar. Here, W is settling iteration by a scalar depending on the known number ζ.

From Definition 9.1, considering that $x_i(t,k)$ is the position of the ith agent, the goals of this study are as follows:

(i) For each agent i to achieve the finite-iteration tracking for a position $x_i(t,k)$.

(ii) To ensure $|x_i(t,k) - x_j(t,k)| \ge K$, for $i = 1, 2, \cdots, n,\ t \in \{0, 1, \cdots, T\}$, $k \in \{0, 1, 2, \cdots\}$. K is the safe preset distance between agents, which is a given constant. Additionally, the reference trajectory $x_i^*(t)$ is defined to satisfy that $|x_i^*(t) - x_j^*(t)| \ge K,\ i, j = 1, 2, \cdots, n,\ i \ne j$.

(iii) To ensure that the number of settling iterations, W, is kept as low as possible.

Remark 9.1 The aforementioned three goals of this study are more practical than the explanation of finite-iteration tracking of discrete systems in Xiong et al. (2020). Reasons include: (1) A safe preset distance K is defined such that $|x_i(t,k) - x_j(t,k)| \ge K$, for $i = 1, 2, \cdots, n,\ t \in \{0, 1, \cdots, T\},\ k \in \{0, 1, 2, \cdots\}$. (2) The tracking with finite-iterations W is kept to a minimum or a suitable predetermined value.

Assumption 9.1 The two requirements have been met in the whole chapter.

(1) The initial state of each iteration remains the same, i.e.,

$$x_i(0,k) = x_i^0, \tag{9.3}$$

where $i = 1, 2, \cdots, n,\ \forall\, k \in \{0, 1, 2, \cdots\}$, x_i^0 is a constant and satisfies that $|x_i^0 - x_j^0| > K, i, j = 1, 2, \cdots, n,\ i \ne j$.

(2) The position $x_i(t,k)$ and the reference trajectory $x_i^*(t)$ of the ith agent are all bounded, which means, $|x_i(t,k)| < M$ and $|x_i^*(t)| < M$, where $M > 0$ is a positive number, $i = 1, 2, \cdots, n,\ t \in \{0, 1, \cdots, T\}, k \in \{0, 1, 2, \cdots\}$.

9.3 ILC Scheme Under FlexRay Protocol

In this part, we propose the ILC scheme under FlexRay communication protocol.

(1) The iterative steps $\{0, 1, \cdots\}$ are divided into several periodic intervals, which means $\{0, 1, 2, \cdots\}$ is grouped into $\{0, 1, \cdots, \varpi - 1\}$, $\{\varpi, \varpi + 1, \cdots, 2\varpi - 1\}$, $\{2\varpi, 2\varpi + 1, \cdots, 3\varpi - 1\}, \cdots$. Here, the periodic iteration ϖ is a positive integer, and $\varpi = \varpi_1 + \varpi_2$, $\varpi_i\ (i = 1, 2)$ are both positive integers. In each periodic interval, $[l\varpi, l\varpi + \varpi_1 - 1]$ is the static segment, and $[l\varpi + \varpi_1, (l+1)\varpi - 1]$ is the dynamic segment, $l = 0, 1, \cdots$.

(2) In the static segment $[l\varpi, l\varpi + \varpi_1 - 1]$ $(l = 0, 1, 2, \cdots)$, there are two goals: The first one is to achieve finite-iteration tracking and the second one is to guarantee that $|x_i(t,k) - x_j(t,k)| \geq K$, for $i = 1, 2, \cdots, n,\ t \in \{0, 1, \cdots, T\},\ k \in \{0, 1, 2, \cdots\}$. The controllers $u_i(t,k)\ (i = 1, 2, \cdots, n)$ are designed into two parts.

Part 2A: In the static segment $[l\varpi, l\varpi + \varpi_1 - 1],\ l = 0, 1, 2, \cdots$,

$$u_i(t, k+1) = u_i(t,k) + \Delta u_i(t,k), \tag{9.4}$$

with

$$\Delta u_i(t,k) = \alpha_i \Delta x_i(t+1, k), \tag{9.5}$$

where α_i satisfies $-1 < \alpha_i < 0, i = 1, 2, \cdots, n$ which is a given constant.

Part 2B: Define $R\ (R > K)$ as the distance at which an agent may locate the positions of other agents. When $|x_i(t,k) - x_j(t,k)| \leq R$, the controllers $u_i(t,k)$ and $u_j(t,k)$ are activated to satisfy the following condition. If not, $u_i(t,k)$ and $u_j(t,k)$ have no change.

$$\begin{cases} \text{controllers have no change}, & \text{if } |x_i(t,k) - x_j(t,k)| > R, \\ \left|u_i(t,k) - u_j(t,k)\right| \geq \tau, & \text{otherwise} \end{cases} \tag{9.6}$$

where $i, j = 1, 2, \cdots, n,\ i \neq j, t \in \{0, 1, \cdots, T\},\ k \in \{0, 1, 2, \cdots\}$, and

$$\tau = \frac{2(R^2 - K^2)Mn \max\limits_{1 \leq i,j \leq n} |a_{ij}|}{|x_i(t,k) - x_j(t,k)|^2 - K^2} + K.$$

Remark 9.2 To prevent the next scenario, the distance R should be selected here to be properly larger than the safe default distance K, which means for some $t \in \{0, 1, \cdots, T\}$ and some $k \in \{1, 2, \cdots\}$, one has $|x_i(t,k) - x_j(t,k)| > R$, but $|x_i(t+$

$1, k) - x_j(t+1, k)| < K$. This contradictory case is, therefore, not discussed in this chapter.

(3) In the dynamic segment $[l\varpi + \varpi_1, (l+1)\varpi - 1]$ $(l = 0, 1, 2, \cdots)$, there are two goals: Making the settling iteration W as minimal as possible is the primary goal. Here, note that the settling iteration W is selected in all dynamic segments (see Remark 9.6 for details). The second goal is to guarantee that $|x_i(t,k) - x_j(t,k)| \geq K$, for $i = 1, 2, \cdots, n,\ t \in \{0, 1, \cdots, T\},\ k \in \{0, 1, 2, \cdots\}$. The controllers $u_i(t,k)$ $(i = 1, 2, \cdots, n)$ are also designed into two parts.

Part 3A: In the dynamic segment $[l\varpi + \varpi_1, (l+1)\varpi - 1]$ $(l = 0, 1, 2, \cdots)$,

$$u_i(t, k+1) = u_i(t,k) + \Delta\widehat{u}_i(t,k), \tag{9.7}$$

with

$$\Delta\widehat{u}_i(t,k) = \begin{cases} \beta_i \Delta x_i(t+1, k), & if\ |1+\alpha_i| \geq \max\limits_{1\leq j\leq n} |1+\alpha_j|, \\ 0, & \text{otherwise}, \end{cases} \tag{9.8}$$

where β_i is a given constant, which will be designed later in Theorem 9.1.

Part 3B: In this part, the controllers $u_i(t,k)$ are designed to be the same as that in (9.6).

Remark 9.3 In the learning control schemes (9.4)–(9.8), further explanations are provided:

(1) There is no conflict in schemes (9.4)–(9.5) (or (9.7)–(9.8)) and (9.6). The controller $u_i(t,k)$ $(i = 1, 2, \cdots, n)$ is first obtained in schemes (9.4)–(9.5) and then checked in scheme (9.6). If the controllers $u_i(t,k)$ and $u_j(t,k)$ are activated in the scheme (9.6), the two controllers are updated according to the scheme (9.6). Otherwise, the two controllers are not updated and remain acquired from schemes (9.4)–(9.5).

(2) The controller $u_i(t,k)$ is only updated when $|1+\alpha_i| \geq \max\limits_{1\leq j\leq n} |1+\alpha_j|$ in the control scheme of 9.8. The triggered condition of updating the controller $u_i(t,k)$ is the inequality $|1+\alpha_i| \geq \max\limits_{1\leq j\leq n} |1+\alpha_j|$. The inequality $|1+\alpha_i| \geq \max\limits_{1\leq j\leq n} |1+\alpha_j|$ means that the convergent speed of tracking agent j is quicker than that of agent i $(1 \leq j \leq n,\ j \neq i)$ at time step t and iterative step k.

Remark 9.4 This remark serves as an explanation of how the event-triggered controller (9.8) in the dynamic segment can offer more bandwidth and payload via an example.

Assume that there is a small network with six nodes (see Fig. 9.1). In one scenario, each agent has communication channel with a 10mb/s initial bandwidth. Each agent receives the control data during the static segment in its ordinary practice. Additionally, in the dynamic segment, if the inequality $|1+\alpha_i| \geq \max\limits_{1\leq j\leq n} |1+\alpha_j|$ holds when $i = 2,\ 4,\ 5$, which means the tracking convergent speeds of agents 2, 4, and

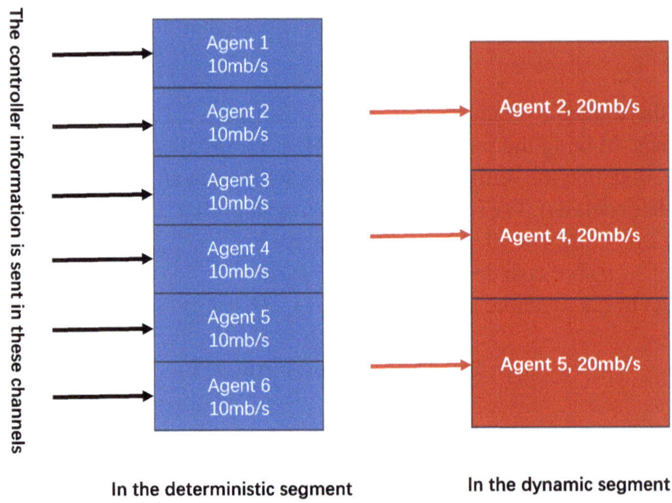

Fig. 9.1 The bandwidth allocation of the information transmission in channels

5 are slower than those of agents 1, 3, and 6. In order to update their controllers and improve their convergence speeds and achieve faster tracking, agents 2, 4, and 5 will have additional 10 mb/s bandwidth, which will increase to 20 mb/s, whereas agents 1, 3, and 6 do not. Additionally, the control mechanism (9.8) is versatile in its design. In other words, if an agent in the static segment completes the specified finite-iteration tracking, its extra 10mb/s of bandwidth can be distributed to another agent in the dynamic segment. As a result, the dynamic segment's event-triggered control system (9.8) can offer more bandwidth and payload.

Remark 9.5 The suggested ILC scheme has the following advantages over conventional ILC schemes (Meng and Moore 2016; Xiong et al. 2016): (1) Greater bandwidth and payload can be offered by the dynamic segment's event-triggered controller (9.8). (2) The settling iterations W of the finite-iteration tracking can be adjusted to its smallest setting, which functions as a practical need in genuine applications.

9.4 Main Results

In this section, by creating suitable control schemes for the static and dynamic segments, respectively, we aim to accomplish the three goals in Section 9.3.

Theorem 9.1 *Consider model (9.1) with the learning control schemes (9.4)–(9.8) in the periodic iteration intervals* $[l\varpi_1, (l+1)\varpi]$ $(l = 0, 1, 2, \cdots)$. *With Assumption 9.1, if parameters* α_i *and* β_i $(i = 1, 2, \cdots, n)$ *in the controllers (9.4), (9.5), (9.7), and (9.8) are designed to satisfy that*

$$-1 < \alpha_i < 0, \tag{9.9}$$

and

$$-1 < \beta_i < 0, \tag{9.10}$$

then the following three objectives will be obtained in the static and dynamic segments: (1) to achieve the finite-iteration tracking; (2) to guarantee that $|x_i(t,k) - x_j(t,k)| \geq K$, *for* $i = 1, 2, \cdots, n$, $t \in \{1, 2, \cdots, T\}$, $k \in \{0, 1, 2, \cdots\}$; *(3) To find a small number of setting iteration W. The settling iteration W can be taken as* $W = (\lfloor \log_{\Xi}^{t} \rfloor + 2)\varpi$, *where* $\Xi = \max\limits_{1 \leq i \leq n} (1 + \alpha_i)^{\varpi_1}$ *and* $t = \frac{1}{\Gamma} \frac{e^{-\zeta}}{\max\limits_{1 \leq i \leq n} (1+\alpha_i)^{\varpi_1}}$.

Proof In order to complete the proof, this theorem will be divided into Corollaries 9.1 and 9.2. □

Remark 9.6 To achieve the third objective, we need to guarantee a small value of setting iteration W. The setting iteration is selected related to α_i and α_j in the static segment of agent i and j, respectively. Therefore, based on the following four processes, W is chosen.

(1) When using Definition 9.1, notice that if the settling iteration W is small, all agents must get at their desired tracking virtually at the same time.

(2) Find agent i with the inequality $|1 + \alpha_i| \geq \max\limits_{1 \leq j \leq n} |1 + \alpha_j|$, which means the convergent speed of tracking of agent j is quicker than that of agent i $(1 \leq j \leq n,\ j \neq i)$.

(3) The communication bandwidth of agent i is added and the communication bandwidth of agent j is cancelled.

(4) Compute $l \geq \log_{\Xi}^{t}$ in (9.29) and $W = (\lfloor \log_{\Xi}^{t} \rfloor + 2)\varpi$, where $\lfloor \log_{\Xi}^{t} \rfloor$ denotes the largest integer part of the number $\log_{\Xi}^{t}$.

Remark 9.7 To minimize the number of updated iteration steps, the ILC event-triggered method is also explored in Xiong et al. (2016). However, the important information on hierarchical distinctions is ignored in Xiong et al. (2016). Thus, to demonstrate this hierarchical difference in this chapter, the iterative steps $\{0, 1, 2, \cdots\}$ are divided into some periodic intervals, and every periodic interval is divided into two sub-segments: the static and dynamic segments. The settling iterations W with the periodic ILC are lower than those with the traditional ILC. The example that follows provides a more thorough explanation.

Corollary 9.1 *Consider model (9.1) with the learning control schemes (9.4)–(9.6) in the static segment* $[l\varpi, l\varpi + \varpi_1 - 1]$ $(l = 0, 1, 2, \cdots)$. *Under Assumption 9.1, the parameters* α_i $(i = 1, 2, \cdots, n)$ *in controllers (9.4) and (9.5) are designed to satisfy*

$$-1 < \alpha_i < 0, \tag{9.11}$$

then the following two goals can be achieved in this static segment: (1) accomplishing the finite-iteration tracking; (2) guaranteeing that $|x_i(t,k) - x_j(t,k)| \geq K$, *for* $i = 1, 2, \cdots, n$, $t \in \{1, 2, \cdots, T\}$, $k \in \{0, 1, 2, \cdots\}$.

Proof The proof is divided into two parts.

(1) Finite-iteration tracking is considered in this section. Define that $\Delta x(t,k) = (\Delta x_1(t,k), \cdots, \Delta x_n(t,k))^{\mathrm{T}}$, according to (9.1), (9.4), and (9.5), one has

$$\begin{aligned} &\Delta x(t+1,k+1) - \Delta x(t+1,k) \\ &= x(t+1,k+1) - x(t+1,k) \\ &= Ax(t,k+1) - Ax(t,k) + u(t,k+1) - u(t,k), \end{aligned} \tag{9.12}$$

for $[l\varpi, l\varpi + \varpi_1 - 1]$ $(l = 0, 1, 2, \cdots)$, $t \in \{0, 1, \cdots, T\}$. Define that $e(t,k) = (e_1(t,k), \cdots, e_n(t,k))^{\mathrm{T}} \triangleq x(t,k+1) - x(t,k)$, from (9.4) and (9.5), one obtains

$$e(t+1,k) = Ae(t,k) + \alpha\Delta x(t+1,k), \tag{9.13}$$

where matrix $\alpha = \mathrm{diag}(\alpha_1, \alpha_2, \cdots, \alpha_n)$. According to Assumption 9.1, one knows $e(0,k) = 0$. From (9.13), it is not difficult to know

$$e(t,k) = \sum_{i=0}^{t-1} A^i \alpha \Delta x(t-i,k), \tag{9.14}$$

for $t \in \{1, \cdots, T\}$, $l\varpi \le k \le l\varpi + \varpi_1 - 1$, $l = 0, 1, 2, \cdots$. Combining (9.12), (9.13), and (9.14), one can obtain

$$\begin{aligned} &\Delta x(t+1,k+1) \\ &= \Delta x(t+1,k) + Ae(t,k) + \alpha\Delta x(t+1,k) \\ &= \Delta x(t+1,k) + \sum_{i=0}^{t-1} A^{i+1}\alpha\Delta x(t-i,k) + \alpha\Delta x(t+1,k), \end{aligned} \tag{9.15}$$

where $l\varpi \le k \le l\varpi + \varpi_1 - 1$, $l = 0, 1, 2, \cdots$.

Let $\Delta X(k) = (\Delta x^{\mathrm{T}}(1,k), \cdots, \Delta x^{\mathrm{T}}(T,k))^{\mathrm{T}}$, one obtains

$$\Delta X(k+1) = \Pi_1 \Delta X(k), \tag{9.16}$$

where $l\varpi \le k \le l\varpi + \varpi_1 - 1$, $l = 0, 1, 2, \cdots$. Matrix Π_1 is given in (9.17), and I_n is the n-dimensional identity matrix. Also, $*_{ij} \in \mathbb{R}^n$ $(i = 2, \cdots, T,\ j = 1, \cdots, T-1)$ can be obtained from (9.15) and (9.16), which are not explained here because the subsequent method does not call for them. Therefore, from (9.11), one has that $\rho(\Pi_1) < 1$. That is, $\Delta X(k) \to \mathbf{0}$ as $l \to +\infty$. As a consequence, system tracking over finite-iteration (9.1) is achievable. Given that it is dependent on the controllers in the static segment $[l\varpi, l\varpi + \varpi_1 - 1]$, $l = 0, 1, 2, \cdots$, the settling iteration W will be calculated using the following corollary:

$$\Pi_1 = \begin{pmatrix} \alpha + I_n & O & O & \cdots & O & O \\ *_{21} & \alpha + I_n & O & \cdots & O & O \\ *_{31} & *_{32} & \alpha + I_n & \cdots & O & O \\ \vdots & \vdots & \vdots & \cdots & \vdots & \vdots \\ *_{T1} & *_{T2} & *_{T3} & \cdots & *_{T,T-1} & \alpha + I_n \end{pmatrix}. \tag{9.17}$$

(2) To verify the condition $|x_i(t,k) - x_j(t,k)| \geq K$, for $i = 1, 2, \cdots, n,\ t \in \{1, 2, \cdots, T\},\ k \in \{0, 1, 2, \cdots\}$.

According to Assumption 9.1, one has that $|x_i(0,0) - x_j(0,0)| > K$, for $i = 1, 2, \cdots, n,\ i \neq j$. Then, fix $k = 0$, for $t = 1$, then check if $|x_i(0,0) - x_j(0,0)| > R$ is fulfilled, for $i = 1, 2, \cdots, n,\ i \neq j$. One has

(a) If $K < |x_i(0,0) - x_j(0,0)| \leq R$, from model (9.1) and the control scheme (9.6), one can obtain

$$\begin{aligned} |x_i(1,0) - x_j(1,0)| &= \left| \sum_{s=1}^{n} a_{is} x_s(0,0) - \sum_{s=1}^{n} a_{js} x_s(0,0) + u_i(0,0) - u_j(0,0) \right| \\ &\geq \left| u_i(0,0) - u_j(0,0) \right| - \left| \sum_{s=1}^{n} a_{is} x_s(0,0) \right| - \left| \sum_{s=1}^{n} a_{js} x_s(0,0) \right| \\ &\geq \frac{2(R^2 - K^2)Mn \max\limits_{1 \leq i,j \leq n} |a_{ij}|}{|x_i(t,k) - x_j(t,k)|^2 - K^2} + K - 2Mn \max_{1 \leq i,j \leq n} |a_{ij}| \\ &\geq K. \end{aligned} \tag{9.18}$$

(b) If $|x_i(0,0) - x_j(0,0)| > R$ is fulfilled, according to Remark 9.2, one has that $|x_i(1,0) - x_j(1,0)| > K$ is also fulfilled. Repeating the above process in $t \in \{2, 3, \cdots, T\}$, one has that $|x_i(t,0) - x_j(t,0)| \geq K$, for $i = 1, 2, \cdots, n$. Note that, from Assumption 9.1, one also has that $|x_i(0,k) - x_j(0,k)| > K$, for $i = 1, 2, \cdots, n,\ i \neq j,\ \forall k \in \{0, 1, 2, \cdots\}$. Then, we prove that $|x_i(1,1) - x_j(1,1)| \geq K$. Calculate $|x_i(0,1) - x_j(0,1)|$, one has

(a) If $K < |x_i(0,1) - x_j(0,1)| \leq R$, similar to (9.18), one also can find

$$\begin{aligned} \left|x_i(1,1) - x_j(1,1)\right| &= \left| \sum_{s=1}^{n} a_{is} x_s(0,1) - \sum_{s=1}^{n} a_{js} x_s(0,1) + u_i(0,1) - u_j(0,1) \right| \\ &\geq \left| u_i(0,1) - u_j(0,1) \right| - \left| \sum_{s=1}^{n} a_{is} x_s(0,1) \right| - \left| \sum_{s=1}^{n} a_{js} x_s(0,1) \right| \\ &\geq \frac{2(R^2 - K^2)Mn \max\limits_{1 \leq i,j \leq n} |a_{ij}|}{|x_i(t,k) - x_j(t,k)|^2 - K^2} + K - 2Mn \max_{1 \leq i,j \leq n} |a_{ij}| \\ &\geq K. \end{aligned} \tag{9.19}$$

(b) If $|x_i(0,1) - x_j(0,1)| > R$ is fulfilled, according to Remark 9.2, one has that $|x_i(1,1) - x_j(1,1)| > K$ is also fulfilled. As a result, one has that $|x_i(1,1) - x_j(1,1)| > K$. Repeating the above process in $k \in \{1, 2, \cdots\}$, one has that $|x_i(t,k) - x_j(t,k)| \geq K$, for $i = 1, 2, \cdots, n,\ t \in \{1, 2, \cdots, T\}$. Hence, the two objectives (1) and (2) have been reached in this static segment. The proof is completed. □

Remark 9.8 In Part B of Corollary 9.1, the proof process and computation are finished. Firstly, one has to prove that $|x_i(t,0) - x_j(t,0)| \geq K$ ($\forall k \in \{0, 1, 2, \cdots\}$) for two cases: $|x_i(0,0) - x_j(0,0)| > R$ and $K < |x_i(0,0) - x_j(0,0)| \leq R$. Then, one has to prove that, for some $k \in \{0, 1, 2, \cdots\}$, $|x_i(t,k) - x_j(t,k)| \geq K$, ($\forall t \in \{1, \cdots, T\}$) for two cases: $|x_i(0,0) - x_j(0,0)| > R$ and $K < |x_i(0,0) - x_j(0,0)| \leq R$.

Corollary 9.2 *Consider model (9.1) with the learning control schemes (9.6)–(9.8) in the dynamic segment* $[l\varpi + \varpi_1, (l+1)\varpi - 1]$ $(l = 0, 1, 2, \cdots)$. *With Assumption 9.1, the parameters* β_i $(i = 1, 2, \cdots, n)$ *in controllers (9.7) and (9.8) are designed to satisfy*

$$-1 < \beta_i < 0, \tag{9.20}$$

then the following goals will be obtained in this dynamic segment: (1) select a settling iteration W as small as feasible; (2) to guarantee that $|x_i(t,k) - x_j(t,k)| \geq K$ *, for* $i = 1, 2, \cdots, n,\ t \in \{1, 2, \cdots, T\},\ k \in \{0, 1, 2, \cdots\}$.

Proof The proof is also divided into two parts.

(1) To find a suitable procedure to make the setting iteration W as small as possible in the dynamic segments. In the schemes (9.7)–(9.8), one has to find the corresponding nodes needed to be located. If there is only a node satisfying $|1+\alpha_i| \geq \max\limits_{1\leq j\leq n} |1+\alpha_j|$, the whole bandwidth is given to node i to increase the tracking controller. If there are two or more nodes satisfying $|1+\alpha_i| \geq \max\limits_{1\leq j\leq n} |1+\alpha_j|$, the whole bandwidth is divided into multiple parts, which are separately given to these nodes in order to add their tracking controllers.

Let $\Delta x(t,k) = (\Delta x_1(t,k), \cdots, \Delta x_n(t,k))^{\mathrm{T}}$, $e(t,k) = (e_1(t,k), \cdots, e_n(t,k))^{\mathrm{T}} \triangleq x(t,k+1) - x(t,k)$. According to (9.1), (9.7), and (9.8), similar to Corollary 9.1, one has

$$e(t+1,k) = Ae(t,k) + \beta \Delta x(t+1,k), \tag{9.21}$$

where matrix $\beta = \mathrm{diag}(\beta_1, \cdots, \beta_n)$. If node i $(i \in \{1, \cdots, n\})$ is a controlled node in this segment, $\beta_i \neq 0$. Otherwise, $\beta_i = 0$.

Let $\Delta X(k) = (\Delta x^{\mathrm{T}}(1,k), \cdots, \Delta x^{\mathrm{T}}(T,k))^{\mathrm{T}}$, similar to Corollary 9.1, one obtains

$$\Delta X(k+1) = \Pi_2 \Delta X(k), \tag{9.22}$$

where $l\varpi + \varpi_1 \leq k \leq (l+1)\varpi - 1$ $(l = 0, 1, 2, \cdots)$. Matrix Π_2 is given in (9.23), and $\widehat{*}_{ij} \in \mathbb{R}^n$ $(i = 2, \cdots, T,\ j = 1, \cdots, T-1)$ can be obtained from (9.22) and (9.23), which are not covered in depth here since the subsequent process does not include any requirements.

$$\Pi_2 = \begin{pmatrix} \beta + I_n & O & O & \cdots & O & O \\ \widehat{*}_{21} & \beta + I_n & O & \cdots & O & O \\ \widehat{*}_{31} & \widehat{*}_{32} & \beta + I_n & \cdots & O & O \\ \vdots & \vdots & \vdots & \cdots & \vdots & \vdots \\ \widehat{*}_{T1} & \widehat{*}_{T2} & \widehat{*}_{T3} & \cdots & \widehat{*}_{T,T-1} & \beta + I_n \end{pmatrix}. \tag{9.23}$$

Combining (9.16) and (9.22), one has that
(1) If $l\varpi \leq k \leq l\varpi + \varpi_1 - 1$,

$$\Delta X(k) = (\Pi_1)^{k-l\varpi}\left[(\Pi_2)^{\varpi-\varpi_1}(\Pi_1)^{\varpi_1}\right]^l \Delta X(0), \tag{9.24}$$

where $l = 0, 1, 2, \cdots$.

(2) If $l\varpi + \varpi_1 \leq k \leq (l+1)\varpi - 1$,

$$\Delta X(k) = (\Pi_2)^{k-l\varpi-\varpi_1}(\Pi_1)^{\varpi_1}\left[(\Pi_2)^{\varpi-\varpi_1}(\Pi_1)^{\varpi_1}\right]^l \Delta X(0), \tag{9.25}$$

where $l = 0, 1, \cdots$.

In the l periodic intervals, assume that nodes $i_1,\ i_2, \cdots,\ i_s$ are picked to be controlled and node i_j $(j = 1, 2, \cdots, s)$ has been controlled for l_j times in the dynamic segment. According to (9.16), (9.17), and (9.22)–(9.25), one has
(i) If $l\varpi \leq k \leq l\varpi + \varpi_1 - 1$,

$$\begin{aligned} \left\|\Delta X(k)\right\| &= \left\|(\Pi_1)^{k-l\varpi}\left[(\Pi_2)^{\varpi-\varpi_1}(\Pi_1)^{\varpi_1}\right]^l \Delta X(0)\right\| \\ &\leq \left\|(\Pi_1)^{k-l\varpi}\left[(\Pi_2)^{\varpi-\varpi_1}(\Pi_1)^{\varpi_1}\right]^l\right\| \cdot \left\|\Delta X(0)\right\| \\ &\leq \Upsilon_1(l)\left\|\Delta X(0)\right\|, \end{aligned} \tag{9.26}$$

where

$$\begin{aligned} \Upsilon_1(l) = &\max_{1\leq i\leq n}(1+\alpha_i)^{k-l\varpi} \times \\ &\max\left\{\max_{1\leq i\leq n, i\neq i_j}(1+\alpha_i)^{\varpi_1 l},\ \max_{1\leq j\leq s}(1+\beta_{i_j})^{l_j(\varpi-\varpi_1)}(1+\alpha_{i_j})^{\varpi_1 l}\right\}. \end{aligned}$$

(ii) If $l\varpi + \varpi_1 \leq k \leq (l+1)\varpi - 1$,

$$\begin{aligned}\left\|\Delta X(k)\right\| &= (\Pi_2)^{k-l\varpi-\varpi_1}(\Pi_1)^{\varpi_1}\left[(\Pi_2)^{\varpi-\varpi_1}(\Pi_1)^{\varpi_1}\right]^l \Delta X(0)\\ &\le \left\|(\Pi_2)^{k-l\varpi-\varpi_1}(\Pi_1)^{\varpi_1}\left[(\Pi_2)^{\varpi-\varpi_1}(\Pi_1)^{\varpi_1}\right]^l\right\| \cdot \|\Delta X(0)\|\\ &\le \Upsilon_2(l)\left\|\Delta X(0)\right\|, \end{aligned} \tag{9.27}$$

where

$$\begin{aligned}\Upsilon_2(l) = &\max_{1\le i\le n}(1+\alpha_i)^{\varpi_1} \times\\ &\max\left\{\max_{1\le i\le n, i\ne i_j}(1+\alpha_i)^{\varpi_1 l}, \max_{1\le j\le s}(1+\beta_{i_j})^{l_j(\varpi-\varpi_1)}(1+\alpha_{i_j})^{\varpi_1 l}\right\}.\end{aligned}$$

Note that $\Upsilon_2(l) \le \Upsilon_1(l)$, one knows that (9.26) must be fulfilled if (9.27) is fulfilled. In the following, only inequality (9.27) is examined for the purpose of simplicity.

In $\Upsilon_2(l)$, the purpose of the term $(1+\beta_{i_j})^{l_j(\varpi-\varpi_1)}$ is to make the following inequality true

$$\begin{aligned}&\max\left\{\max_{1\le i\le n, i\ne i_j}(1+\alpha_i)^{\varpi_1 l}, \max_{1\le j\le s}(1+\beta_{i_j})^{l_j(\varpi-\varpi_1)}(1+\alpha_{i_j})^{\varpi_1 l}\right\}\\ &< \max_{1\le i\le n}(1+\alpha_i)^{\varpi_1 l}.\end{aligned}$$

Without loss of generality, let

$$\begin{aligned}&\Gamma \max_{1\le i\le n}(1+\alpha_i)^{\varpi_1 l} \triangleq\\ &\max\left\{\max_{1\le i\le n, i\ne i_j}(1+\alpha_i)^{\varpi_1 l}, \max_{1\le j\le s}(1+\beta_{i_j})^{l_j(\varpi-\varpi_1)}(1+\alpha_{i_j})^{\varpi_1 l}\right\},\end{aligned}$$

where $0 < \Gamma < 1$. From (9.27) and Definition 9.1, one has $\Upsilon_2(l) \le e^{-\zeta}$. That is,

$$\max_{1\le i\le n}(1+\alpha_i)^{\varpi_1 l} \le \frac{1}{\Gamma}\frac{e^{-\zeta}}{\max_{1\le i\le n}(1+\alpha_i)^{\varpi_1}}. \tag{9.28}$$

In terms of (9.28), it is not hard to get

$$l \ge \log_{\Xi}^{\Theta}, \tag{9.29}$$

where $\Xi = \max_{1\le i\le n}(1+\alpha_i)^{\varpi_1}$ and $\Theta = \frac{1}{\Gamma}\frac{e^{-\zeta}}{\max_{1\le i\le n}(1+\alpha_i)^{\varpi_1}}$.

The settling iteration W may be taken as $W = (\lfloor \log_{\Xi}^{\Theta} \rfloor + 1)\varpi$. Here, since $0 < 1+\alpha_i < 1$, one has that the term $\lfloor \log_{\Xi}^{\Theta} \rfloor$ is small if the positive constant Γ is small. As a consequence, the controllers (9.7) and (9.8) can make the settling iteration W to be smallest possible.

(2) To verify the condition $|x_i(\theta, k) - x_j(t, k)| \geq K$, for $i = 1, 2, \cdots, n$, $t \in \{1, 2, \cdots, T\}$, $k \in \{0, 1, 2, \cdots\}$. In this part, the detailed proof is the same as that in Corollary 9.1, which is omitted here.

As a result, the two goals have been achieved in this dynamic segment. And the proof is completed. □

9.5 Simulation Example

Example 9.1 Consider the following discrete system with eight agents

$$x(t+1, k) = Ax(t, k) + u(t, k). \tag{9.30}$$

In (9.30), $x(t, k) = (x_1(t, k), x_2(t, k), \cdots, x_8(t, k))^{\mathrm{T}} \in \mathbb{R}^8$, and time $T = 10$. Define the initial state $x(0, k) = (-1.2, -0.4, 0.4, 1.2, 2, -2, -3, 3)^{\mathrm{T}}$, $\forall\ k \in \{1, 2, \cdots\}$, and the desired reference trajectory $x^*(t) = (0.1 \sin t - 0.5, 0.1 \cos t + 0.5, 0.1 \sin t + 1.5, 0.1 \cos t + 2.5, 0.1 \sin t + 3.5, 0.1 \sin t - 3.5, 0.1 \sin t - 2.5, 0.1 \cos t - 1.5)^{\mathrm{T}}$, $t \in \{0, 1, \cdots, 10\}$. And the matrix

$$A = \begin{pmatrix} -0.6 & 0 & 0 & 0 & 0 & 0 & 0 & 0 \\ 2 & -0.4 & 0 & 0 & 0 & 0 & 0 & 0 \\ 1 & -1.5 & -0.9 & 0 & 0 & 0 & 0 & 0 \\ 0.5 & 0 & 0.4 & -0.8 & 0 & 0 & 0 & 0 \\ 0 & 1 & 1 & 0 & -0.6 & 0 & 0 & 0 \\ 0.9 & 0 & -1.9 & 0 & 0.8 & -0.4 & 0 & 0 \\ 0 & 1.5 & 0 & 2.1 & 0 & 2.8 & -0.5 & 0 \\ 0 & 0 & -1.6 & -2.5 & 0 & 0 & 3.5 & -0.7 \end{pmatrix}.$$

Also, define the safe preset distance $K = 0.5$ and the detective distance $R = 0.8$. Define the iterations $\varpi = 11$ and $\varpi_1 = 9$, which means that the iterative controllers are the equal in the periodic interval $[l\varpi, (l+1)\varpi]$, $l = 0, 1, 2, \cdots$. In the learning control schemes (9.4)–(9.8), the static segment is $[l\varpi, l\varpi + \varpi_1 - 1]$ and the dynamic segment is $[l\varpi + \varpi_1, (l+1)\varpi - 1]$, $l = 0, 1, 2, \cdots$. The parameters $\alpha_i = -\frac{1}{10}$ and $\beta_i = -\frac{4}{5}$, $i = 1, 2, \cdots, 8$. In (9.28), let $\varsigma = 30$, the settling iteration W can be calculated as $W = 502$. The trajectory of the tracking error $\Delta x(t, k)$ without any control scheme can be seen in Fig. 9.2a. With the control schemes (9.4)–(9.8), the trajectory of the tracking error $\Delta x(t, k)$ with the iteration step $k = 300$ can be seen in Fig. 9.2b. According to the control schemes (9.4)–(9.8), the trajectory of the tracking error $\Delta x(t, k)$ with the iteration step $k = 500$ is shown in Fig. 9.2c. From Fig. 9.2c, the finite-iteration tracking of the system (9.30) is shown.

In addition, Fig. 9.3 presents the comparison of $\|\Delta X(k)\|$ to prove that the settling iteration W can be a small value. In Fig. 9.3a, the red line is the value of $\|\Delta X(k)\|$ with our control schemes (9.4)–(9.8). The blue line is the value of $\|\Delta X(k)\|$ with control schemes (9.4)–(9.6) in the periodic segment $[l\varpi, (l+1)\varpi]$, $l = 0, 1, 2, \cdots$. That is,

Fig. 9.2 The trajectory of the tracking error $\Delta x(t, k)$ in system (9.30) without and with controllers

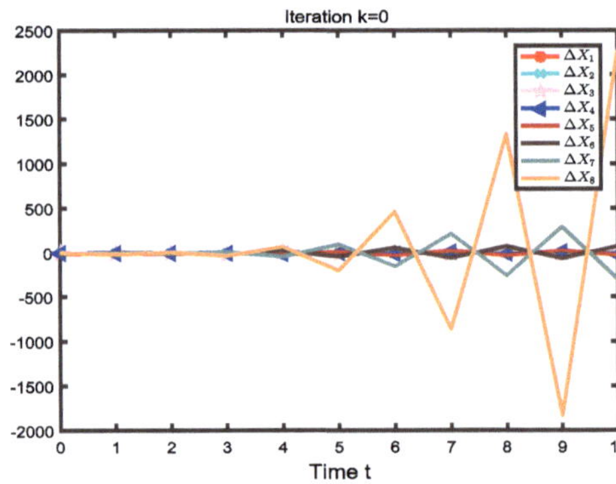

(a) The trajectory of the tracking error $\Delta x(t, k)$ in system (9.30) without any controllers.

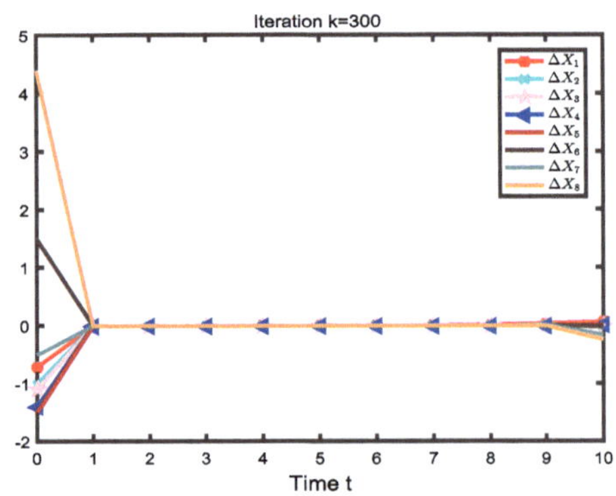

(b) The trajectory of the tracking error $\Delta x(t, k)$ in system (9.30) with control schemes (9.4)-(9.8) at $k = 300$.

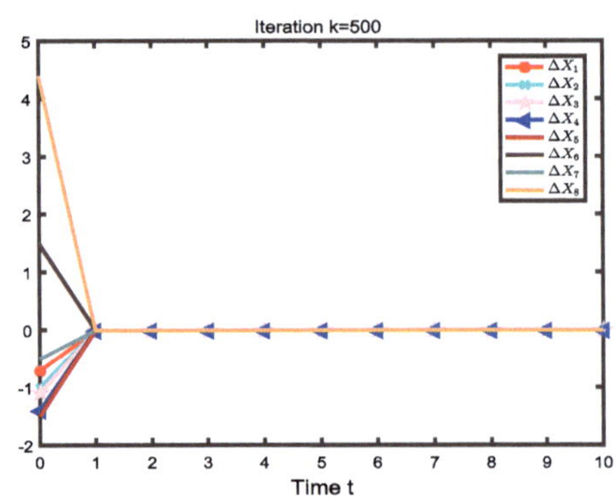

(c) The trajectory of the tracking error $\Delta x(t, k)$ in system (9.30) with control schemes (9.4)-(9.8) at $k = 500$.

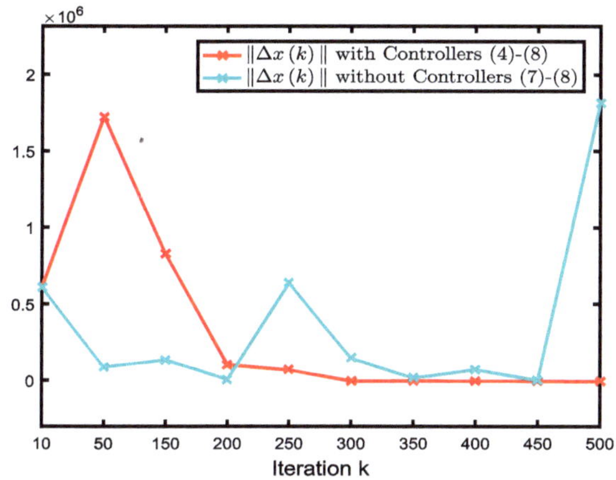

(a) The comparison of $\|\Delta X(k)\|$ with control schemes (9.4)-(9.8) and $\|\Delta X(k)\|$ with control schemes (9.4)-(9.6) at different iteration ks in system (9.30).

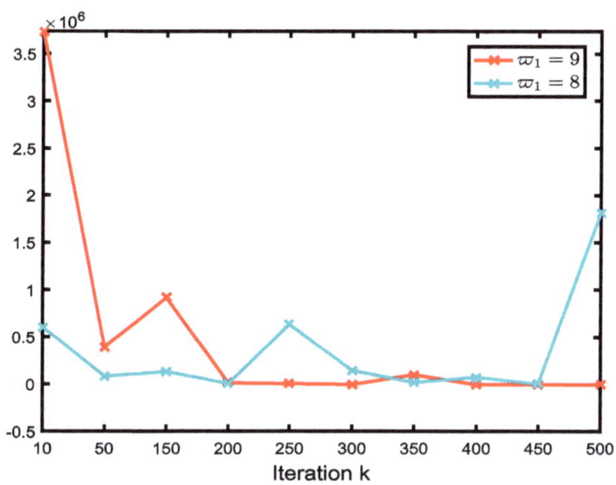

(b) With control schemes (9.4)-(9.8), the comparison of $\|\Delta X(k)\|$ at $\varpi_1 = 9$ and $\varpi_1 = 8$ in system (9.30).

Fig. 9.3 The comparison of $\|\Delta X(k)\|$ in system (9.30) with different control schemes

only the control schemes (9.4)–(9.6) are implemented in the whole periodic segment $[l\varpi, (l+1)\varpi]$. Clearly, $\|\Delta X(k)\|$ with control schemes (9.4)–(9.8) is much smaller than that with control schemes (9.4)–(9.6). Figure 9.3b presents the comparison of $\|\Delta X(k)\|$ in two different ϖ_1s following the same control schemes (9.4)–(9.8). Table 9.1 shows the minimum distances between the agents $x_i(t, k)$ $(i = 1, 2, \cdots, 8)$. It is obvious that the distance between any two agents is always greater than the safe preset distance $K = 0.5$.

Table 9.1 The minimum distances

k	50	100	200	300	400	500
The minimum distance between agents	0.8646	0.8675	0.7717	0.8060	0.8675	0.8675

From Fig. 9.3, one can conclude that: (1) the control schemes we provided (9.4)–(9.8) (considering the static and dynamic segments) are better than the traditional ILC schemes (Meng and Moore 2016; Xiong et al. 2016) (only considering the static segment). According to Fig. 9.3a, it is easy to find the value of $\|\Delta X(k)\|$ with traditional ILC schemes which is significantly larger than that with our ILC schemes at the iteration $W = 500$. As a consequence, for a given ζ, the settling iteration W with our ILC schemes is much smaller than that with the traditional ILC schemes. That is, the convergence rate of the system (9.1) with our ILC schemes is faster than that with the traditional ILC schemes. (2) In Fig. 9.3b, it is easy to find that the value of $\|\Delta X(k)\|$ at $\varpi_1 = 9$ is bigger than that at $\varpi_1 = 8$, which means that the dynamic segment should be appropriately chosen to guarantee that the settling iteration W is small.

9.6 Summary

In this chapter, a novel type of periodic ILC scheme has been developed in order to explore the finite-iteration tracking of network systems. The iterative steps have been divided into multiple periodic intervals in the proposed ILC scheme. Meanwhile, every periodic interval has been divided into two segments: the static and dynamic segments. The following goals have been achieved with our ILC schemes: (1) Each agent has access to finite-iteration tracking. (2) Agents are separated by a safe distance. (3) The settling iterations of the finite-iteration tracking are kept within a limited value that has been predetermined. Thus, the design has the benefit of reducing the communication channel bandwidth load, which enhances the efficiency of finite-iteration tracking. Finally, the outcomes of the simulations have demonstrated how well the design works.

References

Bu XH, Yu QX, Hou ZS, Qian W (2017) Model free adaptive iterative learning consensus tracking control for a class of nonlinear multiagent systems. IEEE Trans Syst Man Cybern Syst. 49(4):677–686

Chi RH, Hou ZS, Jin ST, Huang B (2018) Computationally efficient data-driven higher order optimal iterative learning control. IEEE Trans Neur Netw Learn Syst 29(12):5971–5980

He W, Dong YT, Sun CY (2015) Adaptive neural impedance control of a robotic manipulator with input saturation. IEEE Trans Syst Man Cybern Syst 46(3):334–344

He W, Meng TT, He XY, Sun CY (2019) Iterative learning control for a flapping wing micro aerial vehicle under distributed disturbances. IEEE Trans Cybern 49(4):1524–1535

Jin X (2019) Nonrepetitive leader-follower formation tracking for multiagent systems with LOS range and angle constraints using iterative learning control. IEEE Trans Cybern 49(5):1748–1758

Liu JL, Yin TT, Yue D, Karimi HR, Cao JD (2021) Event-based secure leader-following consensus control for multiagent systems with multiple cyber attacks. IEEE Trans Cybern 51(1):162–173

Meng TT, He W (2017) Iterative learning control of a robotic arm experiment platform with input constraint. IEEE Trans Ind Electron 65(1):664–672

Meng D, Moore KL (2016) Learning to cooperate: networks of formation agents with switching topologies. Automatica 64:278–293

Meng DY, Moore KL (2020) Contraction mapping-based robust convergence of iterative learning control with uncertain, locally Lipschitz nonlinearity. IEEE Trans Syst Man Cybern Syst 50(2):442–454

Murvay P, Groza B (2020) Efficient physical layer key agreement for flexray networks. IEEE Trans Veh Technol 69(9):9767–9780

Park I, Sunwoo M (2011) FlexRay network parameter optimization method for automotive applications. IEEE Trans Ind Electron 58(4):1449–1459

Schmidt K, Schmidt E (2008) Message scheduling for the FlexRay protocol: the static segment. IEEE Trans Veh Technol 58(5):2170–2179

Shen D, Chen HF (2012) Iterative learning control for large scale nonlinear systems with observation noise. Automatica 48(3):577–582

Shen D, Zhang C, Xu Y (2017) Two updating schemes of iterative learning control for networked control systems with random data dropouts. Inform Sci 381:352–370

Tang Y, Zhang DD, Ho DWC, Qian F (2019) Tracking control of a class of cyber-physical systems via a flexray communication network. IEEE Trans Cybern 49(4):1186–1199

Thomopoulos R, Destercke S, Charnomordic B, Johnson I, Abécassis J (2013) An iterative approach to build relevant ontology-aware data-driven models. Inform Sci 221:452–472

Xiong WJ, Yu XH, Chen Y, Gao J (2016) Quantized iterative learning consensus tracking of digital networks with limited information communication. IEEE Trans Neur Netw Learn Syst 28(6):1473–1480

Xiong WJ, Yu XH, Patel R, Yu WW (2016) Iterative learning control for discrete-time systems with event-triggered transmission strategy and quantization. Automatica 72:84–91

Xiong WJ, Xu L, Huang TW, Yu XH, Liu YH (2020) Finite-Iteration tracking of singular coupled systems based on learning control with packet losses. IEEE Trans Syst Man Cybern Syst 50(1):245–255

Yang RN, Zheng WX (2019) Output-based event-triggered predictive control for networked control systems. IEEE Trans Ind Electron 67(12):10631–10640

Yang RN, Zheng WX, Yu YR (2020) Event-triggered sliding mode control of discrete-time two-dimensional systems in Roesser model. Automatica 114:108813

Yu QX, Hou ZS, Bu XH, Yu QF (2019) RBFNN-based data-driven predictive iterative learning control for nonaffine nonlinear systems. IEEE Trans Neur Netw Learn Syst 31(4):1170–1182

Zeng HB, Natale MD, Ghosal A, Sangiovanni-Vincentelli A (2011) Schedule optimization of time-triggered systems communicating over the FlexRay static segment. IEEE Trans Ind Informat 7(1):1–17

Zhang WL, Karimi HR, Tang Y (2020) Distributed tracking for discrete-time multiagent networks via an ultrafast control protocol. IEEE Trans Syst Man Cybern Syst 51(12):7542–7552

Chapter 10
Multi-layered Sampled-Data Tracking Under Cooperative–Antagonistic Interactions

10.1 Introduction

Nowadays, ILC has been widely applied in a variety of fields, such as robotic manipulators (Frénay and Verleysen 2016; Li et al. 2016), data-driven control (Chi et al. 2015, 2016; Janssens et al. 2013), nonlinear systems (Liu et al. 2016; Wei et al. 2015, 2016), linear systems (Bu and Hou 2018; Bu et al. 2019), multi-agent systems (Meng and Moore 2016; Xiong et al. 2018), and rapid thermal processing (Lee et al. 2001; Yang et al. 2003). In Bu and Hou (2018), a new learning algorithm is presented through the application of a projection recognition algorithm and a deterministic equivalence principle. In Xiong et al. (2018), unmeasurable state information is considered, and iterative asynchronous controllers are studied in order to achieve the tracking of T-S fuzzy discrete systems. In Liu et al. (2016), Peng et al. (2014), Peng et al. (2017), a control technique based on neural network learning is applied to achieve the stability of nonlinear systems.

In many networks, such as social networks, transportation networks, and resource allocation networks, it is important to note that nodes are often in competition with one another and cooperate with each other. In a social network, for example, a mutual individual always maintains both friendly and hostile relations with other individuals, and this can generate cooperative and antagonistic groups. Likewise, in an economic system, cooperative and competitive relations between firms arise naturally, especially in producing the same kind of commodity or providing the same type of services. Moreover, in the biological system of a human being, interactions between genes are cooperative and antagonistic in the formation of certain behaviors, decisions, and habits. Of course, cooperative and adversarial relationships can usually be seen among the nodes of many real-world networks. In most existing ILC results, nodes are still assumed to cooperate. Clearly, there is a need to consider different types of interaction relationships between nodes to reflect real-world learning issues. Inspired by Altafini (2013), Meng et al. (2016), Sun et al. (2017), Meng (2018), the interactions between cooperation and antagonism will be considered in the designed

W. Xiong et al., *Iterative Learning Control for Network Systems Under Constrained Information Communication*, Intelligent Control and Learning Systems 12,
https://doi.org/10.1007/978-981-97-0926-7_10

learning controllers. Furthermore, in the ILC results, it is not practical to receive messages in time from the last iteration because of the limited bandwidth of the information storage. In some cases, the messages received may be from the previous learning iteration. Therefore, in this chapter, the state information of these schemes is updated in light of the information of previous learning iterations but not just dependent on the last iteration. As a result, this new type of learning scheme has an accurate prediction and is more consistent with the real situation.

In recent years, researchers have found that many real networks take on different layers of structure. In some real networks, multiple structures are described by different types of links that have multifarious representations or dimensional relationships in diverse subsystems. In information networks, for example, agents may receive information from different social networks, such as Facebook, Twitter, WeChat, and so on. In other words, nodes in the network layer can receive information from other nodes in different layers. Therefore, in the iterative learning schemes, it makes sense to consider multiple layers of connectivity and describe different types of interactions among different neighbors in each layer. According to Boccaletti et al. (2014), De Domenico et al. (2016), He et al. (2017a, b), another kind of learning controller will be considered to be cooperative–antagonistic and multi-layered, which has not been considered in existing ILC results. In addition, sampled-data control (Fridman 2010; Kumar et al. 2018; Li et al. 2014; Manivannan et al. 2018; Wu et al. 2016) is a better control method in line with the rapid development of the modern mechanical industry. The control method greatly reduces the amount of transmitted information among nodes and improves the effective usage of bandwidth. Hence, inspired by Boccaletti et al. (2014), De Domenico et al. (2016), He et al. (2017a, b), Fridman (2010), Wu et al. (2016), Li et al. (2014), the second objective of this chapter is to design another kind of sampled-data multi-layered iterative learning scheme with cooperative–antagonistic interactions. Here, the state information of these schemes is updated based on the sampled-data control and the previous iterations. In addition, the multi-layered and sampled-data control approach will lead to some dramatic consequences.

10.2 Problem Formulation

10.2.1 Model Description

In this chapter, we mainly consider the following discrete network with n nodes

$$x_i(t+1,k) = A_i x_i(t,k) + u_i(t,k), \tag{10.1}$$

where $x_i(t,k) \in \mathbb{R}^m$ is the state of the ith node and matrix $A_i \in \mathbb{R}^{m\times m}$ is the internal connection matrix of node i, $i = 1, \ldots, n$. The ith state $x_i(t,k)$ is related to two variables: the time step $t \in \{0, 1, \ldots, T\}$ and the iterative learning step $k \in \{0, 1, 2, \ldots\}$. $u_i(t, k) \in \mathbb{R}^m$ is the controller of the ith node at iteration step k and time t.

It is worth mentioning that the nodes in the network may have cooperative–antagonistic interactions. This, without a doubt, poses challenges in achieving the target of the network system. To tackle this problem, appropriate control strategies to regulate the system must be developed in order to achieve the system's objectives. Thus, let the desired trajectory of $x_i(t, k)$ be $x_i^*(t)$, and the tracking error $e_i(t, k) = x_i(t, k) - x_i^*(t)$, $\forall\, t \in \{0, 1, \ldots, T\}$, $i = 1, \ldots, n$. Then, the target is to design appropriate iterative learning control schemes $u_i(t, k) \in \mathbb{R}^m$ to satisfy that $\lim_{k \to +\infty} x_i(t, k) = x_i^*(t)$, i.e., $\lim_{k \to +\infty} e_i(t, k) = \mathbf{0}$, $i = 1, 2, \ldots, n$.

10.2.2 ILC Scheme Design

To achieve the consensus, the iterative controllers $u_i(t, k)$ are designed according to the cooperative information of all the nodes. Nonetheless, the information received by an agent may come from agents of multi-layers in a network. For instance, when discussing the relationships of the individuals in a virus transmission network, one has to consider the influence of the individuals in a social network because the spread of social networks makes diseases on a virus spread network more likely to break out. As a result, a kind of multi-layered iterative learning scheme with cooperative–antagonistic interactions will be designed as follows:

$$\begin{cases} u_i(t, k) = u_i^0(t), & k = 1, 2, \ldots, \tau, \\ u_i(t, k) = u_i(t, k - \tau) + \sum\limits_{l=1}^{M} \sum\limits_{j=1}^{n} r_l \Big[\tilde{\alpha}_{ij}^l e_j(t + 1, k - \tau(l)) & \\ \quad - |\tilde{\alpha}_{ij}^l| e_i(t + 1, k - \tau(l)) - \epsilon_i^k e_i(t + 1, k - \tau(l)) \Big], & \mu\tau + 1 \le k \le (\mu + 1)\tau, \end{cases} \tag{10.2}$$

where $\mu = 1, 2, \ldots$, the iteration delay $\tau(l)$ of the lth layer satisfies that $\tau(l) \in [\tau, l - (\mu - 1)\tau - 1]$, τ is a positive integer, and $u_i^0(t) \in \mathbb{R}^m$ is known and independent of the iteration step k. The positive integer M is the number of layers in the entire network, $\tilde{\mathcal{A}}^l = (\tilde{\alpha}_{ij}^l)_{n \times n}$ is the adjacency-weighted matrix of the lth layer, $\epsilon^l = \mathrm{diag}(\epsilon_1^l, \ldots, \epsilon_n^l) \in \mathbb{R}^{n \times n}$, and ϵ_i^l $(i = 1, \ldots, n,\ l = 1, \ldots, M)$ is a nonnegative number and will be given later. Also, $r_l \ge 0$ $(l = 1, \ldots, M)$ is a control parameter and will be given later. From (10.2), one has to pay attention to some key points:

(1) The information received by the ith controller $u_i(t, k)$ at the kth iteration step comes from the previous τ iteration steps and M layers of the whole network.

(2) The relationship between cooperation and competition can be expressed by the positive or negative sign of $\tilde{\alpha}_{ij}$.

Thus, define matrix $\Upsilon^l = (\Upsilon^l_{ij})_{n\times n}$ as

$$\Upsilon^l_{ij} = \begin{cases} l^l_{ii} + \epsilon^l_i, & j = i, \\ l^l_{ij}, & j \neq i, \end{cases}$$

where $\mathcal{L}^l = (l^l_{ij})_{n\times n}$ is the Laplacian of the lth layer, and it has a similar definition as (1.1). Since the nodes in the signed digraph $\tilde{\mathcal{G}}$ may be cooperative or competitive, the row sum of the Laplacian $\mathcal{L}^l$ in the lth layer may be zero or not. This physical property of the Laplacian $\mathcal{L}^l$ is important and will be used in Theorem 10.1. Then (10.2) can be written as

$$\begin{cases} u_i(t,k) = u^0_i(t), & k = 1, 2, \ldots, \tau, \\ u_i(t,k) = u_i(t, k-\tau) - \sum\limits_{l=1}^{M}\sum\limits_{j=1}^{n} r_l \Upsilon^l_{ij} e_j(t+1, k-\tau(l)), & \mu\tau + 1 \leq k \leq (\mu+1)\tau, \end{cases} \tag{10.3}$$

where $\mu = 1, 2, \ldots$. If the iteration delay $\tau(l)$ is 1, (10.3) can be written as

$$\begin{cases} u_i(t,k) = u^0_i(t), & k = 1, 2, \ldots, \tau, \\ u_i(t,k) = u_i(t, k-\tau) - \sum\limits_{l=1}^{M}\sum\limits_{j=1}^{n} r_l \Upsilon^l_{ij} e_j(t+1, k-1), & k = 2, 3, \ldots. \end{cases} \tag{10.4}$$

Furthermore, just as mentioned in the introduction, because the sampled-data control strategy is more reasonable than persistent control strategy, another type of sampled-data multi-layered iterative learning schemes with cooperative–antagonistic interactions is considered as

$$\begin{cases} u_i(t,k) = u^0_i(t), & k = 1, 2, \ldots, \tau, \\ u_i(t,k) = u_i(t, k-\tau) - \sum\limits_{l=1}^{M}\sum\limits_{j=1}^{n} r_l \Upsilon^l_{ij} e_j(t+1, k-\tau(l)), & \mu\tau + 1 \leq k \leq (\mu+1)\tau, \\ e_j(t,k) = e_j(t, k_\nu), & k_\nu \leq k < k_{\nu+1}, \end{cases} \tag{10.5}$$

where $\mu = 1, 2, \ldots$, $\{k_\nu : \nu = 1, 2, \ldots\}$ is the sampled-data iteration sequence, and $0 < k_1 < k_2 < \cdots$.

In the iterative schemes (10.3) and (10.5), the controller $u_i(t,k)$ is adjusted based on the tracking error $e_i(t, k-\tau(l))$. As a result, by repeating the same training according to the tracking error, the desired tracking can be achieved quickly. Therefore, the iterative learning scheme is an efficient control algorithm for tracking discrete networks.

Remark 10.1 In the iterative schemes (10.3) and (10.5), the cooperative–antagonistic interaction relationship is presented in the weighted matrix $\tilde{\mathcal{A}}^l = (\tilde{\alpha}^l_{ij})_{n\times n}$. Here, if the interaction of nodes i and j is cooperative (or competitive), one has $\tilde{\alpha}_{ij} > 0$ (or < 0), where $i \neq j,\ i, j \in \{1, \ldots, n\}$. In addition, the received information of the learning schemes come from different M layers and is reflected in the summation symbol "$\sum_{l=1}^{M}$".

Remark 10.2 In the iterative schemes (10.3) and (10.5), the iteration delay $\tau(l)$ of the lth layer is defined as satisfying that $\tau(l) \in [\tau, k - (\mu - 1)\tau - 1]$. Due to the limitation of communication and energy, the iterative delay $\tau(l)$ always exists in the process of information transmission of sensor networks, so the definition of iterative delay $\tau(l)$ is necessary. In addition, the boundedness of $\tau(l)$ means that if the iteration delay $\tau(k)$ is either too large, the information is regarded as missing.

Assumption 10.1 The initial state in each iteration is assumed as

$$x_i(0, k) = x_i^0 \cdot \big(\underbrace{1, 1, \ldots, 1}_{n}\big)^{\mathrm{T}}, \tag{10.6}$$

where $i = 1, 2, \ldots, n$, $\forall\, k \in \{1, 2, \ldots\}$, x_0^i $(i = 1, 2, \ldots, n)$ is a constant.

10.3 Main Results

In this section, our purpose is to discuss the tracking problem of the system (10.1) with control schemes (10.3) and (10.5).

10.3.1 Convergence Analysis with Cooperative–Antagonistic ILC Scheme

Theorem 10.1 *Consider the system (10.1) with the iterative control scheme (10.3). Under Assumption 10.1, in the interval $t \in [1, T]$, the tracking of the system (10.1) can be obtained if matrices $\mathcal{L}^l$, the nonnegative scalars r_l, and ϵ^l in the iterative scheme (10.3) satisfy that*

$$0 < 1 - \sum_{l=1}^{M}\sum_{j=1}^{n} r_l \Upsilon_{ij}^l < 1. \tag{10.7}$$

Here, if there are at least two antagonistic nodes in the network, (10.7) can be satisfied even though all ϵ_i^l $(i = 1, 2, \ldots, n,\ l = 1, 2, \ldots, M)$ are zero, but if there are all cooperative nodes in the network, (10.7) can be satisfied only if at least one of ϵ_i^l $(i = 1, 2, \ldots, n,\ l = 1, 2, \ldots, M)$ is not zero.

Remark 10.3 Υ_{ij}^l in (10.7) is explained here. If there are at least two opposite nodes in the network, the row sum of at least one row in the Laplacian $\mathcal{L}^l$ of the

lth layer is greater than zero. Then, you can choose suitable nonnegative scalars r_l $(l = 1, \dots, M)$ to meet the condition (10.7) even if all ϵ_i^l $(i = 1, \dots, n,\ l = 1, \dots, M)$ are zero. In contrast, if all nodes in the network are cooperative and all ϵ_i^l $(i = 1, \dots, n,\ l = 1, \dots, M)$ are zero, one only can get $\sum_{l=1}^{M} \sum_{j=1}^{n} r_l \Upsilon_{ij}^l = 0$ for any nonnegative scalar r_l. That is to say, condition (10.7) can be satisfied only if at least one of ϵ_i^l $(i = 1, \dots, n,\ l = 1, \dots, M)$ is not zero.

Proof From systems (10.1) and (10.3), one obtains the following:

(I) When the iterative learning repetition $1 \le k \le \tau - 1, \forall t \in \{1, 2, \dots, T - 1\}$, from (10.2) and Assumption 10.1, one has

$$\begin{aligned} & e_i(t+1, k+1) - e_i(t+1, k) \\ &= x_i(t+1, k+1) - x_i(t+1, k) \\ &= A_i(x_i(t, k+1) - x_i(t, k)) + u_i(t, k+1) - u_i(t, k) \\ &= A_i(x_i(t, k+1) - x_i(t, k)) \\ &= A_i^2(x_i(t-1, k+1) - x_i(t-1, k)) \\ &= A_i^{t+1}(x_i(0, k+1) - x_i(0, k)) = \mathbf{0}. \end{aligned} \tag{10.8}$$

From (10.8), one has

$$e_i(t, k+1) = e_i(t, k), \tag{10.9}$$

for $1 \le k \le \tau - 1$, $t \in \{1, 2, \dots, T\}$, and $i \in \{1, 2, \dots, n\}$.

(II) When the iterative learning step $\tau + 1 \le k \le 2\tau$, $\forall\, t \in \{1, 2, \dots, T - 1\}$, from (10.1) and (10.3), one has

$$\begin{aligned} & e_i(t+1, k) - e_i(t+1, k-\tau) \\ &= x_i(t+1, k) - x_i(t+1, k-\tau) \\ &= A_i(x_i(t, k) - x_i(t, k-\tau)) + u_i(t, k) - u_i(t, k-\tau) \\ &= A_i(e_i(t, k) - e_i(t, k-\tau)) - \sum_{l=1}^{M} \sum_{j=1}^{n} r_l \Upsilon_{ij}^l e_j(t+1, k-\tau(l)), \end{aligned} \tag{10.10}$$

where $\tau(l) \in [\tau, k - (\mu - 1)\tau - 1]$, $l = 1, 2, \dots, M$. Combining (10.9) and (10.10), one also obtains

$$e_i(t, k+1) = e_i(t, k), \tag{10.11}$$

for $\tau + 1 \le k \le 2\tau - 1$, $t \in \{1, 2, \dots, T\}$, and $i \in \{1, 2, \dots, n\}$. Define $\theta_i(t, k - \tau) \triangleq e_i(t, k) - e_i(t, k - \tau)$, from (10.10), one can obtain that

$$\theta_i(t, k-\tau) = A_i \theta_i(t-1, k-\tau) - \sum_{l=1}^{M} \sum_{j=1}^{n} r_l \Upsilon_{ij}^l e_j(t, k-\tau(l)). \tag{10.12}$$

From Assumption 10.1, one has that $\theta_i(0, k-\tau) = \mathbf{0}$, for $\tau+1 \le k \le 2\tau$. Then, from (10.12), one can get that

$$\theta_i(t, k-\tau) = -\sum_{p=0}^{t-1}\sum_{l=1}^{M}\sum_{j=1}^{n} r_l A_i^p \Upsilon_{ij}^l e_j(t-p, k-\tau(l)). \tag{10.13}$$

Then, one knows

$$e_i(t+1, k) = e_i(t+1, k-\tau) - \sum_{p=0}^{t-1}\sum_{l=1}^{M}\sum_{j=1}^{n} r_l A_i^{p+1} \Upsilon_{ij}^l e_j(t-p, k-\tau(l))$$
$$-\sum_{l=1}^{M}\sum_{j=1}^{n} r_l \Upsilon_{ij}^l e_j(t+1, k-\tau(l)). \tag{10.14}$$

Let $E_{i,k} = (e_i^{\mathrm{T}}(1, k), \ldots, e_i^{\mathrm{T}}(T, k))^{\mathrm{T}}$ and $E_{i,k-\tau} = (e_i^{\mathrm{T}}(1, k-\tau), \ldots, e_i^{\mathrm{T}}(T, k-\tau))^{\mathrm{T}}$, one immediately has

$$E_{i,k} = \Pi_1 E_{i,k-\tau}, \tag{10.15}$$

where $i \in \{1, 2, \ldots, n\}$. Π_1 is shown in (10.16), $\Omega \triangleq 1 - \sum_{l=1}^{M}\sum_{j=1}^{n} r_l \Upsilon_{ij}^l$, and I_m is the m-dimensional identity matrix.

$$\Pi_1 = \begin{pmatrix} \Omega * I_m & O & O & \cdots & O & O \\ *_{21} & \Omega * I_m & O & \cdots & O & O \\ *_{31} & *_{32} & \Omega * I_m & \cdots & O & O \\ \vdots & \vdots & \vdots & \cdots & \vdots & \vdots \\ *_{T1} & *_{T2} & *_{T3} & \cdots & *_{T,T-1} & \Omega * I_m \end{pmatrix}, \tag{10.16}$$

$*_{ij} \in \mathbb{R}^m$ $(i = 2, \ldots, T,\ \ j = 1, \ldots, T-1)$ can be obtained from (10.13) and (10.14), which are not shown here because they will not be discussed below. Then, one has

$$\begin{cases} e_i(t, k+1) = e_i(t, k), & \tau+1 \le k \le 2\tau - 1, \\ E_{i,k} = \Pi_1 E_{i,k-\tau}, & \tau+1 \le k \le 2\tau, \end{cases} \tag{10.17}$$

where $t \in \{1, 2, \ldots, T\}, i \in \{1, 2, \ldots, n\}$. Repeating an alike process of Part II), one can get that

$$\begin{cases} e_i(t, k+1) = e_i(t, k), & \mu\tau+1 \le k \le (\mu+1)\tau - 1, \\ E_{i,k} = \Pi_\mu E_{i,k-\tau}, & \mu\tau+1 \le k \le (\mu+1)\tau, \end{cases} \tag{10.18}$$

where $t \in \{1, 2, \ldots, T\}$, $i \in \{1, 2, \ldots, n\}$, $\mu = 2, 3, \ldots,$ Π_μ is a lower-triangular matrix and has the same diagonal elements with Π_1.

According to (10.7), (10.17), and (10.18), one has that $E_{i,k} \to \mathbf{0}$ as $k \to +\infty$, for $i \in \{1, 2, \ldots, n\}$. That is to say, $e_i(t, k) \to \mathbf{0}$ as $k \to +\infty$, for $t \in \{1, 2, \ldots, T\}$, $i \in \{1, 2, \ldots, n\}$. Therefore, one can solve the tracking problem of system (10.1).

In addition, from Remark 10.3, we know that if there are at least two opposing nodes in the network, even if all ϵ_i^l $(i = 1, 2, \ldots, n,\ l = 1, 2, \ldots, M)$ are zero, (10.7) can be satisfied. However, if all cooperative nodes exist in the network, (10.7) can be satisfied only if at least one of ϵ_i^l $(i = 1, 2, \ldots, n,\ l = 1, 2, \ldots, M)$ is not zero. □

Corollary 10.1 *Consider the system (10.1) with the iterative scheme (10.4). Under Assumption 10.1, in the interval* $t \in [1, T]$, *the tracking of the system (10.1) can be obtained if matrices* $\mathcal{L}^l$, *the nonnegative scalars* r_l *and* ϵ^l *in the iterative scheme (10.4) satisfy that*

$$0 < 1 - \sum_{l=1}^{M}\sum_{j=1}^{n} r_l \Upsilon_{ij}^l < 1. \tag{10.19}$$

Here, if there are at least two antagonistic nodes in the network, (10.19) can be satisfied even if all ϵ_i^l $(i = 1, \ldots, n,\ l = 1, \ldots, M)$ *are zero. However, if there are all cooperative nodes in the network, (10.19) can be satisfied only if at least one of* ϵ_i^l $(i = 1, \ldots, n,\ l = 1, \ldots, M)$ *is not zero.*

Proof Similar to Theorem 10.1, the result in Corollary 10.1 is also easy to get. The detailed proof is omitted here. □

Remark 10.4 From Theorem 10.1 and Corollary 10.1, one can see that the obtained conditions (10.7) and (10.19) in Theorem 10.1 and Corollary 10.1 are the same even if the iteration delay $\tau(l)$ exists in (10.3). That means that the iteration delay $\tau(l)$ does not affect the final tracking of the system (10.1), but it may affect the convergent rate of the tracking of the system (10.1).

Remark 10.5 Comparing with Wei et al. (2015, 2016), the advantages of iterative scheme (10.3) can be summarized as follows:

(1) the cooperative–antagonistic interactions are introduced for the first time in the schemes to present the competition relationship among the nodes.

(2) the iterative scheme (10.3) is designed to have various structures, which give a description of different types of links with multi-dimensional relationships in different subsystems of a multi-layer network.

In the following, we shall discuss the tracking of the system (10.1) by using the sampled-data iterative scheme (10.5). For the convenience of proof, the iteration sequence of sampled data is defined as satisfying that

$$\max_{\nu=1}^{+\infty}(k_{\nu+1} - k_\nu) < \tau, \quad \min_{\nu=1}^{+\infty}(k_{\nu+2} - k_\nu) > \tau, \tag{10.20}$$

where $\max_{\nu=1}^{+\infty} x_i$ denotes the maximal value of all x_i, and $\min_{\nu=1}^{+\infty} x_i$ denotes the minimal value of all x_i.

Remark 10.6 In (10.20), the inequality $\max_{\nu=1}^{+\infty}(k_{\nu+1} - k_\nu) < \tau$ indicates that the sampling length of the sampled-data iterations is not large. The inequality $\min_{\nu=1}^{+\infty}(k_{\nu+2} - k_\nu) > \tau$ means that the sampling length of the sampled-data iterations cannot be very small. If the sampling length is large, the validity of the sampling data is very low. However, if the sampling length is too small, the benefit of the sampled-data control is hard to show.

10.3.2 Convergence Analysis with Cooperative–Antagonistic Sampled-Data ILC Scheme

Theorem 10.2 *Consider the system (10.1) with the iterative scheme (10.5). Under Assumption 10.1 and condition (10.20), in the interval $t \in [1, T]$, the tracking of the system (10.1) can be obtained if matrices $\mathcal{L}^l$, the nonnegative scalars r_k and ϵ^l in the iterative scheme (10.5) satisfy that*

$$0 < 1 - \sum_{l=1}^{M}\sum_{j=1}^{n} r_l \Upsilon_{ij}^l < 1. \tag{10.21}$$

Here, if there are at least two antagonistic nodes in the network, (10.21) is satisfied even if all ϵ_i^l $(i = 1, \ldots, n,\ l = 1, \ldots, M)$ are zero. However, if all cooperative nodes exist in the network, (10.21) can be satisfied only if at least one of ϵ_i^l $(i = 1, \ldots, n,\ l = 1, \ldots, M)$ is not zero.

Proof Assume that the sampled-data iteration $\mu_1\tau + 1 \le k_1 \le (\mu_1 + 1)\tau$, $\mu_1 > 0$ is a positive integer. From (10.18), when $\mu < \mu_1$, one has

$$\begin{cases} e_i(t, k+1) = e_i(t, k), & \mu\tau + 1 \le k \le (\mu + 1)\tau - 1, \\ E_{i,k} = \Pi_\mu E_{i,k-\tau}, & \mu\tau + 1 \le k \le (\mu + 1)\tau, \end{cases} \tag{10.22}$$

and

$$\begin{cases} e_i(t, k+1) = e_i(t, k), & \mu_1\tau + 1 \le k \le k_1 - 1, \\ E_{i,k} = \Pi_{\mu_1} E_{i,k-\tau}, & \mu_1\tau + 1 \le k \le k_1. \end{cases} \tag{10.23}$$

According to (10.20), there are three possibilities for the sampled-data iterations k_1, k_2, and k_3:

(1) $\mu_1\tau + 1 \le k_1, k_2 \le (\mu_1 + 1)\tau$, and $(\mu_1 + 1)\tau + 1 \le k_3 \le (\mu_1 + 2)\tau$;

(2) $\mu_1\tau + 1 \le k_1 \le (\mu_1 + 1)\tau$, and $(\mu_1 + 1)\tau + 1 \le k_2, k_3 \le (\mu_1 + 2)\tau$;

(3) $\mu_1\tau + 1 \le k_1 \le (\mu_1 + 1)\tau$, $(\mu_1 + 1)\tau + 1 \le k_2 \le (\mu_1 + 2)\tau$, and $(\mu_1 + 2)\tau + 1 \le k_3 \le (\mu_1 + 3)\tau$.

For the three possibilities, one can have the following:

(1) When $\mu_1\tau + 1 \le k_1, k_2 \le (\mu_1 + 1)\tau$ and $(\mu_1 + 1)\tau + 1 \le k_3 \le (\mu_1 + 2)\tau$, from (10.18), one has

$$\begin{cases} e_i(t, k+1) = e_i(t, k), & k_1 \le k < k_3 - 1, \\ E_{i,k} = \Pi_{\mu_1} E_{i,k-\tau}, & k_1 \le k < k_3, \end{cases} \tag{10.24}$$

and

$$\begin{cases} e_i(t, k+1) = e_i(t, k), & k = k_3, \\ E_{i,k} = \Pi_{\mu_1+1} E_{i,k-\tau}, & k = k_3. \end{cases} \tag{10.25}$$

(2) When $\mu_1\tau + 1 \le k_1 \le (\mu_1 + 1)\tau$ and $(\mu_1 + 1)\tau + 1 \le k_2, k_3 \le (\mu_1 + 2)\tau$, from (10.18), one has

$$\begin{cases} e_i(t, k+1) = e_i(t, k), & k_1 \le k < k_2 - 1, \\ E_{i,k} = \Pi_{\mu_1} E_{i,k-\tau}, & k_1 \le k < k_2, \end{cases} \tag{10.26}$$

and

$$\begin{cases} e_i(t, k+1) = e_i(t, k), & k_2 \le k \le k_3 - 1, \\ E_{i,k} = \Pi_{\mu_1+1} E_{i,l-\tau}, & k_2 \le k \le k_3. \end{cases} \tag{10.27}$$

(3) When $\mu_1\tau + 1 \le k_1 \le (\mu_1 + 1)\tau$, $(\mu_1 + 1)\tau + 1 \le k_2 \le (\mu_1 + 2)\tau$ and $(\mu_1 + 2)\tau + 1 \le k_3 \le (\mu_1 + 3)\tau$, from (10.18), one has

$$\begin{cases} e_i(t, k+1) = e_i(t, k), & k_1 \le k < k_2 - 1, \\ E_{i,k} = \Pi_{\mu_1} E_{i,k-\tau}, & k_1 \le k < k_2, \end{cases} \tag{10.28}$$

and

$$\begin{cases} e_i(t, k+1) = e_i(t, k), & k_2 \le k < k_3 - 1, \\ E_{i,k} = \Pi_{\mu_1+1} E_{i,k-\tau}, & k_2 \le k < k_3. \end{cases} \tag{10.29}$$

and

$$\begin{cases} e_i(t, k+1) = e_i(t, k), & k = k_3, \\ E_{i,k} = \Pi_{\mu_1+2} E_{i,k-\tau}, & k = k_3. \end{cases} \tag{10.30}$$

From (10.24)–(10.30), one obtains

$$\begin{cases} e_i(t, k+1) = e_i(t, k), \quad k_i \le k < k_{i+1} - 1, \ i = 1, 2, \\ E_{i,k_2} = E_{i,k_1} \ \ or \ \ \Pi_{\mu_1+1} E_{i,k_1}, \\ E_{i,k_3} = \Pi_{\mu_1+1} E_{i,k_1} \ \ or \ \ \Pi_{\mu_1+2} \Pi_{\mu_1+1} E_{i,k_1}. \end{cases} \tag{10.31}$$

Define $\mu_{\varpi}\tau + 1 \leq k_{\varpi} \leq (\mu_{\varpi} + 1)\tau$, $\varpi \in \{1, 2, \ldots\}$. Repeating the above process, from (10.31), one gets

$$\begin{cases} e_i(t, k+1) = e_i(t, k), \quad k_i \leq k < k_{i+1} - 1, \; i = 1, 2, \ldots, \\ E_{i,k_{\varpi+1}} = E_{i,k_{\varpi}} \;\; or \;\; \Pi_{\mu_{\varpi}+1} E_{i,k_{\varpi}}, \\ E_{i,k_{\varpi+2}} = \Pi_{\mu_{\varpi}+1} E_{i,k_{\varpi}} \;\; or \;\; \Pi_{\mu_{\varpi}+2} \Pi_{\mu_{\varpi}+1} E_{i,k_{\varpi}}. \end{cases} \tag{10.32}$$

Similar to Theorem 10.1, $\Pi_{\mu_{\varpi}}$ is a lower-triangular matrix and has the same diagonal elements with Π_1 of (10.16). According to (10.21) and (10.32), one has that $E_{i,k_{\mu_{\varpi}}} \to \mathbf{0}$ as $\varpi \to +\infty$, for $i \in \{1, \ldots, n\}$. That is, $e_i(t, k) \to \mathbf{0}$ as $k \to +\infty$, for $t \in \{1, \ldots, T\}$, $i \in \{1, 2, \ldots, n\}$. Therefore, one can solve the tracking problem of the system (10.1).

Similar to Theorem 10.1, we also get that if there are at least two antagonistic nodes in the network, even if all ϵ_i^l $(i = 1, \ldots, n, \; l = 1, \ldots, M)$ are zero, (10.21) can be satisfied. However, if all cooperative nodes exist in the network, (10.21) can be satisfied only if at least one of ϵ_i^l $(i = 1, \ldots, n, \; l = 1, \ldots, M)$ is not zero. □

Remark 10.7 Compared with Meng et al. (2016), Sun et al. (2017), Meng (2018), the control schemes (10.3) and (10.5) have the following advantages:

(1) Cooperative–antagonistic interactions among nodes are considered for the first time in the ILC. In addition, the conditions (10.7) and (10.21) obtained in Theorems 10.1 and 10.2 are also easy to check.

(2) The sampled-data method is proposed for the first time in the scheme (10.5). This control method greatly reduces the amount of transmitted information among nodes and improves the effective utilization of bandwidth. However, due to the slow updating frequency of information in the sampled-data scheme, the convergence rate of the system using the sampled-data control method is slower.

10.4 Simulation Example

In the section, an example is shown to explain the effectiveness of the results in Theorems 10.1 and 10.2.

Example 10.1 Consider the following discrete-time network system with five nodes

$$x_i(t+1, k) = A_i y_i(t, k) + u_i(t, k), \tag{10.33}$$

where $x_i(t, k) = (x_{i1}(t, k), x_{i2}(t, k), x_{i3}(t, k), x_{i4}(t, k))^{\mathrm{T}} \in \mathbb{R}^4$, and the time $T = 10$. Here, $i = 1, 2, 3, 4, 5$, i.e., the number of nodes $n = 5$. The interval connection matrix

$$A_i = \begin{pmatrix} -0.6 & 0 & 0 & 0 \\ 1 & -0.5 & 0 & 0 \\ 0 & 0.5 & -0.75 & 0 \\ 0.5 & 0 & 0.4 & -0.8 \end{pmatrix},$$

for $i = 1, 2, 3, 4, 5$. Define the state $x_i(0, k) = 0.1 \cdot i \cdot (1, 1, 1, 1)^{\mathrm{T}}$ and the desired reference trajectory $x_i^*(t) = 1/10 \cdot (\sin t, \cos t, 1, -1)^{\mathrm{T}}$, $i = 1, 2, 3, 4, 5$, $t \in \{0, 1, 2, \ldots, 10\}$.

The multi-layered ILC scheme (10.3) with cooperative–antagonistic interactions is designed as

$$\begin{cases} u_i(t,k) = u_i^0(t), & k = 1, 2, \ldots, \tau, \\ u_i(t,k) = u_i(t, k-\tau) - \sum\limits_{l=1}^{M} \sum\limits_{j=1}^{n} r_l \Upsilon_{ij}^l e_j(t+1, k-\tau(l)), & \mu\tau + 1 \le k \le (\mu+1)\tau, \end{cases} \tag{10.34}$$

where the number of layers $M = 4$ and $\tau = 3$. Then, the iteration delay $\tau(l)$ of the lth ($l \in \{1, 2, 3, 4\}$) layer satisfies that $3 \le \tau(l) \le k - 3(\mu - 1) - 1$ when $3\mu + 1 \le k \le 3(\mu + 1)$, $\mu = 1, 2, \ldots$. Also, the known $u_i^0(t)$ is defined as $u_i^0(t) = (-\sin(t), \sin(t), -\cos(t), \cos(t))^{\mathrm{T}} \in \mathbb{R}^4$. The adjacency-weighted matrix $\tilde{\mathcal{A}}^l = (\tilde{\alpha}_{ij}^l)_{5\times 5}$ of the lth layer is assumed to have antagonistic nodes and is defined as follows:

$$\tilde{\mathcal{A}}^l = (\tilde{\alpha}_{ij}^l)_{5\times 5} = \begin{pmatrix} 0 & 0.2*l & -0.2*l & 0 & 0 \\ 0.1*l & 0 & -0.5 & 0 & 0 \\ 0.9 & -0.2*l & 0 & 0 & 0 \\ -0.5 & 1 & 0.2*l & 0 & 0 \\ 0.5 & 0 & 0.4*l & -0.8 & 0 \end{pmatrix}.$$

According to Theorem 10.1, we define the control parameter $r_l = 0.1$, $\epsilon^l = O_{5\times 5}$, where $l = 1, \ldots, 4$. There are antagonistic nodes in the network from the expression of matrix $\tilde{\mathcal{A}}^l$. Therefore, we can define $\epsilon^l = O_{5\times 5}$ $(l = 1, 2, \ldots, 4)$ to meet the condition (10.7).

Furthermore, the multi-layered ILC scheme (10.5) for sampled data with cooperative–antagonistic interactions is considered as

$$\begin{cases} u_i(t,k) = u_i^0(t), & k = 1, 2, \ldots, \tau, \\ u_i(t,k) = u_i(t, k-\tau) - \sum\limits_{l=1}^{M} \sum\limits_{j=1}^{n} r_l \Upsilon_{ij}^l e_j(t+1, k-\tau(l)), & \mu\tau + 1 \le k \le (\mu+1)\tau, \\ e_j(t,k) = e_j(t, k_\nu), & when\ k_\nu \le k < k_{\nu+1}, \end{cases} \tag{10.35}$$

where the adjacency-weighted matrix $\tilde{\mathcal{A}}^l = (\tilde{\alpha}_{ij}^l)_{5\times 5}$ of the lth layer is assumed to have no opposing nodes and is defined as

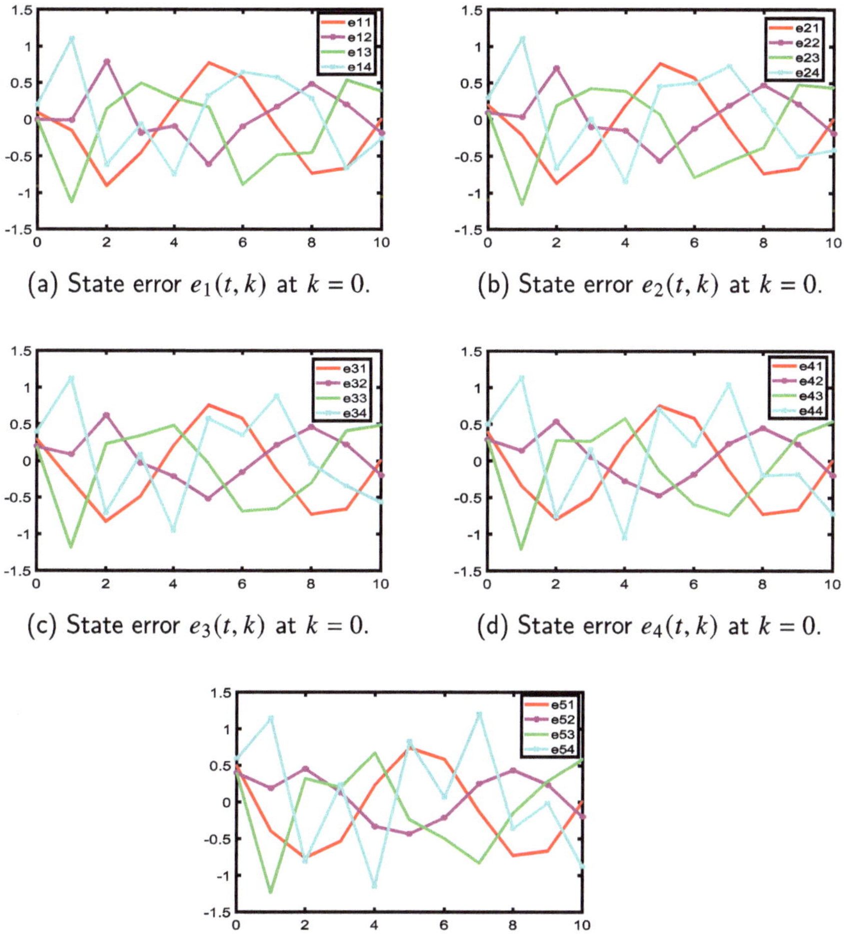

(a) State error $e_1(t,k)$ at $k=0$.

(b) State error $e_2(t,k)$ at $k=0$.

(c) State error $e_3(t,k)$ at $k=0$.

(d) State error $e_4(t,k)$ at $k=0$.

(e) State error $e_5(t,k)$ at $k=0$.

Fig. 10.1 The trajectories of state error $e_i(t,k)$ of the system (10.33) without controllers, $i = 1, 2, 3, 4, 5$

$$\tilde{\mathcal{A}}^l = (\tilde{\alpha}^l_{ij})_{5\times 5} = \begin{pmatrix} 0 & 0.2*l & 0.2*l & 0 & 0 \\ 0.1*l & 0 & 0.5 & 0 & 0 \\ 0.9 & 0.2*l & 0 & 0 & 0 \\ 0.5 & 1 & 0.2*l & 0 & 0 \\ 0.5 & 0 & 0.4*l & 0.8 & 0 \end{pmatrix}.$$

Let ϵ^l =diag(1, 0, 0, 0, 0) $\in \mathbb{R}^{5\times 5}$, and the other parameters are the same as those in (10.34). Moreover, according to the condition (10.20), the sampled-data iteration sequence $\{k_\nu : \nu = 1, 2, \ldots\}$ is defined as $k_1 = 4$, $k_{\nu+1} - k_\nu = 2$, $\nu = 1, 2, \ldots$. There are no antagonistic nodes in the network from the expression of matrix $\tilde{\mathcal{A}}^l$.

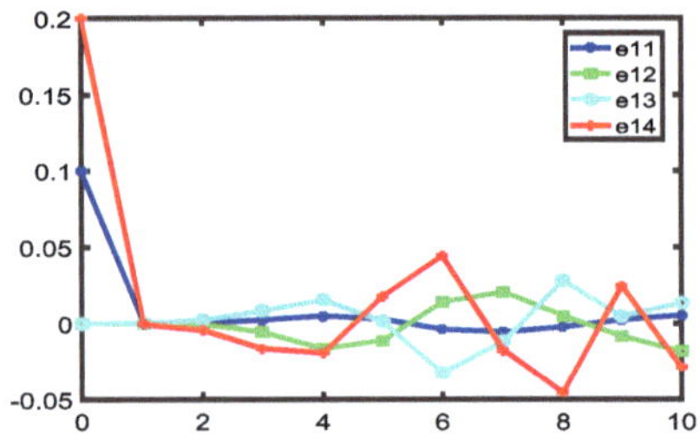

(a) State error $e_1(t, k)$ at $k = 100$.

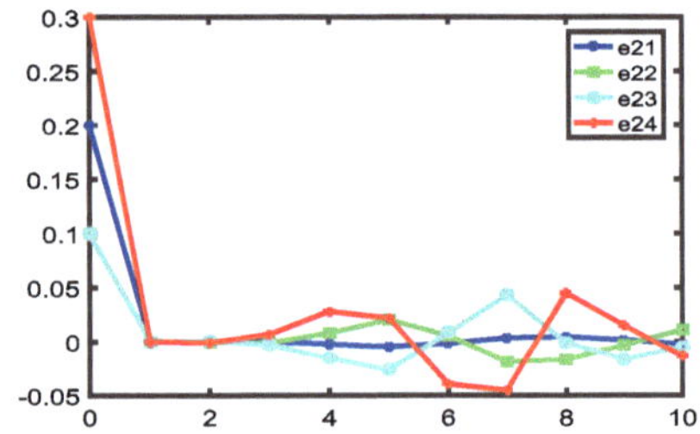

(b) State error $e_2(t, k)$ at $k = 100$.

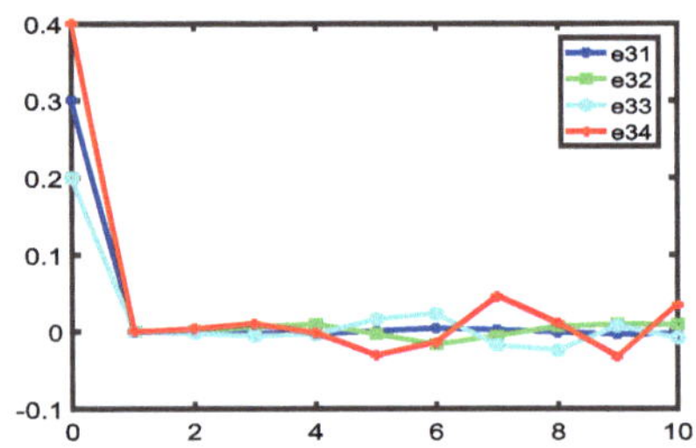

(c) State error $e_3(t, k)$ at $k = 100$.

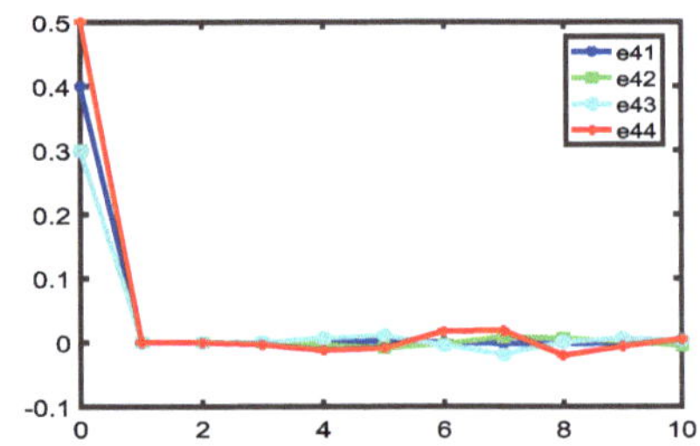

(d) State error $e_4(t, k)$ at $k = 100$.

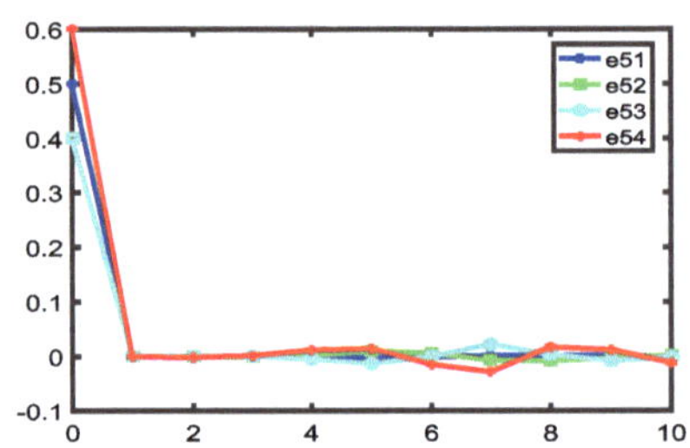

(e) State error $e_5(t, k)$ at $k = 100$.

Fig. 10.2 The trajectories of state error $e_i(t, k)$ of the system (10.33) with controllers (10.34) at $k = 100$, $i = 1, 2, 3, 4, 5$

Therefore, we must define $\epsilon^l = \text{diag}(1, 0, 0, 0, 0)$ $(l = 1, 2, \ldots, 4)$ to meet the condition (10.21). With the above data, the conditions in Theorems 10.1 and 10.2 can be satisfied. That is to say, the finite-iteration tracking of the system (10.33) with learning schemes (10.34) and (10.35) can be obtained from Theorems 10.1–10.2.

The corresponding error traces can be shown in Figs. 10.1, 10.2, 10.3, and 10.4. The state error $e_i(t, k)$ $(i = 1, 2, 3, 4, 5)$ of the system (10.33) with schemes (10.34) is shown in Fig. 10.2. From Fig. 10.3, we can see that the tracking of the system (10.33) is realized by schemes (10.35) at the iteration $k = 400$. The state error $e_i(t, k)$ $(i = 1, 2, 3, 4, 5)$ of the system (10.33) with schemes (10.35) is shown in

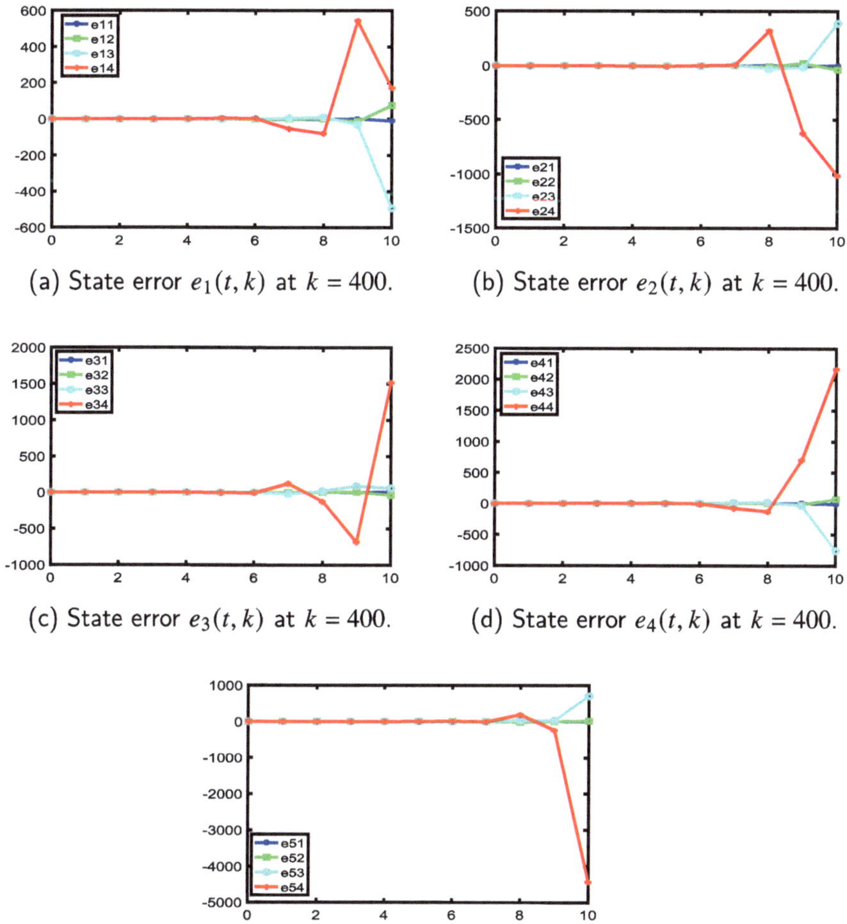

(a) State error $e_1(t, k)$ at $k = 400$.

(b) State error $e_2(t, k)$ at $k = 400$.

(c) State error $e_3(t, k)$ at $k = 400$.

(d) State error $e_4(t, k)$ at $k = 400$.

(e) State error $e_5(t, k)$ at $k = 400$.

Fig. 10.3 The trajectories of state error $e_i(t, k)$ of the system (10.33) with controllers (10.35) at $k = 400$, $i = 1, 2, 3, 4, 5$

Figs. 10.3 and 10.4. Similarly, from Fig. 10.4, the tracking of the system (10.33) is realized by schemes (10.35) at the iteration $k = 1000$. From Figs. 10.2 and 10.3, one has that the convergence rate of the system (10.33) with schemes (10.34) and (10.35) is slow. Moreover, the convergence rate of the system (10.33) with scheme (10.34) is faster than that with scheme (10.35). The reasons are

(1) The schemes are updated according to the previous learning iterations but not just dependent on the last iteration, then it slows the convergence rate of the system (10.33).

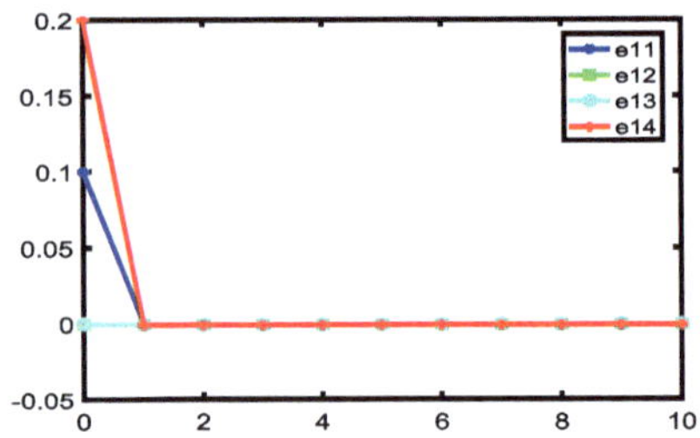

(a) State error $e_1(t, k)$ at $k = 1000$.

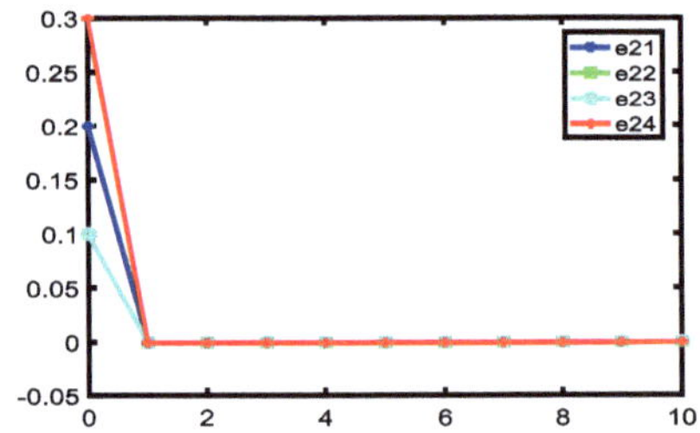

(b) State error $e_2(t, k)$ at $k = 1000$.

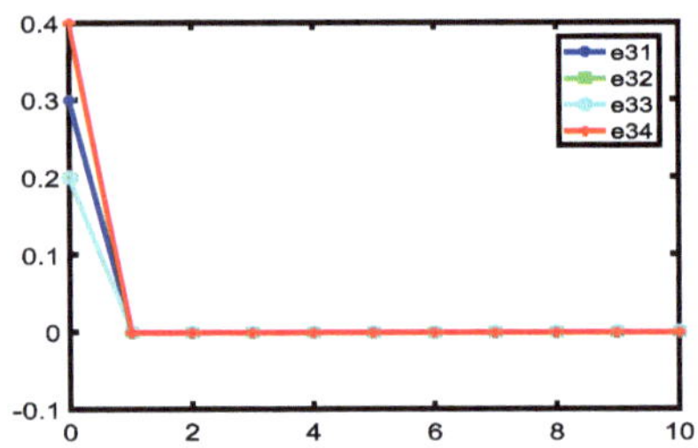

(c) State error $e_3(t, k)$ at $k = 1000$.

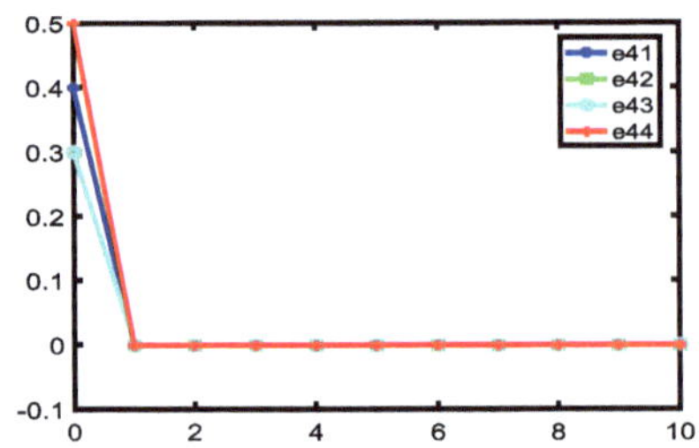

(d) State error $e_4(t, k)$ at $k = 1000$.

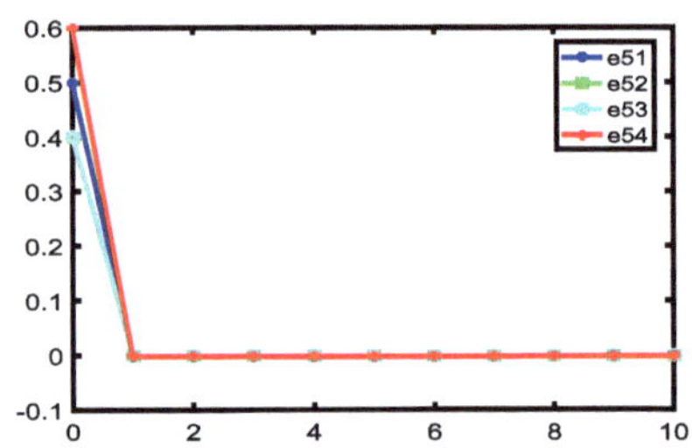

(e) State error $e_5(t, k)$ at $k = 1000$.

Fig. 10.4 The trajectories of state error $e_i(t, k)$ of the system (10.33) with controllers (10.35) at $k = 1000$, $i = 1, 2, 3, 4, 5$

(2) The update frequency of information with the sampled-data scheme (10.35) is slower than that with scheme (10.34), then the convergence rate of system (10.33) with scheme (10.35) is slower than that with scheme (10.34).

In addition, the comparison between two iterative control schemes (10.34) and (10.35) is shown in Fig. 10.5. Here, only the comparison of the first node is addressed for control schemes (10.34) and (10.35). In Fig. 10.5, u_{1i} means the ith component of the first node in the control scheme (10.34), and $\overline{u}_{1i}$ denotes the ith component of the first node in the control scheme (10.35), $i = 1, 2, 3, 4$. From Fig. 10.5, it can be noted that the control variables $u_{1i}(t, k)$ $(i = 1, 2, 3, 4)$ in (10.34) and (10.35) are different

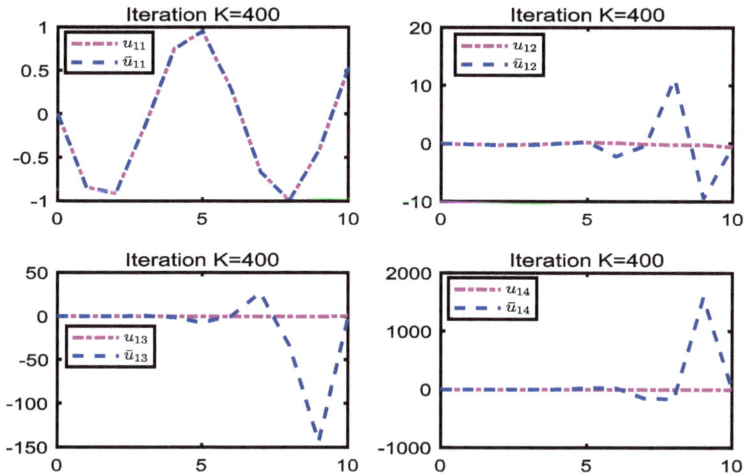

(a) The comparison of $u_{1i}(t,k)$ $(i=1,2,3,4)$ at $k=400$.

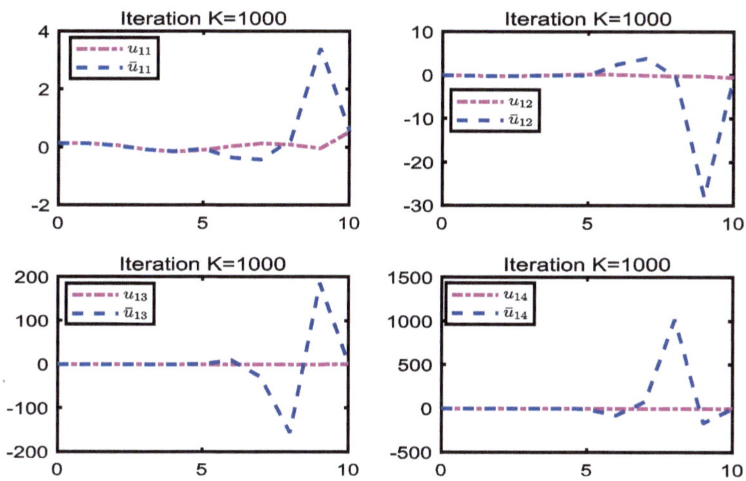

(b) The comparison of $u_{1i}(t,k)$ $(i=1,2,3,4)$ at $k=1000$.

Fig. 10.5 The comparisons of $u_{1i}(t,k)$ $(i=1,2,3,4)$ of controllers (10.34) and (10.35) at $k=400$ and $k=1000$

at $k = 400$ and $k = 1000$. These simulations have shown that the convergence rate of the system (10.33) with the scheme (10.34) is different from the scheme (10.35).

10.5 Summary

The tracking problem of discrete systems has been discussed in this chapter by designing two kinds of multi-layered iterative learning schemes with cooperative–antagonistic interactions. The state information of these schemes has been updated according to previous learning iterations, but it does not only depend on the last iteration. In addition, the iterative learning schemes have been designed to be multi-layered and have cooperative–antagonistic interactions. Finally, a simulation example has been shown to explain the effectiveness of the proposed criteria.

References

Altafini C (2013) Consensus problems on networks with antagonistic interactions. IEEE Trans Autom Control 58(4):935–946

Boccaletti S, Bianconi G, Criado R, Del Genio CI, Gómez-Gardenes J, Romance M, Sendina-Nadal I, Wang Z, Zanin M (2014) The structure and dynamics of multilayer networks. Phys Rep 544(1):1–122

Bu XH, Hou ZS (2018) Adaptive iterative learning control for linear systems with binary-valued observations. IEEE Trans Neural Netw Learn Syst 29(1):232–237

Bu XH, Yu QX, Hou ZS, Qian W (2019) Model free adaptive iterative learning consensus tracking control for a class of nonlinear multiagent systems. IEEE Trans Syst Man Cybern Syst 49(4):677–686

Chi RH, Liu Y, Hou ZS, Jin ST (2015) Data-driven terminal iterative learning control with high-order learning law for a class of nonlinear discrete-time multiple-input-multiple output systems. IET Control Theory Appl 9(7):1075–1082

Chi RH, Huang B, Wang DW, Zhang RK, Feng YJ (2016) Data-driven optimal terminal iterative learning control with initial value dynamic compensation. IET Control Theory Appl 10(12):1357–1364

De Domenico M, Granell C, Porter MA, Arenas A (2016) The physics of spreading processes in multilayer networks. Nat Phys 12(10):901–906

Frénay B, Verleysen M (2016) Reinforced extreme learning machines for fast robust regression in the presence of outliers. IEEE Trans Cybern. 46(12):3351–3363

Fridman E (2010) A refined input delay approach to sampled-data control. Automatica 46(2):421–427

He WL, Xu ZW, Du WL, Chen GR, Kubota N, Qian F (2017) Synchronization control in multiplex networks of nonlinear multi-agent systems. Chaos 27(12):123104

He WL, Chen GR, Han QL, Du WL, Cao JD, Qian F (2017) Multiagent systems on multilayer networks: synchronization analysis and network design. IEEE Trans Syst Man Cybern Syst 47(7):1655–1667

Janssens P, Pipeleers G, Swevers J (2013) A data-driven constrained norm-optimal iterative learning control framework for LTI systems. IEEE Trans Control Syst Tech 21(2):546–551

Kumar SV, Raja R, Anthoni SM, Cao JD, Tu ZW (2018) Robust finite-time non-fragile sampled-data control for TS fuzzy flexible spacecraft model with stochastic actuator faults. Appl Math Comput 321:483–497
Lee KS, Lee J, Chin I, Choi J, Lee JH (2001) Control of wafer temperature uniformity in rapid thermal processing using an optimal iterative learning control technique. Ind Eng Chem Res 40(7):1661–1672
Li HY, Jing XJ, Lam HK, Shi P (2014) Fuzzy sampled-data control for uncertain vehicle suspension systems. IEEE Trans Cybern 44(7):1111–1126
Li XF, Ren QY, Xu JX (2016) Precise speed tracking control of a robotic fish via iterative learning control. IEEE Trans Ind Electron 63(4):2221–2228
Liu YJ, Jing Li, Tong SC, Chen CP (2016) Neural network control-based adaptive learning design for nonlinear systems with full-state constraints. IEEE Trans Neural Netw Learn Syst 27(7):1562–1571
Manivannan R, Samidurai R, Cao JD, Perc M (2018) Design of resilient reliable dissipativity control for systems with actuator faults and probabilistic time-delay signals via sampled-data approach. IEEE Trans Syst Man Cybern Syst 50(11):4243–4255
Meng DY (2018) Dynamic distributed control for networks with cooperative-antagonistic interactions. IEEE Trans Autom Control 63(8):2311–2326
Meng DY, Moore KL (2016) Learning to cooperate: networks of formation agents with switching topologies. Automatica 64:278–293
Meng DY, Jia YM, Du JP (2016) Finite-time consensus for multiagent systems with cooperative and antagonistic interactions. IEEE Trans Neural Netw Learn Syst 27(4):762–770
Peng ZH, Wang D, Zhang HW, Sun G (2014) Distributed neural network control for adaptive synchronization of uncertain dynamical multiagent systems. IEEE Trans Neural Netw Learn Syst 25(8):1508–1519
Peng ZH, Wang D, Wang J (2017) Predictor-based neural dynamic surface control for uncertain nonlinear systems in strict-feedback form. IEEE Trans Neural Netw Learn Syst 28(9):2156–2167
Sun C, Hu GQ, Xie LH (2017) Controllability of multiagent networks with antagonistic interactions. IEEE Trans Autom Control 62(10):5457–5462
Wei QL, Liu DR, Yang X (2015) Infinite horizon self-learning optimal control of nonaffine discrete-time nonlinear systems. IEEE Trans Neural Netw Learn Syst 26(4):866–879
Wei QL, Liu DR, Lin HQ (2016) Value iteration adaptive dynamic programming for optimal control of discrete-time nonlinear systems. IEEE Trans Cybern 46(3):840–853
Wu YQ, Su HY, Shi P, Shu Z, Wu ZG (2016) Consensus of multiagent systems using aperiodic sampled-data control. IEEE Trans Cybern 46(9):2132–2143
Xiong WJ, Xu L, Ho DWC, Cao JD, Huang TW (2018) Synchronous and asynchronous iterative learning strategies of TS fuzzy systems with measurable and unmeasurable state information. IEEE Trans Fuzzy Syst 26(5):3042–3053
Yang DR, Lee KS, Ahn HJ, Lee JH (2003) Experimental application of a quadratic optimal iterative learning control method for control of wafer temperature uniformity in rapid thermal processing. IEEE Trans Semicond Manuf 16(1):36–44

Chapter 11
Stability of Multi-layer Supply Chain Networks with Constraints

11.1 Introduction

A supply chain network (SCN) comprises factories, distributors, wholesalers, and retailers (Alessandri et al. 2011; Dubey et al. 2019; Ha and Tong 2008; Moyaux et al. 2007; Riddalls and Bennett 2002). The network's providers work together and coordinate along the entire chain to generate raw and processed materials, transform these raw materials into the necessary goods, and distribute these products to customers (Chan et al. 2018; Fu et al. 2019; Goodarzian et al. 2020; Nagurney et al. 2014; Yu and Solvang 2020). To address the management issues of SCNs, for instance, a distributed model predictive control method is suggested in Fu et al. (2020). Here, each agent relies on this technique to select the regionally advantageous tactic and limit information sharing with neighbors. In de Albuquerque et al. (2013), stochastic reward nets are used in conjunction with mathematical optimization to assess the global warming potential and energetic supply chain indices.

Inventory (Roy et al. 2019; Tancrez et al. 2012) is one of the main issues in a SCN. There are safety inventory and expected inventory values (He et al. 2020; Zepeda et al. 2016) in a real manufacturing and order process. To prevent inventory backlogs or product scarcity losses, businesses will employ a variety of manufacturing or ordering tactics. As a result, the study of SCN inventory control issues has huge theoretical implications. However, there is a lot of unpredictability that could impact a SCN's inventory. For instance, a retailer's orders are very different from the demands of end consumers; wholesalers' orders may exhibit a strong oscillation due to various market hot spots and demand; and a factory's production schedule may be unstable due to factors like a shortage of raw materials, the COVID-19 epidemic, etc.

As a result, it is crucial to consider and debate the equilibrium analysis of a supply chain with uncertainties (see Chan et al. 2018; Liu et al. 2019; Tang et al. 2020 and the references therein). Two important concerns about the inventory of SCNs, nonetheless, remain unanswered in this body of literature: (Q_1) How may the convergence rates of the ideal inventory, or equilibrium, be increased while reducing costs? (Q_2)

W. Xiong et al., *Iterative Learning Control for Network Systems Under Constrained Information Communication*, Intelligent Control and Learning Systems 12,
https://doi.org/10.1007/978-981-97-0926-7_11

How can the restricted control resource be preserved while the control scheme is being designed? A new learning control system is created in this study to address the aforementioned two concerns. It is motivated by the references (Wan et al. 2017; Wen et al. 2020; Zhang et al. 2017; Zhu et al. 2019) prior to addressing some state-related limitations, a multi-layer supply chain network with three tiers (production layer, distribution layer, and sales layer) will be constructed. Define the states' optimum stock levels and treat the fluctuating amounts of products sent between different nodes as control inputs. After that, create the correct control inputs (Tang et al. 2019; Wei et al. 2018; Xiong et al. 2020) to fulfill all limitations and achieve the convergence of the optimum inventory levels. Here, a comparison between our control approach and the conventional ones will be addressed in the context of a given scenario in order to highlight the benefits of our control strategy. And the illustration gives the potential causes. Our technique makes two contributions: (i) It can conserve scarce control resources. (ii) It can increase all nodes' rates of convergence. The aforementioned queries can be addressed as a consequence.

11.2 Problem Formulation

11.2.1 Model Description

The production layer (PL), distribution layer (DL), and sales layer (SL) comprise a multi-layer SCN. Assume the network as a whole is a directed graph with $N-1$ nodes. Each node in this system consists of a collection of buffers that contain M various categories of products. Additionally, node 0 is a made-up node that serves as the initial source at PL, and node N serves as the ultimate source at SL. The collection of indexes denoting the various items accessible at node i, $i \in \{1, \dots, N-1\}$ is known as $D(i)$ ($D(i) \subseteq M$). Figure 11.1 shows the three-layer supply chain network diagram. The discrete-time model of a SCN with multiple layers is as follows:

$$\begin{cases} x_{il}(t+1) = x_{il}(t) + u_{0il}(t-\tau_{0il}) - \sum\limits_{j \in O_l(i)} u_{ijl}(t), \; i \in PL, \;\; l \in D(i), \\ x_{il}(t+1) = x_{il}(t) + \sum\limits_{j \in I_l(i)} u_{jil}(t-\tau_{jil}) - \sum\limits_{j \in O_l(i)} u_{ijl}(t), \quad i \in DL, \; l \in D(i), \\ x_{il}(t+1) = x_{il}(t) + \sum\limits_{j \in I_l(i)} u_{jil}(t-\tau_{jil}) - u_{iNl}(t), \quad i \in SL, \; l \in D(i), \\ y_{il}(t+1) = y_{il}(t) - u_{iNl}(t) + \omega_{il}(t), \quad i \in SL, \; l \in D(i), \end{cases} \tag{11.1}$$

where the time stamp is $t = 0, 1, 2 \dots$, and the inventory level of node i is $x_{il}(t)$ which contains goods of type l at time t, $i = 1, \dots, N-1$, $l \in D(i)$. $u_{ijl}(t)$ is the variable quantity of good l transmitted from node i to node j at time t, $j = 1, \dots, N-1$, $l \in D(i) \cap D(j)$, $i \in I(j)$, $I(j) = \{i : there\ is\ a\ transmission\ from\ node\ i\ to\ node\ j\}$. τ_{ijl} is the delay in the transmission of good l from node

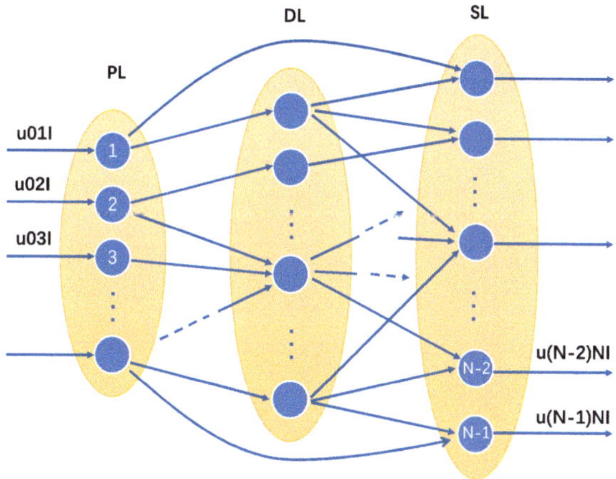

Fig. 11.1 A diagram of a multi-layer supply chain network with 3 layers

i to node j, $j \in \{1, \ldots, N-1\}$, $l \in D(i) \cap D(j)$, $i \in I(j)$. $\omega_{il}(t)$ is the external need of good l at node i, $i \in SL$, $l \in D(i)$. Moreover, $y_{il}(t)$ is the backlogged demand of good l at node i, $i \in SL$, $l \in D(i)$. $I_l(i) = \{j : j \in I(i),\ and\ l \in D(i) \cap D(j)\}$. $O_l(i) = \{j : j \in O(i),\ and\ l \in D(i) \cap D(j)\}$, where $O(j) = \{i : there\ is\ a\ transmission\ from\ node\ j\ to\ node\ i\}$.

11.2.2 Constraints of States and Objectives

To increase the viability and application of the model (11.1), some constraints on its states will be described in this section.

A safety stock is typically kept on hand to satisfy unforeseen requests in order to avoid a stock shortage (i.e., a node or more nodes do not have sufficient inventory levels to meet the external requirements of other nodes). As a result of the constrained inventory capacity, each node has a maximum amount of storage. Hence, the state variable $x_il(t)$ should have an upper constraint. Consequently, one must

$$\underline{x}_{il} \leq x_{il}(t) \leq \overline{x}_{il},\ l \in D(i), \tag{11.2}$$

where $i = 1, \ldots, N-1$, $t = 0, 1 \ldots$, $\underline{x}_{il}$ and $\overline{x}_{il}$ $(0 < \underline{x}_{il} < \overline{x}_{il})$ are positive scalars. Without a doubt, the backlog of goods should also be nonnegative. Thus, one is constrained by the following:

$$y_{il}(t) \geq 0,\ i \in SL,\ l \in D(i), \tag{11.3}$$

where $t = 1, 2, 3 \ldots$.

Each $u_{ijl}(t)$ is viewed as the control input for the entire research. Let x_{il}^* represent the optimal inventory level for the state $x_{il}(t)$. One of our key goals in the following is to develop acceptable control inputs $u_{ijl}(t)$ to accomplish two goals: (1) The variables' aforementioned restrictions are all met. (2) The x_{il}^* ideal inventory's convergence is attained. This is,

$$\left|x_{il}(t) - x_{il}^*\right| < \theta, \tag{11.4}$$

$i = 1, \ldots, N-1$, $\forall\, t > K$, where K is a known constant and θ is a specified tiny positive number. In addition, an assumption is needed in the following.

Assumption 11.1 The parameters are assumed as follows:

$$\underline{x}_{il} \le x_{il}(0) \le \overline{x}_{il},\ i = 1, \ldots, N-1, \tag{11.5}$$

$$y_{il}(0) > 0,\ i \in SL, \tag{11.6}$$

$$\underline{x}_{il} + \vartheta_{il} < x_{il}^* < \overline{x}_{il} - \vartheta_{il},\ i = 1, 2, \ldots, N-1, \tag{11.7}$$

$$\omega_{il}(k) \ge \underline{\omega}_{il} > 0,\ i \in SL,\ k = 0, 1, 2\ldots, \tag{11.8}$$

where $l \in D(i)$, $\underline{\omega}_{il}$, ϑ_{il} are tiny positive constants, and $\vartheta_{il} < \frac{1}{2}(\overline{x}_{il} - \underline{x}_{il})$.

11.3 Control Strategy with Limitations

In this section, the objective is to design appropriate control inputs $u_{ijl}(t)$ to guarantee the limitations (11.2) and (11.3).

I. Discuss $y_{il}(t)$. Suppose $u_{iNl}(0) = 0$, which denotes that there is initially no control. Design the following when $t = 1, 2, 3\ldots$

$$u_{iNl}(t) = \delta_{iNl}(t) \cdot \left|x_{il}(t) - x_{il}^*\right|,\ i = 1, \ldots, N-1, \tag{11.9}$$

with

$$\begin{cases} 0 \le \delta_{iNl}(t) < \begin{cases} \frac{\underline{\omega}_{il}}{|x_{il}(t) - x_{il}^*|},\ if\ y_{il}(t) \le \varepsilon_{il}, \\ \frac{y_{il}(t) + \underline{\omega}_{il}}{|x_{il}(t) - x_{il}^*|},\ if\ y_{il}(t) > \varepsilon_{il}, \end{cases} ,\ if\ x_{il}(t) \ne x_{il}^*, \\ \delta_{iNl}(t) = 0, \qquad\qquad if\ x_{il}(t) = x_{il}^*, \end{cases} \tag{11.10}$$

where a tiny value, $\varepsilon_{il} > 0$, can be thought of as the $y_{il}(t)$ threshold. One has the following property while using the control schemes (11.9) and (11.10).

Remark 11.1 If $y_{il}(t) > \varepsilon_{il}$, then the range of $\delta_{iNl}(t)$ is large, as shown in (11.10). This is logical given that a larger backlog in excellent inventories could result in a larger amount of good l being moved out. As a result, when $y_{il}(t)$ is larger, the alternate range of $\delta_{iNl}(t)$ is greater.

Property 11.1 Think about the variable states $y_{il}(t)$ in the model from (11.1). If the control input$u_{iNl}(t)$ is created as formula (11.9) with (11.10), $i \in SL$, $l \in D(i)$, $t = 1, 2\ldots$, the limitation (11.3) holds under Assumption 11.1.

Proof Inductive mathematics will be used to demonstrate this property. According to Assumption 11.1, one has $y_{il}(0) > 0$. Based on the model (11.1), one has $y_{il}(1) = y_{il}(0) + \omega_{il}(0) \geq y_{il}(0) + \underline{\omega}_{il} > 0$. Assume that, for some $t \in \{2, 3, 4\ldots\}$, one has $y_{il}(t) > 0$. Then, one only needs to prove that $y_{il}(t+1) > 0$.

(1) If $0 < y_{il}(t) \leq \varepsilon_{il}$, one can obtain

$$\begin{aligned} & y_{il}(t+1) \\ & = y_{il}(t) - u_{iNl}(t) + \omega_{il}(t) \\ & = y_{il}(t) - \delta_{iNl}(t) \cdot \left|x_{il}(t) - x^*_{il}\right| + \omega_{il}(t) \\ & > y_{il}(t) - \frac{\underline{\omega}_{il}}{\left|x_{il}(t) - x^*_{il}\right|} \cdot \left|x_{il}(t) - x^*_{il}\right| + \omega_{il}(t) \\ & \geq y_{il}(t) > 0. \end{aligned} \tag{11.11}$$

(2) If $y_{il}(t) > \varepsilon_{il}$, one obtains

$$\begin{aligned} & y_{il}(t+1) \\ & = y_{il}(t) - u_{iNl}(t) + \omega_{il}(t) \\ & = y_{il}(t) - \delta_{iNl}(t) \cdot \left|x_{il}(t) - x^*_{il}\right| + \omega_{il}(t) \\ & > y_{il}(t) - \frac{y_{il}(t) + \underline{\omega}_{il}}{\left|x_{il}(t) - x^*_{il}\right|} \cdot \left|x_{il}(t) - x^*_{il}\right| + \omega_{il}(t) \\ & = \omega_{il}(t) - \underline{\omega}_{il} \geq 0. \end{aligned} \tag{11.12}$$

As a result, one has $y_{il}(t+1) > 0$. The mathematical induction suggests that one has that $y_{il}(t) > 0$, for $\forall\, t \in \{1, 2, 3\ldots\}$. That is, the limitation (11.3) holds. □

II. Discuss $x_{il}(t)$. Let $u_{ijl}(0) = 0$ $(i = 0, 1, \ldots, N-1,\ j = 1, \ldots, N)$, which also means that in the beginning, there is no control.

(1) When $i \in PL$, $l \in D(i)$, design that

$$u_{ijl}(t) = \delta_{ijl}(t) \cdot \left|x_{il}(t) - x^*_{il}\right|, \tag{11.13}$$

where $i = 0, 1, \ldots, N-1$, $j = 1, \ldots, N-1$, $\delta_{ijl}(t) \geq 0$, and

$$\begin{cases} \Xi(t) > 0,\ \Pi(t) < \overline{x}_{il} - x^*_{il}, & if\ \underline{x}_{il} \leq x_{il}(t) < x^*_{il}, \\ \Xi(t) = 0,\ \Pi(t) = 0, & if\ x_{il}(k) = x^*_{il}, \\ \Pi(t) < 0,\ \Xi(t) > \underline{x}_{il} - x^*_{il}, & if\ x^*_{il} < x_{il}(t) \leq \overline{x}_{il}, \end{cases} \tag{11.14}$$

where

$$\Xi(t) = \delta_{0il}(t-\tau_{0il}) \min_{i \in PL} \left| x_{il}(t) - x_{il}^* \right| - \sum_{j \in O_l(i)} \delta_{ijl}(t) \max_{i \in PL} \left| x_{il}(t) - x_{il}^* \right|,$$

$$\Pi(t) = \delta_{0il}(t-\tau_{0il}) \max_{i \in PL} \left| x_{il}(t) - x_{il}^* \right| - \sum_{j \in O_l(i)} \delta_{ijl}(t) \min_{i \in PL} \left| x_{il}(t) - x_{il}^* \right|.$$

(2) When $i \in DL,\ l \in D(i)$, $u_{ijl}(t)$ is designed to have the equivalent expression with (11.13) with

$$\begin{cases} \Xi^1(t) > 0,\ \Pi^1(t) < \overline{x}_{il} - x_{il}^*, & if\ \underline{x}_{il} \le x_{il}(t) < x_{il}^*, \\ \Xi^1(t) = 0,\ \Pi^1(t) = 0, & if\ x_{il}(t) = x_{il}^*, \\ \Pi^1(t) < 0,\ \Xi^1(t) > \underline{x}_{il} - x_{il}^*, & if\ x_{il}^* < x_{il}(t) \le \overline{x}_{il}, \end{cases} \tag{11.15}$$

where $i, j = 1, \ldots, N-1$, $\delta_{ijl}(t) \ge 0$, and

$$\Xi^1(t) = \sum_{j \in I_l(i)} \delta_{jil}(t-\tau_{jil}) \min_{i \in DL} \left| x_{il}(t) - x_{il}^* \right| - \sum_{j \in O_l(i)} \delta_{ijl}(t) \max_{i \in DL} \left| x_{il}(t) - x_{il}^* \right|,$$

$$\Pi^1(t) = \sum_{j \in I_l(i)} \delta_{jil}(t-\tau_{jil}) \max_{i \in DL} \left| x_{il}(t) - x_{il}^* \right| - \sum_{j \in O_l(i)} \delta_{ijl}(t) \min_{i \in DL} \left| x_{il}(t) - x_{il}^* \right|.$$

(3) When $i \in SL,\ l \in D(i)$, $u_{ijl}(t)$ is designed to have the equivalent expression with (11.13) with

$$\begin{cases} \Xi^2(t) > 0,\ \Pi^2(t) < \overline{x}_{il} - x_{il}^*, & if\ \underline{x}_{il} \le x_{il}(t) < x_{il}^*, \\ \Xi^2(t) = 0,\ \Pi^2(t) = 0, & if\ x_{il}(t) = x_{il}^*, \\ \Pi^2(t) < 0,\ \Xi^2(t) > \underline{x}_{il} - x_{il}^*, & if\ x_{il}^* < x_{il}(t) \le \overline{x}_{il}, \end{cases} \tag{11.16}$$

where $i = 1, \ldots, N-1,\ j = 1, \ldots, N,\ \delta_{ijl}(t) \ge 0$, and

$$\Xi^2(t) = \sum_{j \in I_l(i)} \delta_{jil}(t-\tau_{jil}) \min_{i \in SL} \left| x_{il}(t) - x_{il}^* \right| - \delta_{iNl}(t) \max_{i \in SL} \left| x_{il}(t) - x_{il}^* \right|,$$

$$\Pi^2(t) = \sum_{j \in I_l(i)} \delta_{jil}(t-\tau_{jil}) \max_{i \in SL} \left| x_{il}(t) - x_{il}^* \right| - \delta_{iNl}(t) \min_{i \in SL} \left| x_{il}(t) - x_{il}^* \right|.$$

Remark 11.2 From (11.13)–(11.16), it is clear that the parameters $\delta_{ijl}(t)$ in the control inputs $u_{ijl}(t)$ are changed to suit various requirements based on the discrepancies between$x_{il}(t)$ and x_{il}^*. Consequently, the ideal inventory level x_{il}^* can quickly converge. As a result, the control scheme (11.13)–(11.16) is referred to as a learning scheme.

Furthermore, one has the following property when using the learning control system (11.13)–(11.16).

Property 11.2 Consider the variable states $x_{il}(t)$ in the model (11.1). When using Assumption 11.1, the limitation (11.2) holds true if the learning controllers $u_{ijl}(t)$ are made to meet the requirements (11.13)–(11.16).

Proof Inductive mathematics will be used to demonstrate this property. When $i \in PL, l \in D(i)$, according to Assumption 11.1, one has $\underline{x}_{il} \leq x_{il}(0) \leq \overline{x}_{il}$. Based on the model (11.1), one has $x_{il}(1) = x_{il}(0)$, which means that $\underline{x}_{il} \leq x_{il}(1) \leq \overline{x}_{il}$. Assume that, for some $t \in \{2, 3, 4\ldots\}$, one has $x_{il}(t) > 0$. Then, one has to prove that $x_{il}(t+1) > 0$.

(1) If $\underline{x}_{il} \leq x_{il}(t) < x^*_{il}$, one can acquire from relation (11.14)

$$\begin{aligned} & x_{il}(t+1) \\ &= x_{il}(t) + u_{0il}(t-\tau_{0il}) - \sum_{j \in O_l(i)} u_{ijl}(t) \\ &= x_{il}(t) + \delta_{0il}(t-\tau_{0il}) \cdot \left|x_{il}(t) - x^*_{il}\right| - \sum_{j \in O_l(i)} \delta_{ijl}(t) \cdot \left|x_{il}(t) - x^*_{il}\right| \\ &\geq x_{il}(t) + \Xi(t) > x_{il}(t) \geq \underline{x}_{il}. \end{aligned} \tag{11.17}$$

Also,

$$\begin{aligned} & x_{il}(t+1) \\ &= x_{il}(t) + \delta_{0il}(t-\tau_{0il}) \cdot \left|x_{il}(t) - x^*_{il}\right| - \sum_{j \in O_l(i)} \delta_{ijl}(t) \cdot \left|x_{il}(t) - x^*_{il}\right| \\ &\leq x_{il}(t) + \Pi(k) < x_{il}(t) + \overline{x}_{il} - x^*_{il} < \overline{x}_{il}. \end{aligned} \tag{11.18}$$

(2) If $x_{il}(t) = x^*_{il}$, one can obtain from (11.14)

$$x_{il}(t+1) = x_{il}(t) = x^*_{il}. \tag{11.19}$$

(3) If $x^*_{il} < x_{il}(t) \leq \overline{x}_{il}$, one can obtain from (11.14)

$$\begin{aligned} & x_{il}(t+1) \\ &= x_{il}(t) + \delta_{0il}(t-\tau_{0il}) \cdot \left|x_{il}(t) - x^*_{il}\right| - \sum_{j \in O_l(i)} \delta_{ijl}(t) \cdot \left|x_{il}(t) - x^*_{il}\right| \\ &\leq x_{il}(t) + \Pi(t) < x_{il}(t) \leq \overline{x}_{il}. \end{aligned} \tag{11.20}$$

Also,

$$\begin{aligned} & x_{il}(t+1) \\ &= x_{il}(t) + \delta_{0il}(t-\tau_{0il}) \cdot \left|x_{il}(t) - x^*_{il}\right| - \sum_{j \in O_l(i)} \delta_{ijl}(t) \cdot \left|x_{il}(t) - x^*_{il}\right| \\ &\geq x_{il}(t) + \Xi(t) > x_{il}(t) + \underline{x}_{il} - x^*_{il} \geq \underline{x}_{il}. \end{aligned} \tag{11.21}$$

As a result, one has $\underline{x}_{il} \leq x_{il}(t+1) \leq \overline{x}_{il}$, for $i \in PL, l \in D(i)$. The mathematical induction suggests that one has that $\underline{x}_{il} \leq x_{il}(t) \leq \overline{x}_{il}$, for $t \in \{1, 2, \ldots\}, i \in PL$,

$l \in D(i)$. That is, the limitation (11.2) holds for $i \in PL, l \in D(i)$. Correspondingly, using a similar analysis as above, one can also obtain that the limitation (11.2) holds for $i \in DL$, $l \in D(i)$ and $i \in SL$, $l \in D(i)$. Hence, the limitation (11.2) holds for all variable states $x_{il}(t)$ in the model (11.1). □

11.4 Main Results

The major target of this section is to discuss the convergence of state variables $x_{il}(t)$ by developing suitable control inputs $u_{ijl}(t)$. To get the limitation (11.2), the control inputs $u_{ijl}(t)$ can only be changed based on the requirements (11.9)–(11.10) and (11.13)–(11.16). Additionally, a comparison between our control method and the conventional ones will be discussed in the following example in order to highlight the benefits of our approach.

11.4.1 Learning Control Scheme Analysis

The time sequence $\{1, 2, \ldots\}$ is divided into periodic parts $\{1, \ldots, K_1\}$, $\{K_1 + 1, K_1 + 2, \ldots, 2K_1\}$, $\{2K_1 + 1, 2K_1 + 2, \ldots, 3K_1\}, \ldots$. Here, $K_1 = K_2 + K_3$ and K_i $(i = 1, 2, 3)$ are all positive integers. Based on conditions (11.9)–(11.10) and (11.13)–(11.16), the learning control inputs $u_{ijl}(t)$ are designed in the following.

Part 1: When $t \in (mK_1, mK_1 + K_2]$ $(m \in \{0, 1, 2, \ldots\})$, the learning control inputs $u_{ijl}(t)$ are adjusted as follows:

(1) When $i \in PL, l \in D(i)$, design that

$$\begin{cases} \Pi(t) < 0, \ \Xi(t) > \max\{2x_{il}^* - 2x_{il}(t), \ \underline{x}_{il} - x_{il}^*\}, & if \ x_{il}^* < x_{il}(t) \le \overline{x}_{il}, \\ \Xi(t) > 0, \ \Pi(t) < \min\{2x_{il}^* - 2x_{il}(t), \ \overline{x}_{il} - x_{il}^*\}, & if \ \underline{x}_{il} \le x_{il}(t) < x_{il}^*, \\ \Xi(t) = 0, \ \Pi(t) = 0, & if \ x_{il}(t) = x_{il}^*, \end{cases} \tag{11.22}$$

where $t \in \{2, 3\ldots\}$, $\Xi(t)$ and $\Pi(t)$ have the equivalent expressions as those in (11.14).

(2) When $i \in DL, l \in D(i)$, design that

$$\begin{cases} \Pi^1(t) < 0, \ \Xi^1(t) > \max\{2x_{il}^* - 2x_{il}(t), \ \underline{x}_{il} - x_{il}^*\}, & if \ x_{il}^* < x_{il}(t) \le \overline{x}_{il}, \\ \Xi^1(t) > 0, \ \Pi^1(t) < \min\{2x_{il}^* - 2x_{il}(k), \ \overline{x}_{il} - x_{il}^*\}, & if \ \underline{x}_{il} \le x_{il}(t) < x_{il}^*, \\ \Xi^1(t) = 0, \ \Pi^1(t) = 0, & if \ x_{il}(t) = x_{il}^*, \end{cases} \tag{11.23}$$

where $t \in \{2, 3\ldots\}$, $\Xi^1(t)$ and $\Pi^1(t)$ have the equivalent expressions as those in (11.15).

(3) When $i \in SL, l \in D(i)$, design that

$$\begin{cases} \Pi^2(t) < 0, \ \Xi^2(t) > \max\{2x_{il}^* - 2x_{il}(t), \ \underline{x}_{il} - x_{il}^*\}, & if \ x_{il}^* < x_{il}(t) \le \overline{x}_{il}, \\ \Xi^2(t) > 0, \ \Pi^2(t) < \min\{2x_{il}^* - 2x_{il}(t), \ \overline{x}_{il} - x_{il}^*\}, & if \ \underline{x}_{il} \le x_{il}(t) < x_{il}^*, \\ \Xi^2(t) = 0, \ \Pi^2(t) = 0, & if \ x_{il}(t) = x_{il}^*, \end{cases} \tag{11.24}$$

where $t \in \{2, 3 \ldots\}$, $\Xi^2(t)$ and $\Pi^2(t)$ have the equivalent expressions as those in (11.16).

Part 2: When $t \in (mK_1 + K_2, (m+1)K_1]$ $(m \in \{0, 1, 2, \ldots\})$, one has

(1) When $i \in PL, l \in D(i)$, define $\left|x_{\varpi_1 l}^* - x_{\varpi_1 l}(t)\right| = \max \left|x_{il}^* - x_{il}(t)\right|$, where $\varpi_1 \in PL$. Then, $u_{ijl}(t) = \widehat{\delta}_{ijl}(t) \cdot \left|x_{il}(t) - x_{il}^*\right|$ with $\widehat{\delta}_{ijl}(t)$ satisfying that

$$\begin{cases} \widehat{\delta}_{0il}(t) = 0, \ \widehat{\delta}_{ijl}(t) = 0, & if \ i \ne \varpi_1, \\ \widehat{\delta}_{0il}(t) = \frac{\lambda_1(PL)}{\lambda_1(\varpi_1)} \delta_{0il}(t), \ \widehat{\delta}_{ijl}(t) = \frac{\lambda_1(PL)}{\lambda_1(\varpi_1)} \delta_{ijl}(t), & if \ i = \varpi_1, \end{cases} \tag{11.25}$$

where $\delta_{0il}(t)$, $\delta_{ijl}(t)$ are given in (11.13) and (11.22), $\lambda_1(PL)$ is the number of all nodes in PL, and $\lambda_1(\varpi_1) = \{The \ number \ of \ all \ nodes \ i : i = \varpi_1, \ and \ i \in PL\}$, $i = 0, 1, \ldots, N-1, \ j = 1, \ldots, N-1$. Moreover, we define that

$$\begin{cases} \widehat{\Phi}(t) = x_{il}^* - x_{il}(t), \ if \ x_{il}^* < x_{il}(t) \le \overline{x}_{il}, \ and \ \widehat{\Phi}(t) < x_{il}^* - x_{il}(t), \\ \widehat{\Phi}(t) = x_{il}^* - x_{il}(t), \ if \ \underline{x}_{il} \le x_{il}(t) < x_{il}^*, \ and \ \widehat{\Phi}(t) > x_{il}^* - x_{il}(t), \\ \widehat{\Phi}(k) = 0, \qquad if \ x_{il}(t) = x_{il}^*, \\ \widehat{\delta}_{0il}(t), \ \widehat{\delta}_{ijl}(t) \ are \ the \ same \ as \ those \ in \ (11.25), \ otherwise, \end{cases} \tag{11.26}$$

where $\widehat{\Phi}(t) = \widehat{\delta}_{0il}(t - \tau_{0il}) \cdot \left|x_{il}(t) - x_{il}^*\right| - \sum_{j \in O_l(i)} \widehat{\delta}_{ijl}(t) \cdot \left|x_{il}(t) - x_{il}^*\right|$.

(2) When $i \in DL, l \in D(i)$, define $\left|x_{\varpi_2 l}^* - x_{\varpi_2 l}(t)\right| = \max \left|x_{il}^* - x_{il}(t)\right|$, where $\varpi_2 \in DL$. Then, $u_{ijl}(t) = \widehat{\delta}_{ijl}(t) \cdot \left|x_{il}(t) - x_{il}^*\right|$ with $\widehat{\delta}_{ijl}(t)$ satisfying that

$$\begin{cases} \widehat{\delta}_{jil}(t) = 0, \ \widehat{\delta}_{ijl}(t) = 0, & if \ i \ne \varpi_2, \\ \widehat{\delta}_{jil}(t) = \frac{\lambda_2(DL)}{\lambda_2(\varpi_2)} \delta_{jil}(t), \ \widehat{\delta}_{ijl}(t) = \frac{\lambda_2(DL)}{\lambda_2(\varpi_2)} \delta_{ijl}(t), & if \ i = \varpi_2, \end{cases} \tag{11.27}$$

where $\delta_{jil}(t)$, $\delta_{ijl}(t)$ are given in (11.13) and (11.23), $\lambda_2(DL)$ is the number of all nodes in DL, and $\lambda_2(\varpi_2) = \{The \ number \ of \ all \ nodes \ i : i = \varpi_2, \ and \ i \in DL\}$, $i, j = 1, 2, \ldots, N-1$. Moreover, we define that

$$\begin{cases} \widehat{\Phi}^1(t) = x_{il}^* - x_{il}(t), \ if \ x_{il}^* < x_{il}(t) \le \overline{x}_{il}, \ and \ \widehat{\Phi}^1(t) < x_{il}^* - x_{il}(t), \\ \widehat{\Phi}^1(t) = x_{il}^* - x_{il}(t), \ if \ \underline{x}_{il} \le x_{il}(t) < x_{il}^*, \ and \ \widehat{\Phi}^1(t) > x_{il}^* - x_{il}(t), \\ \widehat{\Phi}^1(t) = 0, \qquad if \ x_{il}(t) = x_{il}^*, \\ \widehat{\delta}_{jil}(t), \ \widehat{\delta}_{ijl}(t) \ are \ the \ same \ as \ those \ in \ (11.27), \ otherwise, \end{cases} \tag{11.28}$$

where $\widehat{\Phi}^1(t) = \sum_{j \in I_l(i)} \widehat{\delta}_{jil}(t - \tau_{jil}) \cdot \left|x_{il}(t) - x_{il}^*\right| - \sum_{j \in O_l(i)} \widehat{\delta}_{ijl}(t) \cdot \left|x_{il}(t) - x_{il}^*\right|$.

(3) When $i \in SL$, $l \in D(i)$, define $\left|x^*_{\varpi_3 l} - x_{\varpi_3 l}(t)\right| = \max \left|x^*_{il} - x_{il}(t)\right|$, where $\varpi_3 \in SL$. Then, $u_{ijl}(t) = \widehat{\delta}_{ijl}(t) \cdot \left|x_{il}(t) - x^*_{il}\right|$ with $\widehat{\delta}_{ijl}(t)$ satisfying that

$$\begin{cases} \widehat{\delta}_{jil}(t) = 0,\ \widehat{\delta}_{iNl}(t) = 0, & if\ i \neq \varpi_3, \\ \widehat{\delta}_{jil}(t) = \frac{\lambda_3(SL)}{\lambda_3(\varpi_3)}\delta_{jil}(t),\ \widehat{\delta}_{iNl}(t) = \frac{\lambda_3(SL)}{\lambda_3(\varpi_3)}\delta_{iNl}(t), & if\ i = \varpi_3, \end{cases} \tag{11.29}$$

where $\delta_{jil}(t)$, $\delta_{iNl}(t)$ are given in (11.13) and (11.24), $\lambda_3(SL)$ is the number of all nodes in SL, and $\lambda_3(\varpi_3) = \{The\ number\ of\ all\ nodes\ i : i = \varpi_3,\ and\ i \in SL\}$, $i = 1, \ldots, N-1$, $j = 1, \ldots, N$. Moreover, we define the following

$$\begin{cases} \widehat{\Phi}^2(t) = x^*_{il} - x_{il}(t),\ if\ x^*_{il} < x_{il}(t) \le \overline{x}_{il},\ \widehat{\Phi}^2(t) < x^*_{il} - x_{il}(t), \\ \widehat{\Phi}^2(t) = x^*_{il} - x_{il}(t),\ if\ \underline{x}_{il} \le x_{il}(t) < x^*_{il},\ \widehat{\Phi}^2(t) > x^*_{il} - x_{il}(t), \\ \widehat{\Phi}^2(t) = 0, \qquad if\ x_{il}(t) = x^*_{il}, \\ \widehat{\delta}_{jil}(t),\ \widehat{\delta}_{iNl}(t)\ are\ the\ same\ as\ those\ in\ (11.29),\ otherwise, \end{cases} \tag{11.30}$$

where $\widehat{\Phi}^2(t) = \sum\limits_{j \in I_l(i)} \widehat{\delta}_{jil}(t - \tau_{jil}) \cdot \left|x_{il}(t) - x^*_{il}\right| - \widehat{\delta}_{iNl}(t) \cdot \left|x_{il}(t) - x^*_{il}\right|$.

Remark 11.3 In (11.25), when $i \in PL$, $l \in D(i)$, if $i \neq \varpi_1$, $\widehat{\delta}_{0il}(t) = 0$, and $\widehat{\delta}_{ijl}(t) = 0$ means that there are no control inputs in node i. When $i = \varpi_1$, $\widehat{\delta}_{0il}(t) = \frac{\lambda_1(PL)}{\lambda_1(\varpi_1)}\delta_{0il}(t)$, and $\widehat{\delta}_{ijl}(t) = \frac{\lambda_1(PL)}{\lambda_1(\varpi_1)}\delta_{ijl}(t)$ means that there are control inputs in node i, and the controllers of node i at t ($t \in (mK_1 + K_2, (m+1)K_1]$) are $\frac{\lambda_1(PL)}{\lambda_1(\varpi_1)}$ times of that at $t = mK_1 + K_2$ ($m \in \{0, 1, 2, \ldots\}$). As a result, those who control the resources of nodes j ($j \neq \varpi_1$) are evenly reassigned to those needed nodes, i.e., nodes i ($i = \varpi_1$). That is, the limited control resource can be saved. Equations (11.27) and (11.29) have similar explanations.

11.4.2 Convergence Analysis

Theorem 11.1 *Discuss System (11.1). With Assumption 11.1, when $t \in (mK_1, mK_1 + K_2]$ ($m \in \{0, 1, 2, \ldots\}$), if the learning control inputs$u_{ijl}(t)$ are created in part 1 of control scheme, the restrictions (11.2) and (11.3) hold, and the convergence of state variables $x_{il}(t)$ can be realized.*

Proof The proof will be completed with two parts.

(1) Prove the limitations (11.2) and (11.3).

According to Property 11.1, one can achieve the limitation (11.3) by satisfying the requirements (11.9) and (11.10). One only needs to discuss the limitation (11.2) in the following two situations, in accordance with Property 11.2 and Conditions (11.22)–(11.24).

Case 1: $x^*_{il} < x_{il}(t) \le \overline{x}_{il}$. In this case, similar to (11.20), one can obtain that $x_{il}(t+1) \le \overline{x}_{il}$. Also, one has from (11.22)

$$
\begin{aligned}
& x_{il}(t+1) \\
& = x_{il}(t) + \delta_{0il}(t-\tau_{0il}) \cdot \left|x_{il}(t) - x_{il}^*\right| - \sum_{j \in O_l(i)} \delta_{ijl}(t) \cdot \left|x_{il}(t) - x_{il}^*\right| \\
& \geq x_{il}(t) + \Xi(t) > x_{il}(t) + \underline{x}_{il} - x_{il}^* > \underline{x}_{il}.
\end{aligned} \tag{11.31}
$$

The mathematical induction suggests that one has that $\underline{x}_{il} \leq x_{il}(t) \leq \overline{x}_{il}$, for $t \in (mK_1, mK_1 + K_2]$, $m \in \{0, 1, \ldots\}$.

Case 2: $\underline{x}_{il} \leq x_{il}(t) < x_{il}^*$. In this case, similar to (11.17), one can obtain that $x_{il}(t+1) \geq \underline{x}_{il}$. Also, one has from (11.22)

$$
\begin{aligned}
& x_{il}(t+1) \\
& = x_{il}(t) + \delta_{0il}(t-\tau_{0il}) \cdot \left|x_{il}(t) - x_{il}^*\right| - \sum_{j \in O_l(i)} \delta_{ijl}(t) \cdot \left|x_{il}(t) - x_{il}^*\right| \\
& \leq x_{il}(t) + \Pi(t) < x_{il}(t) + \overline{x}_{il} - x_{il}^* < \overline{x}_{il}.
\end{aligned} \tag{11.32}
$$

The mathematical induction suggests that one also has that $\underline{x}_{il} \leq x_{il}(t) \leq \overline{x}_{il}$, for $t \in (mK_1, mK_1 + K_2]$, $m \in \{0, 1, \ldots\}$. As a result, the limitation (11.2) holds for $i \in PL,\ l \in D(i)$. Correspondingly, using the similar analysis as above, one can also obtain that the limitation (11.2) holds for $i \in DL$, $l \in D(i)$ and $i \in SL$, $l \in D(i)$. Hence, when $t \in (mK_1, mK_1 + K_2]$ ($m \in \{0, 1, \ldots\}$), the limitation (11.2) holds when the control inputs $u_{ijl}(t)$ satisfy the conditions (11.9)–(11.10), (11.13)–(11.16), and (11.22)–(11.24).

(2) Prove the convergence of state variables $x_{il}(t)$. When $i \in PL,\ l \in D(i)$, according to (11.13) and (11.22), we discuss the convergence of $x_{il}(t)$ in two cases.

Case 1: $x_{il}^* < x_{il}(t) \leq \overline{x}_{il}$. In this case, from (11.22) and (11.31), one has

$$
\begin{aligned}
& x_{il}(t+1) - x_{il}^* \\
& = x_{il}(t) + \delta_{0il}(t-\tau_{0il}) \cdot \left|x_{il}(t) - x_{il}^*\right| - \sum_{j \in O_l(i)} \delta_{ijl}(t) \cdot \left|x_{il}(t) - x_{il}^*\right| - x_{il}^* \\
& \leq x_{il}(t) + \Pi(t) - x_{il}^* < x_{il}(k) - x_{il}^*.
\end{aligned} \tag{11.33}
$$

With (11.22), one has

$$
\begin{aligned}
& x_{il}(t+1) - x_{il}^* \\
& = x_{il}(t) + \delta_{0il}(t-\tau_{0il}) \cdot \left|x_{il}(t) - x_{il}^*\right| - \sum_{j \in O_l(i)} \delta_{ijl}(t) \cdot \left|x_{il}(t) - x_{il}^*\right| - x_{il}^* \\
& \geq x_{il}(t) + \Xi(t) - x_{il}^* > x_{il}(t) + 2x_{il}^* - 2x_{il}(t) - x_{il}^* \\
& = x_{il}^* - x_{il}(t).
\end{aligned} \tag{11.34}
$$

As a result, one can find a constant ς $(0 < \varsigma < 1)$ such that

$$
\left|x_{il}(t+1) - x_{il}^*\right| < \varsigma \left|x_{il}(t) - x_{il}^*\right|. \tag{11.35}
$$

Case 2: $\underline{x}_{il} \le x_{il}(t) < x_{il}^*$. In this case, from (11.22) and (11.32), similar to Case 1, one also can find a constant ς_1 $(0 < \varsigma_1 < 1)$ such that

$$\left|x_{il}(t+1) - x_{il}^*\right| < \varsigma_1 \left|x_{il}(t) - x_{il}^*\right|. \tag{11.36}$$

From (11.35) and (11.36), one can find a constant $\widehat{K}_1$ such that

$$\left|x_{il}(t) - x_{il}^*\right| < \theta, \tag{11.37}$$

$\forall\, t > \widehat{K}_1$, and θ is a given small positive number, $i = 1, \ldots, N-1$. Hence, the convergence of the ideal inventory x_{il}^* is achieved for $i \in PL,\ l \in D(i)$. Accordingly, using an analysis similar to the one described above, it is possible to determine the convergence of $x_{il}(t)$ for $i \in DL,\ l \in D(i)$ and $i \in SL,\ l \in D(i)$. Hence, when $t \in (mK_1, mK_1 + K_2]$ $(m \in \{0, 1, \ldots\})$, the convergence of $x_{il}(t)$ is achieved when the control inputs $u_{ijl}(t)$ satisfy the conditions (11.9)–(11.10), (11.13)–(11.16), and (11.22)–(11.24). □

Theorem 11.2 *Discuss System (11.1). With Assumption 11.1, when* $(t \in (mK_1 + K_2, (m+1)K_1])$ $(m \in \{0, 1, 2, \ldots\})$*, the limitations (11.2) and (11.3) hold, and the convergence of state variables* $x_{il}(t)$ *can be achieved if the learning control inputs* $u_{ijl}(t)$ *are designed in part 2 of the control scheme.*

Proof When $t \in (mK_1 + K_2, (m+1)K_1]$ $(m \in \{0, 1, \ldots\})$, two components will be used to complete the evidence.

(1) Prove the limitations (11.2) and (11.3).

With the conditions (11.9)–(11.10), one can obtain the limitation (11.3) according to Property 11.1. For the limitation (11.2), according to Property 11.2 and conditions (11.22), (11.25) and (11.26), when $i \in PL,\ l \in D(i)$, one only needs to discuss the limitation (11.2) in the following four cases.

Case 1: $x_{il}^* < x_{il}(t) \le \overline{x}_{il}$, and $\widehat{\Phi}^1(t) < x_{il}^* - x_{il}(t)$. In this case, according to (11.22), (11.25) and (11.26), one has $\widehat{\Phi}^1(k) < 0$, which means that $x_{il}(t+1) < x_{il}(t)$. In addition,

$$\begin{aligned}
&x_{il}(t+1)\\
&= x_{il}(t) + \delta_{0il}(t - \tau_{0il}) \cdot \left|x_{il}(t) - x_{il}^*\right| - \sum_{j \in O_l(i)} \delta_{ijl}(t) \cdot \left|x_{il}(t) - x_{il}^*\right|\\
&= x_{il}(t) + x_{il}^* - x_{il}(t) = x_{il}^*.
\end{aligned} \tag{11.38}$$

As a result, one has $\underline{x}_{il} \le x_{il}(t+1) \le \overline{x}_{il}$. According to the mathematical induction, one has that $\underline{x}_{il} \le x_{il}(t) \le \overline{x}_{il}$, for $t \in (mK_1 + K_2, (m+1)K_1]$, $m \in \{0, 1, \ldots\}$.

Case 2: $x_{il}^* < x_{il}(t) \le \overline{x}_{il}$, and $\widehat{\Phi}^1(t) \ge x_{il}^* - x_{il}(t)$. In this case, according to (11.22), (11.25) and (11.26), one also has $\widehat{\Phi}^1(t) < 0$, i.e., $x_{il}(t+1) < x_{il}(t)$. In addition,

$$
\begin{aligned}
& x_{il}(t+1) \\
& = x_{il}(t) + \delta_{0il}(t - \tau_{0il}) \cdot \left| x_{il}(t) - x_{il}^* \right| - \sum_{j \in O_l(i)} \delta_{ijl}(t) \cdot \left| x_{il}(k) - x_{il}^* \right| \\
& \geq x_{il}(t) + x_{il}^* - x_{il}(t) = x_{il}^*.
\end{aligned} \tag{11.39}
$$

As a result, one has $\underline{x}_{il} < x_{il}^* \leq x_{il}(t+1) < x_{il}(t) \leq \overline{x}_{il}$.The mathematical induction indicates that one has that $\underline{x}_{il} \leq x_{il}(t) \leq \overline{x}_{il}$, for $t \in (mK_1 + K_2, (m+1)K_1]$, $m \in \{0, 1, 2, \ldots\}$.

Case 3: $\underline{x}_{il} \leq x_{il}(t) < x_{il}^*$, and $\widehat{\Phi}^1(t) > x_{il}^* - x_{il}(t)$. In this case, according to (11.22), (11.25) and (11.26), one has $\widehat{\Phi}^1(t) > 0$, i.e., $x_{il}(t+1) > x_{il}(t)$. In addition,

$$
\begin{aligned}
& x_{il}(t+1) \\
& = x_{il}(t) + \delta_{0il}(t - \tau_{0il}) \cdot \left| x_{il}(t) - x_{il}^* \right| - \sum_{j \in O_l(i)} \delta_{ijl}(t) \cdot \left| x_{il}(t) - x_{il}^* \right| \\
& = x_{il}(t) + x_{il}^* - x_{il}(t) = x_{il}^*.
\end{aligned} \tag{11.40}
$$

As a result, one has $\underline{x}_{il} \leq x_{il}(t+1) \leq \overline{x}_{il}$. The mathematical induction indicates that one has that $\underline{x}_{il} \leq x_{il}(t) \leq \overline{x}_{il}$, for $t \in (mK_1 + K_2, (m+1)K_1]$, $m \in \{0, 1, \ldots\}$.

Case 4: $\underline{x}_{il} \leq x_{il}(t) < x_{il}^*$, and $\widehat{\Phi}^1(t) \leq x_{il}^* - x_{il}(t)$. In this case, according to (11.22), (11.25) and (11.26), one has $\widehat{\Phi}^1(t) > 0$, i.e., $x_{il}(t+1) > x_{il}(t)$. In addition,

$$
\begin{aligned}
& x_{il}(t+1) \\
& = x_{il}(t) + \delta_{0il}(t - \tau_{0il}) \cdot \left| x_{il}(t) - x_{il}^* \right| - \sum_{j \in O_l(i)} \delta_{ijl}(t) \cdot \left| x_{il}(t) - x_{il}^* \right| \\
& \leq x_{il}(t) + x_{il}^* - x_{il}(t) = x_{il}^*.
\end{aligned} \tag{11.41}
$$

As a result, one has $\underline{x}_{il} \leq x_{il}(t) < x_{il}(k+1) \leq x_{il}^*$. The mathematical induction indicates that one has that $\underline{x}_{il} \leq x_{il}(t) \leq \overline{x}_{il}$, for $t \in (mK_1 + K_2, (m+1)K_1]$, $m \in \{0, 1, \ldots\}$.

As a result, the limitation (11.2) holds for $i \in PL$, $l \in D(i)$. Correspondingly, using a similar analysis as above, one can also obtain that the limitation (11.2) holds for $i \in DL$, $l \in D(i)$ and $i \in SL$, $l \in D(i)$. Hence, when $t \in (mK_1 + K_2, (m+1)K_1]$ ($m \in \{0, 1, \ldots\}$), the limitation (11.2) holds when the control inputs $u_{ijl}(t)$ satisfy the conditions (11.9)–(11.10), (11.22)–(11.24), and (11.25)–(11.30).

(2) Prove the convergence of state variables $x_{il}(t)$. When $i \in PL$, $l \in D(i)$, according to (11.22), (11.25), and (11.26), we discuss the convergence of $x_{il}(t)$ in four cases.

Case 1: $x_{il}^* < x_{il}(t) \leq \overline{x}_{il}$, $\widehat{\Phi}^1(t) < x_{il}^* - x_{il}(t)$. From (11.38), one has

$$
x_{il}(t+1) - x_{il}^* = 0. \tag{11.42}
$$

Case 2: $x_{il}^* < x_{il}(t) \leq \overline{x}_{il}$, $\widehat{\Phi}^1(t) \geq x_{il}^* - x_{il}(t)$. From (11.39), one has

$$0 \le x_{il}(t+1) - x_{il}^* < x_{il}(t) - x_{il}^*. \tag{11.43}$$

Case 3: $\underline{x}_{il} \le x_{il}(t) < x_{il}^*$, $\widehat{\Phi}^1(t) > x_{il}^* - x_{il}(t)$. From (11.40), one has

$$x_{il}(t+1) - x_{il}^* = 0. \tag{11.44}$$

Case 4: $\underline{x}_{il} \le x_{il}(t) < x_{il}^*$, $\widehat{\Phi}^1(t) \le x_{il}^* - x_{il}(t)$. According to (11.41), one has

$$0 \ge x_{il}(t+1) - x_{il}^* > x_{il}(t) - x_{il}^*. \tag{11.45}$$

From (11.42)–(11.45), one can find a constant ς_2 $(0 < \varsigma_2 < 1)$ such that

$$\left|x_{il}(t+1) - x_{il}^*\right| < \varsigma_2 \left|x_{il}(t) - x_{il}^*\right|. \tag{11.46}$$

That is, one can find a constant $\widehat{K}_2$ such that

$$\left|x_{il}(t) - x_{il}^*\right| < \theta, \tag{11.47}$$

$\forall\ t > \widehat{K}_2$ and θ is a given small positive number, $i = 1, 2, \ldots, N-1$. As a result,$i \in PL, l \in D(i)$ achieves the convergence of the ideal inventory x_{il}^*. Accordingly, using an analysis similar to the one described above, it is possible to determine the convergence of $x_{il}(t)$for both $i \in DL,\ l \in D(i)$ and $i \in SL,\ l \in D(i)$. As a result, when $t \in (mK_1 + K_2, (m+1)K_1]$ $(m \in \{0, 1, 2, \ldots\})$, the convergence of $x_{il}(t)$ is achieved if the control inputs $u_{ijl}(t)$ are designed in part 2 of the control scheme. Then, let $\widehat{K} = \max\left\{\widehat{K}_1, \widehat{K}_2\right\}$, combining the two results in Theorems 11.1 and 11.2, one can conclude that the convergence of $x_{il}(t)$ is achieved if the control inputs $u_{ijl}(t)$ are designed in control scheme, $t = 1, 2, 3, \ldots$. □

Remark 11.4 Compared with Tang et al. (2020), Zhang et al. (2017), our control scheme makes a dual contribution: (1) It can save the limited control resource. In this scheme, when $t \in (mK_1 + K_2, (m+1)K_1]$ $(m \in \{0, 1, 2, \ldots\})$, the limited control resources of all nodes are reallocated to those with slow convergence rates, which can use the entire control resources sensibly since some nodes may not need any control resources in the time interval due to their fast convergence rates. (2) It can improve the convergence rates of all nodes. The following example shows the convergence rates of all nodes for our control scheme and the conventional ones. It is clear that the control scheme can increase the convergence rates of all nodes by first accelerating the convergence rates of those nodes that are required.

11.5 Simulation Example

In the section, an example will be shown to demonstrate the effectiveness of Theorems 11.1 and 11.2.

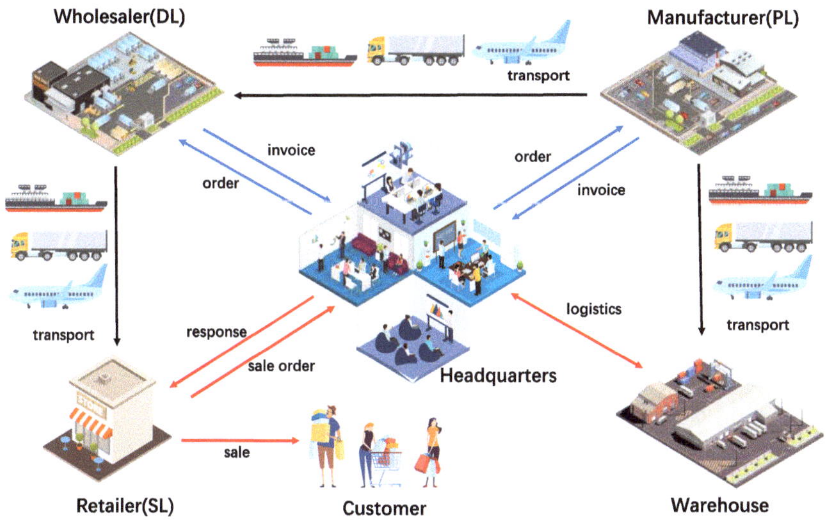

Fig. 11.2 A skeleton of a supply chain network

Example 11.1 A supply chain network's basic structure is depicted in Fig. 11.2. The supply chain network is shown in Fig. 11.2 to have a three-layer structure (see Fig. 11.3). The factory serves as the PL in this network and creates autos as products. Every plant type has a different vehicle model and operating cost. The wholesaler is no longer active. The wholesalers purchase cars from manufacturers and sell them to specific merchants in the wholesale market. Specifically, retailers are on SL. Applying data from Kulkarni and Kulkarni (2020), we choose at random 90 trading periods for three factories (F1,...,F3), four wholesalers (W1,...,W4), and four retailers (R1,...,R4). It's well known that a supply chain operation could have a bullwhip effect (Lee et al. 1997). In order to satisfy the downstream consumers, the upstream supplier constantly maintains an excess of inventory. As a result, there must be a significant amount of safety stock from merchants to the factories, which is a waste of inventory resources and money. Therefore, the upstream suppliers should modify the production strategy to get closer to an optimum inventory in order to save manufacturing expenses and storage costs.

For the model (11.1) of Fig. 11.3, the following list of parameters includes some. The lower inventory bound $\underline{x}_{ij} = 1$, the upper inventory bound $\overline{x}_{ij} = 140$, and the ideal inventory $x^*_{ij} = d_k = 14$. The initial backlog $y_{il}(0)$ of goods satisfy that $y_{il}(0) > 0,\ i \in SL$. And the external need $\omega_{il}(k) = 3$ for $i \in SL,\ t = 0, 1, 2 \ldots$. From Fig. 11.3, for each node i $(i = 1, 2, \ldots, 11)$, one can obtain its $I_l(i)$ and $O_l(i)$. For example, $I_l(5) = \{1, 3\}$, and $O_l(5) = \{8, 11\}$. In (11.10), define $\varepsilon_{il} = 4$, for $i = 1, 2, \ldots, N-1$. In control scheme, $K_1 = 20,\ \ K_2 = 6$, i.e., the time sequence $\{1, 2, 3, \ldots\}$ is divided into periodic parts $\{1, 2, \ldots, 10\}$, $\{11, 12, \ldots, 20\}, \ldots$. When $t \in (10m, 10m + 7]$ $(m \in \{0, 1, 2, \ldots\})$, the controllers $u_{ijl}(t)$ are designed to satisfy conditions (11.14)–(11.16) and (11.22)–(11.24). When

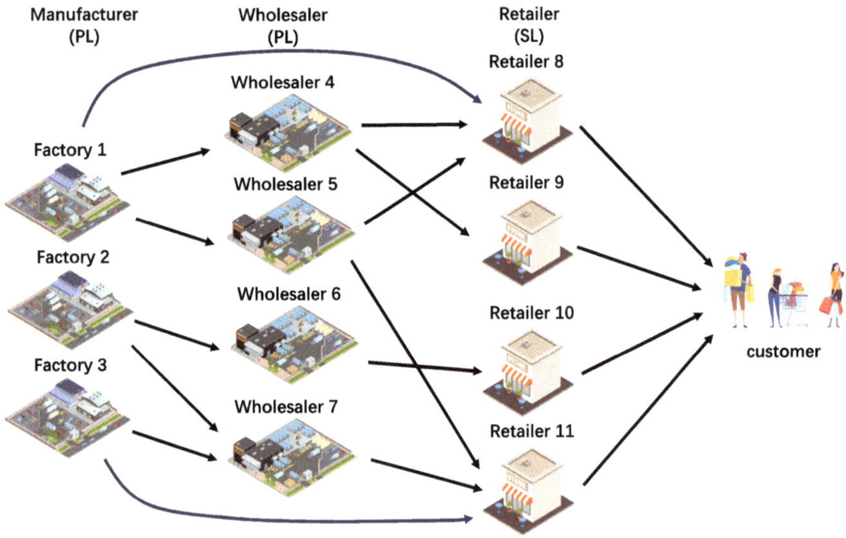

Fig. 11.3 A three-layer structure of a supply chain network

$t \in (10m + 7, 10(m + 1)]$ ($m \in \{0, 1, 2, \ldots\}$), the controllers $u_{ijl}(t)$ are designed to satisfy conditions (11.25)–(11.30). In control scheme, choose appropriate $\delta_{ijl}(t)$ in (11.13) to ensure that conditions (11.14)–(11.16) and (11.22)–(11.30) are satisfied. The trajectories of states $x_{il}(t)$ are presented in Fig. 11.4, and the trajectories of states $y_{il}(t)$ are shown in Fig. 11.5. From Fig. 11.4, one can see that, with our control scheme, all states $x_{il}(t) \in [2, 120]$ and the convergence of the ideal inventory x_{il}^* of states $x_{il}(t)$ can be obtained. From Fig. 11.5, the limitation (11.3) holds when the control input $u_{iNl}(t)$ is designed to satisfy condition (11.10).

Table 11.1 provides a comparison between our approach and a conventional one in order to highlight the benefits of our control scheme. Table 11.1 shows that our control strategy has a faster convergence rate than the conventional one for all states in the system (11.1). Here, the traditional one is a scheme only with (11.22)–(11.24). There are two potential causes: (1) The idle transport resources can be redistributed to the required nodes since certain nodes do not need to be controlled. (2) In order to boost the convergence rate of all nodes, the nodes with slow convergence rates receive extra transport resources to convey commodities, which can hasten their convergence rates. Additionally, we incorporate the major parameters listed in Metters (1997) into a cost function to assess the efficacy of our plan in a genuine supply chain.

In addition, the cost function for each trading time slot C_k is shown below:

$$C_t = \sum_{l \in D(i)} \Big\{ c_l(x_{lt} - x_{lt-1})_+ + h_l(x_{lt} - d_t)_+ + \pi_l(d_t - x_{lt})_+ \\ - V_l\big[d_t - (d_t - x_{lt})_+\big]_+ - \alpha_l V_l\big[(x_{lt-1} - x_{lt})_+ - d_t\big]_+ \Big\}. \tag{11.48}$$

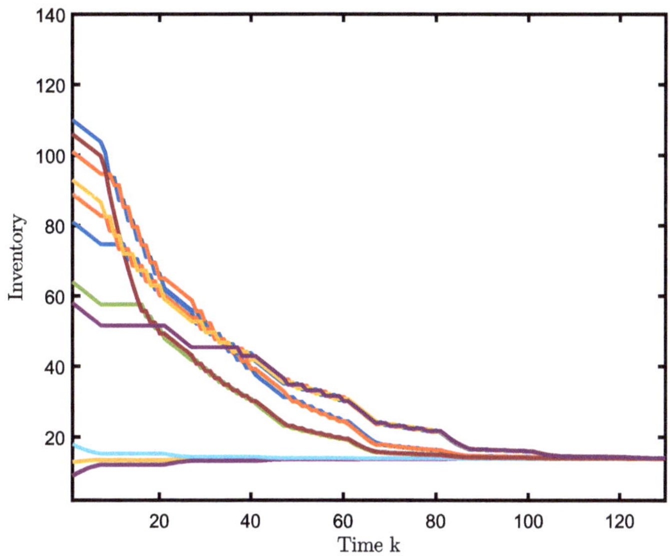

Fig. 11.4 The trajectories of states $x_{il}(k)$ in system (11.1) with control scheme

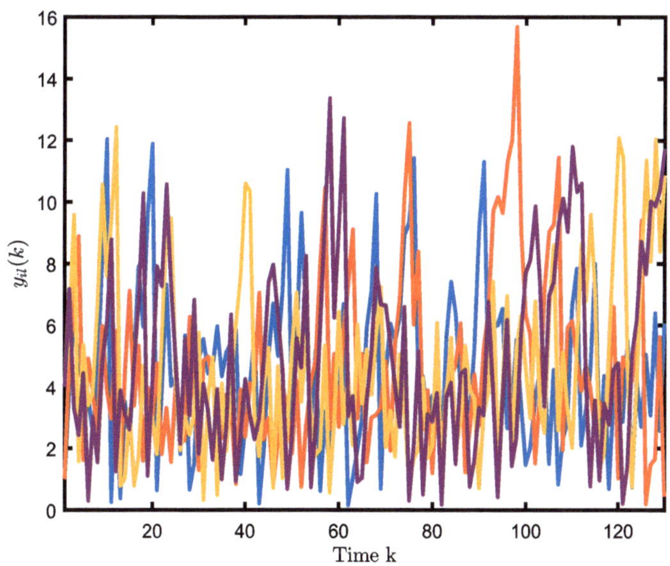

Fig. 11.5 The trajectories of states $y_{il}(k)$ in system (11.1) with control scheme

Table 11.1 The comparison of the value $\sum_{i=1}^{11} |x_{il}(t) - x_{il}^*|$ with different ks between our control scheme and the traditional scheme

k	20	40	60	80	100	120
$\sum_{i=1}^{11} \left\|x_{il}(t) - x_{il}^*\right\|$ in the traditional scheme	432.3	265.98	128.3	30.8	1.247	0.038
$\sum_{i=1}^{11} \left\|x_{il}(t) - x_{il}^*\right\|$ in control scheme	460.8	200.9	100.07	38.42	9.88	2.503

The production cost is defined by the first term of the cost function, where c_l is the cost in dollars of creating product l, and x_{lt} and x_{lt-1} are the quantities of inventory for l in time slots t and $t-1$, respectively. The second term, h_l, which stands for holding cost per unit, denotes the cost of storage. The quantity of demand during period t is d_t.The third component is the penalty component, where π_l represents the financial cost of unmet demand per unit in addition to lost profits. The final term, where V_l denotes the revenue per unit and α_l denotes the discount factor, represents the revenue in the supply chain. In other words, for the entire supply chain, if the operators produce more goods than the market requires, they will not only incur higher storage costs, but also lose money by selling the excess inventory at a loss. Additionally, if the product supply falls short of market demand, the operator will be punished. In (11.48), we assume the producing cost $c = 15500$ for a vehicle, and selling revenue $V = 16000$. The storage cost $h = 120$, unsatisfied penalty $\pi = 1200$, and the pricing discount rate $\alpha = 0.7$.

Here, we use the aggregated inventory produced by our control scheme, the conventional one, and the initial dataset (baseline) to apply the cost function (11.48) and assess the correlation between inventory and cost. Figure 11.6 displays the total cost (income) for each operator in the suggested supply chain. According to Fig. 11.6, it is expensive for supply chain operators to make or buy goods to serve as the starting inventory for the first trade period. Then, compared to the baseline and conventional schemes, our scheme can generate more consistent revenue. It means that if the supply chain operator quickly changes the inventory amount to the ideal inventory, they will earn a significant income.

Remark 11.5 Compared with Chan et al. (2018), Liu et al. (2019), the contribution of this study is: (1) The mentioned model (11.1), which is a three-layer multi-layer SCN, is more general. A comprehensive description of the network transmission relationship between nodes may also be found in the model (11.1). (2) By creating a new control method, our control strategy, with a more consistent quantity of revenue, makes it possible to quickly converge on the ideal inventory levels (see Fig. 11.6).

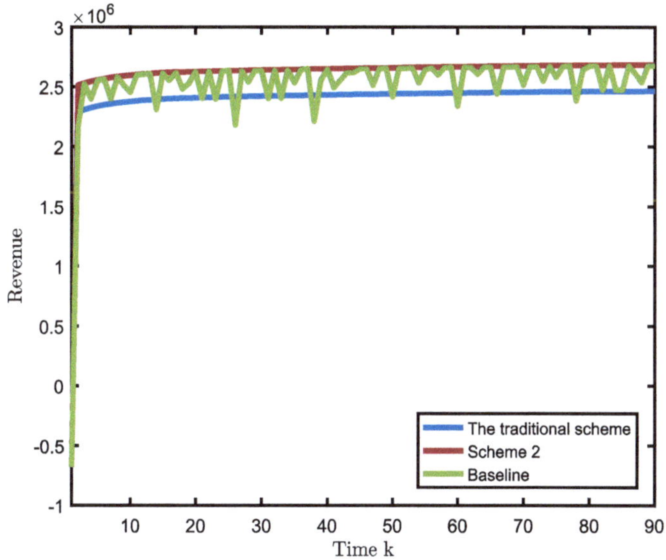

Fig. 11.6 The revenues of the bullwhip effect in system (11.1) with three schemes

11.6 Summary

This chapter has talked about the convergence of a multi-layer SCN with three layers. First, a three-layer multi-layer supply chain network has been constructed, and several state-related limitations have been taken into account. The convergence of the aforementioned SCN was then attained by creating suitable control inputs, which satisfied all state restrictions. Additionally, a new learning control scheme has been implemented in order to accelerate the convergence of the ideal inventory, and the simulation compares its convergence rate with that of conventional schemes. It has been demonstrated that our control scheme converges more quickly than other schemes, and the example addresses some of the potential causes.

References

Alessandri A, Gaggero M, Tonelli F (2011) Min-max and predictive control for the management of distribution in supply chains. IEEE Trans Control Syst Technol 19(5):1075–1089

Chan CK, Zhou Y, Wong KH (2018) A dynamic equilibrium model of the oligopolistic closed-loop supply chain network under uncertain and time-dependent demands. Transp Res E Logist Transp Rev 118:325–354

de Albuquerque GA, Maciel P, Lima RMF, Magnani F (2013) Strategic and tactical evaluation of conflicting environment and business goals in green supply chains. IEEE Trans Syst Man Cybern Syst 43(5):1013–1027

Dubey R, Gunasekaran A, Childe SJ, Papadopoulos T, Blome C, Luo ZW (2019) Antecedents of resilient supply chains: an empirical study. IEEE Trans Eng Manag 66(1):8–19
Fu DF, Zhang HT, Yu Y, Ionescu CM, Aghezzaf EH, Keyser R (2019) A distributed model predictive control strategy for the bullwhip reducing inventory management policy. IEEE Trans Ind Informat 15(2):932–941
Fu DF, Zhang HT, Dutta A, Chen GR (2020) A cooperative distributed model predictive control approach to supply chain management. IEEE Trans Syst Man Cybern Syst 50(12):4894–4904
Goodarzian F, Hosseini-Nasab H, Munuzuri J, Fakhrzad MB (2020) A multi-objective pharmaceutical supply chain network based on a robust fuzzy model: a comparison of meta-heuristics. Appl Soft Comput 92:106331
Ha AY, Tong S (2008) Contracting and information sharing under supply chain competition. Manag Sci 54(4):701–715
He Y, Huang HF, Li D (2020) Inventory and pricing decisions for a dual-channel supply chain with deteriorating products. Oper Res 20(3):1461–1503
Javid AA, Azad N (2010) Incorporating location, routing and inventory decisions in supply chain network design. Transp Res E Logist Transp Rev 46(5):582–597
Kulkarni S, Kulkarni C (2020) Dataset for leveraging blockchain to improve automotive supply chain traceability. https://data.mendeley.com/datasets/n24z7r2z28/2
Lee HL, Padmanabhan V, Whang S (1997) Information distortion in a supply chain: the bullwhip effect. Manag Sci 43(4):546–558
Liu YK, Chen YJ, Yang GQ (2019) Developing multiobjective equilibrium optimization method for sustainable uncertain supply chain planning problems. IEEE Trans Fuzzy Syst 27(5):1037–1051
Mawandiya BK, Jha JK, Thakkar JJ (2020) Optimal production-inventory policy for closed-loop supply chain with remanufacturing under random demand and return. Oper Res 20(3):1623–1664
Metters R (1997) Quantifying the bullwhip effect in supply chains. J Oper Manag 15(2):89–100
Moyaux T, Chaib-draa B, D'Amours S (2007) Information sharing as a coordination mechanism for reducing the bullwhip effect in a supply chain. IEEE Trans Syst Man Cybern C 37(3):396–409
Müller MA, Grüne L (2016) Economic model predictive control without terminal constraints for optimal periodic behavior. Automatica 70:128–139
Nagurney A, Dong J, Zhang D (2002) A supply chain network equilibrium model. Transp Res E Logist Transp Rev 38(5):281–303
Nagurney A, Yu M, Floden J, Nagurney LS (2014) Supply chain network competition in time-sensitive markets. Transp Res E Logist Transp Rev 70:112–127
Riddalls C, Bennett S (2002) Production-inventory system controller design and supply chain dynamics. Int J Syst Sci 33(3):181–195
Roy A, Gilbert SM, Lai GM (2019) The implications of visibility on the use of strategic inventory in a supply chain. Manag Sci 65(4):1752–1767
Tancrez JS, Lange JC, Semal P (2012) A location-inventory model for large three-level supply chains. Transp Res E Logist Transp Rev 48(2):485–502
Tang Y, Zhang DD, Ho DWC, Qian F (2019) Tracking control of a class of cyber-physical systems via a flexray communication network. IEEE Trans Cybern 49(4):1186–1199
Tang Y, Wu XT, Shi P, Qian F (2020) Input-to-state stability for nonlinear systems with stochastic impulses. Automatica 113:108766
Wan Y, Cao JD, Wen GH (2017) Quantized synchronization of chaotic neural networks with scheduled output feedback control. IEEE Trans Neural Netw Learn Syst 28(11):2638–2647
Wei QL, Li BK, Song RZ (2018) Discrete-time stable generalized self-learning optimal control with approximation errors. IEEE Trans Neural Netw Learn Syst 29(4):1226–1238
Wen GH, Wang PJ, Huang TW, Lü JH, Zhang F (2020) Distributed consensus of layered multi-agent systems subject to attacks on edges. IEEE Trans Circuits Syst I 67(9):3152–3162
Xiong WJ, Yu XH, Liu C, Wen GH, Wen SP (2020) Simplifying complex network stability analysis via hierarchical node aggregation and optimal periodic control. IEEE Trans Neural Netw Learn Syst 32(7):3098–3107

Yu H, Solvang WD (2020) A fuzzy-stochastic multi-objective model for sustainable planning of a closed-loop supply chain considering mixed uncertainty and network flexibility. J Clean Prod 266:121702

Zepeda ED, Nyaga GN, Young GJ (2016) Supply chain risk management and hospital inventory: effects of system affiliation. J Oper Manag 44:30–47

Zhang WB, Tang Y, Huang TW, Kurths J (2017) Sampled-data consensus of linear multi-agent systems with packet losses. IEEE Trans Neural Netw Learn Syst 28(11):2516–2527

Zhu SB, Zhou J, Chen GR, Lu JA (2019) Estimating the region of attraction on a complex dynamical network. SIAM J Control Optim 57(2):1189–1208

Chapter 12
Security of Network Systems Under Cyber-Attack

12.1 Introduction

From the existing literature, one knows that the system robustness and stability are fully considered under time delay (Li and Ren 2021; Xiao et al. 2018), model uncertainties (Meng 2019; Meng and Moore 2017), and system switching process (Xue et al. 2022), etc. In addition, network systems are ubiquitous in real-life, and malicious cyber-attacks often occur in the cyberspace and signal transmission process. Malicious attacks might result in data dropout (Bu et al. 2013; Liu and Ruan 2017, 2018) and system failure (Li et al. 2015), etc. How can systems be kept stable and safe from various malicious attacks? This problem attracts attention to the secure control, and a number of security control protocols are proposed. For example, Wan et al. (2017) proposes a security control strategy based on observation and switching topology. In He et al. (2018), the impulsive control method is used to achieve the secure synchronization of systems. Further, the security problems of multiple cyber-attacks are studied by using the cooperative control scheme (Tahoun 2021) and PID control strategy (Zhao et al. 2021). These results provide effective control strategies for achieving system security. However, some sufficient conditions, such as conditions (16a) and (16b) in Ref. Zhao et al. (2021), are a little complicated and hard to check in the real application if the dimension of the model is big. In addition, the combination of multiple cyber-attacks increases the difficulty of the theoretical analysis. Hence, to achieve some simple security conditions of systems with multiple cyber-attacks, we try to find applicable approaches for analyzing and obtaining some concise conditions.

In the past several decades, ILC has been widely studied because of its advantages of simple design and strong adaptability. This control method is applied in bionic robotic fish (Wang et al. 2020), large-scale urban road networks (Ren et al. 2020), batch processes (Zhou et al. 2019), network systems (Fu et al. 2017; Xiong et al. 2022), etc. In contrast to many traditional control approaches, the ILC strategy achieves tracking by using the information from prior knowledge. And it does not

W. Xiong et al., *Iterative Learning Control for Network Systems Under Constrained Information Communication*, Intelligent Control and Learning Systems 12,
https://doi.org/10.1007/978-981-97-0926-7_12

heavily rely on mathematical models. Thus, many ILC-based tracking techniques are proposed, such as robust learning tracking (Meng 2019; Meng and Moore 2017; Meng and Zhang 2021), finite-iteration tracking (Xiong et al. 2020, 2021), cooperative tracking (Deng et al. 2020; Zhang and Meng 2021), etc. It is clear that ILC algorithms provide another effective way to solve high-precision tracking control problems. Until now, the research on ILC still attracts the interest of science and engineering technology communities.

Actually, with the ILC approach, some reasonable control algorithms (Ren et al. 2022; Xiong et al. 2022; Yu et al. 2022) are designed to solve the security problem. For example, two criteria are proposed to analyze the security of discrete nonlinear systems under injection attacks (Xiong et al. 2022). The security problem is considered by data-driven control under the deception and false data injection attacks (Yu et al. 2022). Furthermore, denial-of-service (DoS) attacks are also common in cyberspace, which attracts the interest of researchers (Guo et al. 2021; Yang et al. 2020). However, since there are different forms of DoS attacks, the above methods in Xiong et al. (2022), Yu et al. (2022) are ineffective in dealing with DoS attacks. As a result, our main purpose in this study is to design ILC strategies to discuss system security under random deception and DoS attacks. Note that when deception and DoS attacks occur simultaneously, it is hard to obtain the system's actual output since the attacks might falsify the observations. As a result, a major difficulty is how to design ILC strategies according to the output measurements with attacks. Also, the theoretical analysis is difficult when DoS and deception attacks take place randomly. Thus, it is meaningful to consider the system security problem under random deception and DoS attacks by designing an ILC algorithm. An output estimation formula is proposed under the DoS and deception attacks to achieve system security. Then, the proper ILC algorithms are designed according to the output estimation formula.

12.2 Problem Formulation

For a discrete-time network system, the dynamic of each node is described as

$$\begin{cases} x_i(t+1,k) = Ax_i(t,k) + Bu_i(t,k), \quad t \in \mathcal{T}^-, \\ y_i(t,k) = Cx_i(t,k), \quad t \in \mathcal{T}, \end{cases} \tag{12.1}$$

where k denotes the kth iteration process, $x_i(0,k)$ is the kth initial state. $x_i(t,k) \in \mathbb{R}^n$ is the state vector, $u_i(t,k) \in \mathbb{R}^p$ is the input vector, and $y_i(t,k) \in \mathbb{R}^m$ is the output vector. The coefficient matrices A, B, C, and D in (12.1) are with appropriate dimensions. $\mathcal{T} = \{1,\ldots,T\}$ and $\mathcal{T}^- = \{1,\ldots,T-1\}$ represent time intervals, T is the fixed terminal time.

In this chapter, the iterative learning process under cyber-attacks is shown in Fig. 12.1. It is worth mentioning that security analysis with injection attacks is investigated in Xiong et al. (2022). However, other types of cyber-attacks are common as

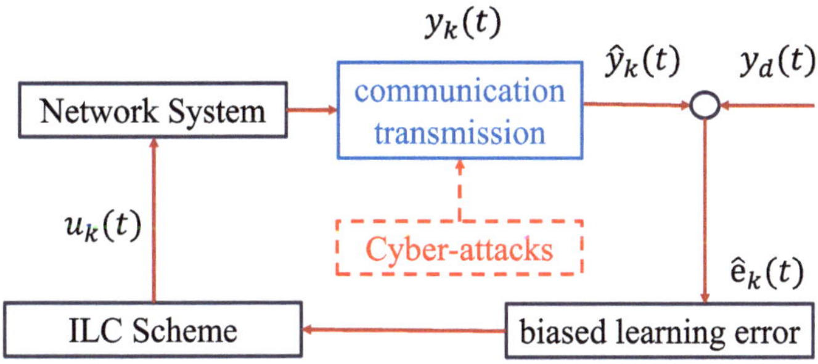

Fig. 12.1 Iterative learning processes under cyber-attacks

well. Perhaps it is more realistic and meaningful to consider multiple attacks at the same time. Inspired by Zhao et al. (2021), we consider the data transmission over the measurement channel. The following equation describes the impact pattern of random deception and DoS attacks:

$$\hat{y}_i(t,k) = y_i(t,k) + \alpha_{i,k}^t \beta_{i,k}^t \mu_i(t,k) + \alpha_{i,k}^t \big(1 - \beta_{i,k}^t\big) \nu_i(t,k), \tag{12.2}$$

where $\hat{y}_i(t,k)$ is the received signal by the observer subject to attacks. $\mu_i(t,k) \in \mathbb{R}^m$ and $\nu_i(t,k) \in \mathbb{R}^m$ denote the deception attack and the DoS attack, respectively. Specifically, the deception attack is described as

$$\mu_i(t,k) = -y_i(t,k) + \xi_i(t,k), \tag{12.3}$$

where $\xi_i(t,k)$ is an arbitrary bounded signal, i.e., $\|\xi_i(t,k)\| \le \xi_{i,b}$, $\xi_{i,b} > 0$. In addition, the DoS attack mode is as follows:

$$\nu_i(t,k) = -y_i(t,k). \tag{12.4}$$

Moreover, Bernoulli stochastic variables $\alpha_{i,k}^t$ and $\beta_{i,k}^t$ in (12.2) are mutually uncorrelated, and satisfy

$$\begin{aligned} &\text{Prob}\big\{\alpha_{i,k}^t = 1\big\} = \mathbb{E}\big(\alpha_{i,k}^t\big) = \bar{\alpha}_i, \ \ 0 \le \bar{\alpha}_i < 1, \\ &\text{Prob}\big\{\beta_{i,k}^t = 1\big\} = \mathbb{E}\big(\beta_{i,k}^t\big) = \bar{\beta}_i, \ \ 0 \le \bar{\beta}_i < 1, \end{aligned}$$

where $\bar{\alpha}$ and $\bar{\beta}$ are two known constants. Moreover, $\alpha_{i,k_1}^{t_1}$ and $\alpha_{i,k_2}^{t_2}$ are mutually independent for all $k_1 \ne k_2$ and $t_1, t_2 \in \mathcal{T}$, and so is $\beta_{i,k}^t$. According to (12.2), (12.3), and (12.4), one obtains

$$\hat{y}_i(t,k) = \big(1 - \alpha_{i,k}^t\big) y_i(t,k) + \alpha_{i,k}^t \beta_{i,k}^t \xi_i(t,k). \tag{12.5}$$

Define the tracking error as $e_i(t,k) = y_d(t) - y_i(t,k)$, where $y_d(t) \in \mathbb{R}^p$ is the desired output or trajectory. The trajectory tracking will be discussed by adjusting the control input with the tracking error. To ensure that systems can achieve the safe tracking target, we first introduce the definition of security in the following.

Definition 12.1 Let the desired security level be specified as $\sigma > 0$. The network system (12.1) is said to achieve the σ-secure in the meaning sense if the inequality $\lim_{k\to\infty} \mathbb{E}(\|e_i(t,k)\|) \le \sigma$ holds, $\forall\, t \in \mathcal{T}$.

Thus, the purpose of this chapter is to analyze the security of the system in accordance with Definition 12.1 and reasonable ILC scheme.

12.3 ILC Scheme Design

In general, the real output series $\{y_i(t,k)\}$ might not be directly obtained and used due to the cyber-attacks. And the incorrect measurement signals are probably applied to adjust control strategies. As a result, the tracking target is difficult to achieve. Fortunately, one can use the desired trajectory or output to compensate for the incorrect output. Therefore, similar to Liu and Ruan (2017, 2018), the available or estimation output is designed as

$$\check{y}_i(t,k) = \left(1 - \alpha_{i,k}^t\right) y_i(t,k) + \alpha_{i,k}^t \beta_{i,k}^t \xi_i(t,k) + \alpha_{i,k}^t y_d(t), \tag{12.6}$$

and the available tracking error is defined as $\check{e}_i(t,k) \triangleq y_d(t) - \check{y}_i(t,k)$.

According to (12.6), for the discrete-time network system (12.1), the ILC strategy is constructed as follows:

$$\begin{aligned} &u_i(t,k+1) \\ &= u_i(t,k) + K \sum_{j\in N_i} \alpha_{ij} \left[\check{y}_j(t+1,k) - \check{y}_i(t+1,k)\right] + \alpha q_i K \check{e}_i(t+1,k), \end{aligned} \tag{12.7}$$

where K is a learning gain matrix, α denotes the learning parameter. $q_i > 0$ indicates that the node v_i is able to access the information of the desired output. In general, the initial input $u_0(t)$ is chosen arbitrarily. The iterative learning strategy (12.7) is similar to (3) in Meng and Zhang (2021). Nevertheless, the difference is that our ILC (12.7) uses the available output $\check{y}_i(t,k)$ rather than the actual output $y_i(t,k)$. It means the impact of cyber-attacks is fully considered.

Remark 12.1 Compared with Definition 2 in Zhao et al. (2021), Definition 12.1 proposes a security concept from the perspective of iteration, which is appropriate for repetitive systems. Moreover, Definition 12.1 reflects the influence of random attacks and can clearly depict the security level with parameter σ.

Let $x(t,k) = (x_1^{\mathrm{T}}(t,k), \ldots, x_N^{\mathrm{T}}(t,k))^{\mathrm{T}} \in \mathbb{R}^{nN}$, $y(t,k) = (y_1^{\mathrm{T}}(t,k), \ldots, y_N^{\mathrm{T}}(t,k))^{\mathrm{T}} \in \mathbb{R}^{mN}$ and $u(t,k) = (u_1^{\mathrm{T}}(t,k), \ldots, u_N^{\mathrm{T}}(t,k))^{\mathrm{T}} \in \mathbb{R}^{pN}$. According to (12.1), one obtains

$$\begin{cases} x(t+1,k) = \tilde{A}x(t,k) + \tilde{B}u(t,k), \ t \in \mathcal{T}^-, \\ y(t,k) = \tilde{C}x(t,k), \ t \in \mathcal{T}, \end{cases} \tag{12.8}$$

where $\tilde{A} = I_N \otimes A$, $\tilde{B} = I_N \otimes B$, and $\tilde{C} = I_N \otimes C$.

Analogously, the estimation output (12.6) can be rewritten as

$$\check{y}(t,k) = \big(I_{mN} - \mathcal{A}_k^t\big)y(t,k) + \mathcal{A}_k^t\mathcal{B}_k^t\xi(t,k) + \mathcal{A}_k^t(\mathbf{1} \otimes y_d(t)), \tag{12.9}$$

where $\check{y}(t,k) = (\check{y}_1^{\mathrm{T}}(t,k), \ldots, \check{y}_N^{\mathrm{T}}(t,k))^{\mathrm{T}}$, $\xi(t,k) = (\xi_1^{\mathrm{T}}(t,k), \ldots, \xi_N^{\mathrm{T}}(t,k))^{\mathrm{T}}$, $\mathcal{A}_k^t = \mathrm{diag}\big(I_m \otimes \alpha_{1,k}^t, \ldots, I_m \otimes \alpha_{N,k}^t\big)$, and $\mathcal{B}_k^t = \mathrm{diag}\big(I_m \otimes \beta_{1,k}^t, \ldots, I_m \otimes \beta_{N,k}^t\big)$. Equation (12.9) depicts the actual measurable influence of two types of attacks on the entire network system. Furthermore, the learning control strategy for the network system (12.8) can be written as

$$u(t,k+1) = u(t,k) + \big[(\mathcal{L}(\mathcal{G}) + \alpha Q) \otimes K\big]\check{e}(t+1,k), \tag{12.10}$$

where $\check{e}(t,k) = (\check{e}_1^{\mathrm{T}}(t,k), \ldots, \check{e}_N^{\mathrm{T}}(t,k))^{\mathrm{T}}$. And (12.10) fully utilizes the available measurement error to train the control scheme, making it more suitable for cyber-attacked systems.

Remark 12.2 In the control strategy (12.7), we assume that each node has the same learning gain matrix K and step parameter α for the sake of analysis. In fact, the learning control parameters for each node can be different. That is, the learning gain parameters of each node are K_i and α_i, $i \in \{1, \ldots, N\}$. Then, control strategy (12.10) becomes

$$u(t,k+1) = u(t,k) + \bar{K}\big[(\mathcal{L}(\mathcal{G}) + \Lambda Q) \otimes I_m\big]\check{e}(t+1,k), \tag{12.11}$$

where $\bar{K} = \mathrm{diag}(K_1, \ldots, K_N)$ and $\Lambda = \mathrm{diag}(\alpha_1, \ldots, \alpha_N)$. At this point, the iterative learning control strategy is only formally different and essentially unchanged.

The following assumption is needed for the theoretical analysis to obtain this target.

Assumption 12.1 Each iteration's initial state does not change, i.e., $x_i(0,k) = x_i(0,k+1) = x_d(0)$.

Definition 12.2 $x = (x_1, \ldots, x_n)^{\mathrm{T}}$ and $y = (y_1, \ldots, y_n)^{\mathrm{T}}$ are two real vectors, then the partial order relation $\prec$ is defined as $x \prec y$ if and only if $x_i \leq y_i$ for all $i \in \{1, \ldots, n\}$.

12.4 Main Results

This section focuses on analyzing the security problem caused by cyber-attacks. The robust convergence and boundedness of system trajectories will be discussed to illustrate the impact of malicious attacks. The results show that our strategy proposes a possible scheme to realize tracking and security.

12.4.1 Robust Convergence Analysis

Theorem 12.1 *Let Assumption 12.1 hold, and the ILC strategy (12.7) is applied to network system 12.1. The σ-secure tracking is achieved, i.e., $\lim_{k\to\infty} \mathbb{E}(\|e(t,k)\|) \leq \sigma$ if one chooses learning matrix K and step parameter α to satisfy the following inequality:*

$$\mathbb{E}\Big(\Big\| I_{mN} - \tilde{C}\tilde{B}\mathcal{K}\big(I_{mN} - \mathcal{A}_k^t\big)\Big\|\Big) = \varrho < 1, \tag{12.12}$$

where $\mathcal{K} = \big[(\mathcal{L}(\mathcal{G}) + \alpha Q) \otimes K\big]$.

Proof According to (12.8), one gets

$$\begin{aligned} e(t,k+1) &= \mathbf{1} \otimes y_d(t) - y(t,k+1) \\ &= \mathbf{1} \otimes y_d(t) - y(t,k) + y(t,k) - y(t,k+1) \\ &= e(t,k) - \tilde{C}[x(t,k+1) - x(t,k)], \end{aligned} \tag{12.13}$$

where $e(t,k+1) = (e_1^{\mathrm{T}}(t,k+1), \ldots, e_N^{\mathrm{T}}(t,k+1))^{\mathrm{T}}$.

Let $\Delta x(t,k) = x(t,k+1) - x(t,k)$, then one has

$$\begin{aligned} \Delta x(t,k) &= \tilde{A}\Delta x(t-1,k) + \tilde{B}[u(t-1,k) - u(t-1,k-1)] \\ &= \tilde{A}\Delta x(t-1,k) + \tilde{B}\big[(\mathcal{L}(\mathcal{G}) + \alpha Q) \otimes K\big]\big[\mathbf{1} \otimes y_d(t) - \check{y}(t,k)\big] \\ &= \tilde{A}\Delta x(t-1,k) + \tilde{B}\mathcal{K}\big[e(t,k) + y(t,k) - \check{y}(t,k)\big]. \end{aligned} \tag{12.14}$$

According to (12.9), it is not difficult to know

$$y(t,k) - \check{y}(t,k) = -\mathcal{A}_k^t\mathcal{B}_k^t\xi(t,k) - \mathcal{A}_k^t e(t,k). \tag{12.15}$$

Then, from (12.14) and (12.15), (12.13) becomes

$$\begin{aligned} e(t,k+1) = {} & \Big[I_{mN} - \tilde{C}\tilde{B}\mathcal{K}\big(I_{mN} - \mathcal{A}_k^t\big)\Big]e(t,k) \\ & - \tilde{C}\tilde{A}\Delta x(t-1,k) + \tilde{C}\tilde{B}\mathcal{K}\mathcal{A}_k^t\mathcal{B}_k^t\xi(t,k). \end{aligned} \tag{12.16}$$

From (12.16), by applying mathematical induction, one can obtain

$$e(t,k+1) = \Big[I_{mN} - \tilde{C}\tilde{B}\mathcal{K}\big(I_{mN} - \mathcal{A}_k^t\big)\Big]e(t,k) - \sum_{i=0}^{t-1}\tilde{C}\tilde{A}^i\tilde{B}\mathcal{K}\big(I_{mN} - \mathcal{A}_k^{t-1-i}\big)e(t-1-i,k) + \sum_{i=0}^{t}\tilde{C}\tilde{A}^i\tilde{B}\mathcal{K}\mathcal{A}_k^{t-i}\mathcal{B}_k^{t-i}\xi(t-i,k). \tag{12.17}$$

Taking norm for both sides of equation (12.17),

$$\big\|e(t,k+1)\big\| \le \big\|I_{mN} - \tilde{C}\tilde{B}\mathcal{K}\big(I_{mN} - \mathcal{A}_k^t\big)\big\| \cdot \big\|e(t,k)\big\| + \sum_{i=0}^{t-1}\big\|\tilde{C}\tilde{A}^i\tilde{B}\mathcal{K}\big\| \cdot \big\|I_{mN} - \mathcal{A}_k^{t-1-i}\big\| \cdot \big\|e(t-1-i,k)\big\| + \sum_{i=0}^{t}\big\|\tilde{C}\tilde{A}^i\tilde{B}\mathcal{K}\big\| \cdot \big\|\mathcal{A}_k^{t-i}\big\| \cdot \big\|\mathcal{B}_k^{t-i}\big\| \cdot \big\|\xi(t-i,k)\big\|. \tag{12.18}$$

Taking the expectation of equation (12.18), one has

$$\mathbb{E}\Big(\big\|e(t,k+1)\big\|\Big) \le \mathbb{E}\Big(\big\|I_{mN} - \tilde{C}\tilde{B}\mathcal{K}\big(I_{mN} - \mathcal{A}_k^t\big)\big\|\Big)\mathbb{E}\Big(\big\|e(t,k)\big\|\Big) + \sum_{i=0}^{t-1}\big\|\tilde{C}\tilde{A}^i\tilde{B}\mathcal{K}\big\|\mathbb{E}\Big(\big\|I_{mN} - \mathcal{A}_k^{t-1-i}\big\|\Big)\mathbb{E}\Big(\big\|e(t-1-i,k)\big\|\Big) + \sum_{i=0}^{t}\big\|\tilde{C}\tilde{A}^i\tilde{B}\mathcal{K}\big\|\mathbb{E}\Big(\big\|\mathcal{A}_k^{t-i}\big\| \cdot \big\|\mathcal{B}_k^{t-i}\big\|\Big)\big\|\xi_b\big\|, \tag{12.19}$$

where $\xi_b = ((\mathbf{1} \otimes \xi_{b,1})^{\mathrm{T}}, \ldots, (\mathbf{1} \otimes \xi_{b,N})^{\mathrm{T}})^{\mathrm{T}}$.

Note that $\mathrm{Prob}\{\alpha_{i,k}^t = 1\} = \mathbb{E}(\alpha_{i,k}^t) = \bar{\alpha}_i$, $\mathrm{Prob}\{\beta_{i,k}^t = 1\} = \mathbb{E}(\beta_{i,k}^t) = \bar{\beta}_i$, $\mathcal{A}_k^t$, and $\mathcal{B}_k^t$ are diagonal matrices. The elements of $\mathcal{A}_k^t$ and $\mathcal{B}_k^t$ take values of 0 or 1, then $\|I_{mN} - \mathcal{A}_k^{t-1-i}\| = 1$ or 0. As a result,

$$\mathrm{Prob}\Big\{\big\|I_{mN} - \mathcal{A}_k^{t-1-i}\big\| = 0\Big\} = \mathrm{Prob}\Big\{I_{mN} = \mathcal{A}_k^{t-1-i}\Big\} = \prod_{i=1}^{N}\bar{\alpha}_i,$$

$$\mathrm{Prob}\Big\{\big\|I_{mN} - \mathcal{A}_k^{t-1-i}\big\| = 1\Big\} = 1 - \mathrm{Prob}\Big\{I_{mN} = \mathcal{A}_k^{t-1-i}\Big\} = 1 - \prod_{i=1}^{N}\bar{\alpha}_i.$$

Analogously,

$$\text{Prob}\Big\{\|\mathcal{A}_k^{t-1-i}\|=0\Big\}=\text{Prob}\Big\{O=\mathcal{A}_k^{t-1-i}\Big\}=\prod_{i=1}^{N}(1-\bar{\alpha}_i),$$
$$\text{Prob}\Big\{\|\mathcal{A}_k^{t-1-i}\|=1\Big\}=1-\text{Prob}\Big\{O=\mathcal{A}_k^{t-1-i}\Big\}=1-\prod_{i=1}^{N}(1-\bar{\alpha}_i),$$

and

$$\text{Prob}\Big\{\|\mathcal{B}_k^{t-i}\|=0\Big\}=\text{Prob}\Big\{O=\mathcal{B}_k^{t-i}\Big\}=\prod_{i=1}^{N}(1-\bar{\beta}_i),$$
$$\text{Prob}\Big\{\|\mathcal{B}_k^{t-i}\|=1\Big\}=1-\text{Prob}\Big\{O=\mathcal{B}_k^{t-i}\Big\}=1-\prod_{i=1}^{N}(1-\bar{\beta}_i).$$

Then, one can obtain that

$$\mathbb{E}\Big(\|I_{mN}-\mathcal{A}_k^{t-1-i}\|\Big)=1\times\Big(1-\prod_{i=1}^{N}\bar{\alpha}_i\Big)+0\times\prod_{i=1}^{N}\bar{\alpha}_i=1-\prod_{i=1}^{N}\bar{\alpha}_i,$$
$$\mathbb{E}\Big(\|\mathcal{A}_k^{t-i}\|\cdot\|\mathcal{B}_k^{t-i}\|\Big)=\Big(1-\prod_{i=1}^{N}(1-\bar{\alpha}_i)\Big)\Big(1-\prod_{i=1}^{N}(1-\bar{\beta}_i)\Big). \tag{12.20}$$

For the convenience of the following discussion, let $w_1=1-\prod_{i=1}^{N}\bar{\alpha}_i$, $w_2=1-\prod_{i=1}^{N}\bar{\beta}_i$, $\bar{w}_1=1-\prod_{i=1}^{N}(1-\bar{\alpha}_i)$ and $\bar{w}_2=1-\prod_{i=1}^{N}(1-\bar{\beta}_i)$. Combing with (12.19), one knows

$$\begin{aligned}\mathbb{E}\Big(\|e(t,k+1)\|\Big)\le\ &\mathbb{E}\Big(\|I_{mN}-\tilde{C}\tilde{B}\mathcal{K}(I_{mN}-\mathcal{A}_k^t)\|\Big)\mathbb{E}\Big(\|e(t,k)\|\Big)\\&+w_1\sum_{i=0}^{t-1}\|\tilde{C}\tilde{A}^i\tilde{B}\mathcal{K}\|\mathbb{E}\Big(\|e(t-1-i,k)\|\Big)\\&+\bar{w}_1\bar{w}_2\sum_{i=0}^{t}\|\tilde{C}\tilde{A}^i\tilde{B}\mathcal{K}\|\,\|\xi_b\|.\end{aligned} \tag{12.21}$$

Let $\mathbb{E}\big(\|I_{mN}-\tilde{C}\tilde{B}\mathcal{K}(I_{mN}-\mathcal{A}_k^t)\|\big)=\varrho$, for $t\in\{0,1,\ldots,T\}$, (12.21) becomes

$$\underbrace{\begin{pmatrix}\mathbb{E}(\|e(0,k+1)\|)\\\mathbb{E}(\|e(1,k+1)\|)\\\vdots\\\mathbb{E}(\|e(T,k+1)\|)\end{pmatrix}}_{\mathbf{E}_{vec}(\|e(k+1)\|)}\prec\Psi\underbrace{\begin{pmatrix}\mathbb{E}(\|e(0,k)\|)\\\mathbb{E}(\|e(1,k)\|)\\\vdots\\\mathbb{E}(\|e(T,k)\|)\end{pmatrix}}_{\mathbf{E}_{vec}(\|e(k)\|)}+\underbrace{\begin{pmatrix}\bar{w}_1\bar{w}_2\|CB\mathcal{K}\|\,\|\xi_b\|\\\bar{w}_1\bar{w}_2\sum_{i=0}^{1}\|CA^iB\mathcal{K}\|\,\|\xi_b\|\\\vdots\\\bar{w}_1\bar{w}_2\sum_{i=0}^{l}\|CA^iB\mathcal{K}\|\,\|\xi_b\|\end{pmatrix}}_{M_{vec}(\|\xi_b\|)}, \tag{12.22}$$

where the matrix Ψ is defined as

$$\Psi = \begin{pmatrix} \varrho & 0 & \cdots 0 \\ w_1\|CAB\mathcal{K}\| & \varrho & \cdots 0 \\ \vdots & \vdots & \cdots \vdots \\ w_1\|CA^T B\mathcal{K}\| & w_1\|CA^{T-1}B\mathcal{K}\| & \cdots \varrho \end{pmatrix}. \tag{12.23}$$

According to the mathematical induction and (12.22), one gets

$$\begin{aligned} \mathbf{E}_{vec}(\|e(k+1)\|) &\prec \Psi\mathbf{E}_{vec}(\|e(k)\|) + M_{vec}(\|\xi_b\|) \\ &\prec \Psi^2\mathbf{E}_{vec}(\|e(k-1)\|) + (\Psi + I_{T+1})M_{vec}(\|\xi_b\|) \\ &\prec \Psi^{k+1}\mathbf{E}_{vec}(\|e(0)\|) + \sum_{j=0}^{k}\Psi^j M_{vec}(\|\xi_b\|). \end{aligned} \tag{12.24}$$

From (12.12), one can know that $\rho(\Psi) < 1$. Thus, $\lim_{k\to+\infty} \Psi^{k+1}\mathbf{E}_{vec}(\|e(0)\|) = \mathbf{0}$, and $\sum_{j=0}^{k}\Psi^j M_{vec}(\|\xi_b\|)$ is absolute convergence. It may be assumed that $\lim_{k\to+\infty} \Psi^k = \bar{\Psi}$, then $\lim_{k\to+\infty} \Psi^k M_{vec}(\|\xi_b\|) = \bar{\Psi}M_{vec}(\|\xi_b\|)$. Hence, $\mathbb{E}(\|e(k)\|) \prec \bar{\Psi}M_{vec}(\|\xi_b\|)$ as $k \to +\infty$. Let σ be the maximum element in vector $\bar{\Psi}M_{vec}(\|\xi_b\|)$, then one obtains $\lim_{k\to+\infty} \mathbb{E}(\|e_i(t,k)\|) \le \sigma$, $i \in \{1, \ldots, N\}$, $t \in \mathcal{T}$. According to Definition 12.1, one knows that the σ-secure tracking is achieved. □

Remark 12.3 When the dynamic equations of each node are heterogeneous, i.e., $A_1 \neq \cdots \neq A_N$, $B_1 \neq \cdots \neq B_N$ and $C_1 \neq \cdots \neq C_N$, the learning control scheme (12.11) can be used to achieve the security target. At this time, $\tilde{B} = \mathrm{diag}(B_1, \ldots, B_N)$ and $\tilde{C} = \mathrm{diag}(C_1, \ldots, C_N)$, a similar result can be obtained by combining the proof of Theorem 12.1.

Remark 12.4 In real application, it is not easy to verify the condition (12.12). To address this issue, one can deflate the inequality in order to obtain an easily verifiable condition as follows:

$$\begin{aligned} &\mathbb{E}\Big(\big\|I_{mN} - \tilde{C}\tilde{B}\mathcal{K}(I_{mN} - \mathcal{A}_k^t)\big\|\Big) \\ &\le \mathbb{E}\Big(\big\|I_{mN} - \tilde{C}\tilde{B}\mathcal{K}\big\| + \big\|\tilde{C}\tilde{B}\mathcal{K}\big\| \cdot \big\|\mathcal{A}_k^t\big\|\Big) \\ &\le \big\|I_{mN} - \tilde{C}\tilde{B}\mathcal{K}\big\| + \big\|\tilde{C}\tilde{B}\mathcal{K}\big\| \cdot \mathbb{E}\Big(\big\|\mathcal{A}_k^t\big\|\Big) \\ &\le \big\|I_{mN} - \tilde{C}\tilde{B}\mathcal{K}\big\| + \big\|\tilde{C}\tilde{B}\mathcal{K}\big\|\Big(1 - \prod_{i=1}^{N}(1-\bar{\alpha}_i)\Big). \end{aligned} \tag{12.25}$$

As a result, the appropriate parameters can be selected to make inequality $\big\|I_{mN} - \tilde{C}\tilde{B}\mathcal{K}\big\| + \big\|\tilde{C}\tilde{B}\mathcal{K}\big\|\big(1 - \prod_{i=1}^{N}(1-\bar{\alpha}_i)\big) < 1$ hold. Then, according to Theorem 12.1, the security of network system can be achieved.

12.4.2 Boundedness of All System Trajectories

In this subsection, we mainly analyze the boundedness of all system trajectories. First, the following lemma is proposed.

Lemma 12.1 (Meng and Moore 2017) *For the following discrete-time systems:*

$$\theta(t+1,k) = \Omega(t,k)\theta(t,k) + \vartheta(t,k),$$

where $t \in \mathcal{S}^- = \{0, \ldots, T-1\}$, $k \in \mathbb{Z}_+$. $\vartheta(t,k) \in \mathbb{R}^n$ *and* $\Omega(t,k) \in \mathbb{R}^{n\times n}$ *are bounded, i.e., the following inequalities hold:*

$$\sup_{k\in\mathbb{Z}_+} \|\vartheta(t,k)\| \le b_\vartheta(t),\ b_\vartheta(t) > 0, \qquad \sup_{k\in\mathbb{Z}_+, t\in\mathcal{S}^-} \|\Omega(t,k)\| \le b_\Omega,\ b_\Omega > 0.$$

For $\forall\ t \in \mathcal{S}^-$, *there exist an iterative sequence* $\{\epsilon_l(t) : l \in \mathbb{Z}_+\}$, $\epsilon_0(t) = 0$, $0 < \epsilon_{l+1}(t) - \epsilon_l(t) \le \tilde{\epsilon}(t)$ *and* $\tilde{\epsilon}_l(t) > 0$ *such that*

$$\left\| \prod_{j=\epsilon_l(t)}^{\epsilon_{l+1}(t)-1} \Omega(t,j) \right\| \le \chi < 1,\ \ l \in \mathbb{Z}_+.$$

Then, for $t \in \mathcal{S}^-$, *the following results hold.*

(1) $\vartheta(t,k) \in \mathbb{R}^n$ *is bounded, i.e.,* $\sup_{k\in\mathbb{Z}_+} \|\theta(t,k)\| \le b_\theta(t)$, $b_\theta(t) \ge 0$. *Furthermore,* $\limsup_{k\to\infty} \|\theta(t,k)\| \le b_{\theta_{\sup}}(t)$, *where* $b_{\theta_{\sup}}(t) \ge 0$ *and* $b_{\theta_{\sup}}(t) < b_\theta(t)$.

(2) $\lim_{k\to\infty} \theta(t,k) = \mathbf{0}$, *if* $\lim_{k\to\infty} \vartheta(t,k) = \mathbf{0}$ *holds.*

Theorem 12.2 *Let Assumption 12.1 hold, ILC strategy (12.7) is applied to network system 12.1. If one chooses learning matrix K and step parameter α satisfy the inequality*

$$\mathbb{E}\Big(\Big\| I_{mN} - \mathcal{K}\big(I_{mN} - \mathcal{A}_k^t\big)\tilde{C}\tilde{B}\Big\|\Big) = \varrho < 1, \tag{12.26}$$

where $\mathcal{K} = \big[(\mathcal{L}(\mathcal{G}) + \alpha Q) \otimes K\big]$. *The state* $x(t,k)$ *and control* $u(t,k)$ *is bounded in the sense of mathematical expectation, i.e.,*

$$\sup_{k\in\mathbb{Z}_+, t\in\mathcal{S}} \mathbb{E}(\|x(t,k)\|) \le B_x, \qquad \sup_{k\in\mathbb{Z}_+, t\in\mathcal{S}} \mathbb{E}(\|u(t,k)\|) \le B_u,$$

where $\mathcal{S} = \{0, \ldots, T\}$, $B_x > 0$ *and* $B_u > 0$ *are two constants. Furthermore, the measurement* $y(t,k)$ *and the real error* $e(t,k)$ *are also bounded in the sense of mathematical expectation, i.e.,*

$$\sup_{k\in\mathbb{Z}_+,t\in\mathcal{S}} \mathbb{E}(\|y(t,k)\|) \le B_y, \quad \sup_{k\in\mathbb{Z}_+,t\in\mathcal{S}} \mathbb{E}(\|e(t,k)\|) \le B_e,$$

where $\mathcal{S} = \{0, \ldots, T\}$, $B_y > 0$ and $B_e > 0$ are two constants.

Proof According to (12.8) and (12.10), one has

$$\begin{aligned} u(t,k+1) &= u(t,k) + \mathcal{K}\check{e}(t+1,k) \\ &= u(t,k) + \mathcal{K}\Big[\big(I_{mN} - \mathcal{A}_k^t\big)(\mathbf{1}\otimes y_d(t+1)) \\ &\quad -\big(I_{mN} - \mathcal{A}_k^t\big)\tilde{C}x(t+1,k) - \mathcal{A}_k^t\mathcal{B}_k^t\xi(t+1,k)\Big] \\ &= \big[I_{mN} - \mathcal{K}\big(I_{mN} - \mathcal{A}_k^t\big)\tilde{C}\tilde{B}\big]u(t,k) + \varphi(t,k), \end{aligned} \tag{12.27}$$

where

$$\varphi(t,k) = \mathcal{K}\Big[\big(I_{mN} - \mathcal{A}_k^t\big)\big(\mathbf{1}\otimes y_d(t+1) - \tilde{C}\tilde{A}x(t,k)\big) - \mathcal{A}_k^t\mathcal{B}_k^t\xi(t+1,k)\Big].$$

Thus, according to (12.27), it is not hard to get

$$\mathbb{E}(\|u(t,k+1)\|) \le \mathbb{E}\Big(\big\|I_{mN} - \mathcal{K}\big(I_{mN} - \mathcal{A}_k^t\big)\tilde{C}\tilde{B}\big\|\Big)\mathbb{E}(\|u(t,k)\|) + \mathbb{E}(\|\varphi(t,k)\|), \tag{12.28}$$

and

$$\mathbb{E}\big(\|\varphi(t,k)\|\big) \le w_1\|\mathcal{K}\|\|\mathbf{1}\otimes y_d(t+1)\| + \bar{w}_1\bar{w}_2\|\xi_b\| + w_1\|\tilde{C}\tilde{A}\|\mathbb{E}(\|x(t,k)\|), \tag{12.29}$$

where w_1, $\bar{w}_1$, and $\bar{w}_2$ are defined in the proof of Theorem 12.1.

Next, the boundedness is demonstrated by applying the method in Meng and Moore (2017).

Step 1. For $t = 0$ and $k \in \mathbb{Z}_+$, one has $\sup_{k\in\mathbb{Z}_+} \mathbb{E}(\|x(0,k)\|) \le B_{x0} \triangleq B_x(0)$. Combining with (12.29), one knows

$$\begin{aligned} \mathbb{E}\big(\|\varphi(0,k)\|\big) &\le w_1\|\mathcal{K}\|\|\mathbf{1}\otimes y_d(1)\| + \bar{w}_1\bar{w}_2\|\xi_b\| + w_1\|\tilde{C}\tilde{A}\|\mathbb{E}(\|x(0,k)\|) \\ &\le w_1\|\mathcal{K}\|B_{y_d} + \bar{w}_1\bar{w}_2\|\xi_b\| + w_1\|\tilde{C}\tilde{A}\|B_x(0), \end{aligned} \tag{12.30}$$

where $B_{y_d} > 0$ denotes the upper bound of desired trajectory $y_d(t)$. Then, according to Lemmas 12.1 and (12.28), it is easy to know that $\sup_{k\in\mathbb{Z}_+} \mathbb{E}(\|u(0,k)\|) \le B_u(0)$.

Step 2. For $t > 0$, assume that $\sup_{k\in\mathbb{Z}_+} \mathbb{E}(\|x(t,k)\|) \le B_x(t)$, $B_x(t) > 0$, and $\sup_{k\in\mathbb{Z}_+} \mathbb{E}(\|u(t,k)\|) \le B_u(t)$, $B_u(t) > 0$. $B_x(t)$ and $B_u(t)$ are bounded. From (12.8), it is not hard to know

$$\begin{aligned}\mathbb{E}\big(\|x(t+1,k)\|\big) &\leq \|\tilde{A}\|\mathbb{E}\big(\|x(t,k)\|\big) + \|\tilde{B}\|\mathbb{E}\big(\|u(t,k)\|\big) \\ &\leq \|\tilde{A}\|B_x(t) + \|\tilde{B}\|B_u(t) \triangleq B_x(t+1), \quad k \in \mathbb{Z}_+. \end{aligned} \tag{12.31}$$

Linking (12.31) and (12.29),

$$\mathbb{E}\big(\|\varphi(t+1,k)\|\big) \leq w_1\|\mathcal{K}\|B_{y_d} + \bar{w}_1\bar{w}_2\|\xi_b\| + w_1\|\tilde{C}\tilde{A}\|B_x(t+1). \tag{12.32}$$

From Lemmas 12.1 and (12.28), one knows that $\mathbb{E}\big(\|u(t+1,k)\|\big) \leq B_u(t+1)$. Furthermore, according to (12.31) and (12.32), the goal is proved by applying the induction method. As a result, let $B_x \triangleq \max_{t\in\mathcal{S}} B_x(t)$ and $B_u \triangleq \max_{t\in\mathcal{S}} B_u(t)$, one can immediately obtain that $\sup_{k\in\mathbb{Z}_+,t\in\mathcal{S}} \mathbb{E}(\|x(t,k)\|) \leq B_x$, $\sup_{k\in\mathbb{Z}_+,t\in\mathcal{S}} \mathbb{E}(\|u(t,k)\|) \leq B_u$. It means the first result in Theorem 12.2 is proved. According to the relationship among $y(t,k)$, $x(t,k)$, and $e(t,k)$, one can also imitate the proof of the first conclusion and obtain the second result in Theorem 12.2. Thus, the proof process is omitted. □

Remark 12.5 Similar to Remark 12.4, the condition (12.26) can be simplified as

$$\begin{aligned}&\mathbb{E}\Big(\Big\|I_{mN} - \mathcal{K}\big(I_{mN} - \mathcal{A}_k^t\big)\tilde{C}\tilde{B}\Big\|\Big) \\ &\leq \big\|I_{mN} - \mathcal{K}\tilde{C}\tilde{B}\big\| + \|\mathcal{K}\|\left(1 - \prod_{i=1}^{N}(1-\bar{\alpha}_i)\right)\big\|\tilde{C}\tilde{B}\big\| < 1.\end{aligned} \tag{12.33}$$

The condition (12.33) is easy to verify and calculate in applications.

Remark 12.6 Theorem 12.2 demonstrates that the designed iterative learning control strategy also ensures that the individual variables in the network system are bounded. In addition, compared with the control strategy in Zhao et al. (2021), our learning control scheme provides a concise and rational approach. Compared with Xiong et al. (2022), this chapter considers the security problem under two more general network attack types. Also, we propose an effective iterative learning control scheme and obtain an easily verifiable sufficiency criterion.

12.5 Simulation Example

In this section, a numerical example is presented to demonstrate the validity of the results in Theorems 12.1 and 12.2.

Example 12.1 Consider a network system with six nodes, and the communication topology is shown in Fig. 12.2. It should be noted that node 0 represents the desired or target trajectory. The directed edge weight is set to 1.

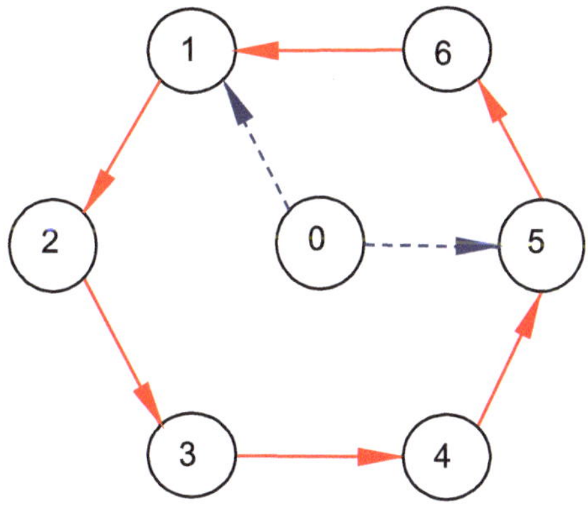

Fig. 12.2 Network communication topology diagram

The dynamical behavior of each node is described by (12.1). Particularly, the matrices are as

$$A = \begin{pmatrix} 0 & 1 \\ 1 & 0 \end{pmatrix}, \quad B = \begin{pmatrix} 0.5 \\ 1 \end{pmatrix}, \quad C = \begin{pmatrix} 1 & 1 \end{pmatrix}.$$

Let the terminal time $T = 10$, and the initial state of each node is as follows:

$$x_1(0,k) = \begin{pmatrix} 0 \\ 0 \end{pmatrix}, \quad x_2(0,k) = \begin{pmatrix} 1 \\ -1 \end{pmatrix}, \quad x_3(0,k) = \begin{pmatrix} 2 \\ -2 \end{pmatrix},$$
$$x_4(0,k) = \begin{pmatrix} 1.5 \\ -1.5 \end{pmatrix}, \quad x_5(0,k) = \begin{pmatrix} 2.5 \\ -2.5 \end{pmatrix}, \quad x_6(0,k) = \begin{pmatrix} 3 \\ -3 \end{pmatrix}.$$

The target trajectory is set as $y_d(t) = 0.2\sin(0.4\pi t), t \in \{0, 1, \ldots, 10\}$. The iterative learning control scheme is described by (12.7), where the learning parameters $K = 0.2$ and $\alpha = 1$. In addition, the probability of network attacks occurring is as follows:

$$\text{Prob}\{\alpha_{1,k}^t = 1\} = 0.5, \ \text{Prob}\{\alpha_{2,k}^t = 1\} = 0.2, \ \text{Prob}\{\alpha_{3,k}^t = 1\} = 0.4,$$
$$\text{Prob}\{\alpha_{4,k}^t = 1\} = 0.3, \ \text{Prob}\{\alpha_{5,k}^t = 1\} = 0.4, \ \text{Prob}\{\alpha_{6,k}^t = 1\} = 0.3.$$

Then, according to Remark 12.4, one knows that $\|I_{mN} - \tilde{C}\tilde{B}\mathcal{K}\| + \|\tilde{C}\tilde{B}\mathcal{K}\|[1 - \prod_{i=1}^{N}(1-\bar{\alpha}_i)] = 0.9812 < 1$. It means that the condition in Theorem 12.1 holds. From Remark 12.5, one knows $\|I_{mN} - \mathcal{K}\tilde{C}\tilde{B}\| + \|\mathcal{K}\|(1 - \prod_{i=1}^{N}(1-\bar{\alpha}_i))\|\tilde{C}\tilde{B}\| = 0.9812 < 1$. It demonstrates the condition in Theorem 12.2 holds. Moreover, let the probability of deception attacks occurs as

$$\text{Prob}\{\beta_{1,k}^t = 1\} = 0.2, \ \text{Prob}\{\beta_{2,k}^t = 1\} = 0.1, \ \text{Prob}\{\beta_{3,k}^t = 1\} = 0.3,$$
$$\text{Prob}\{\beta_{4,k}^t = 1\} = 0.2, \ \text{Prob}\{\beta_{5,k}^t = 1\} = 0.1, \ \text{Prob}\{\beta_{6,k}^t = 1\} = 0.2.$$

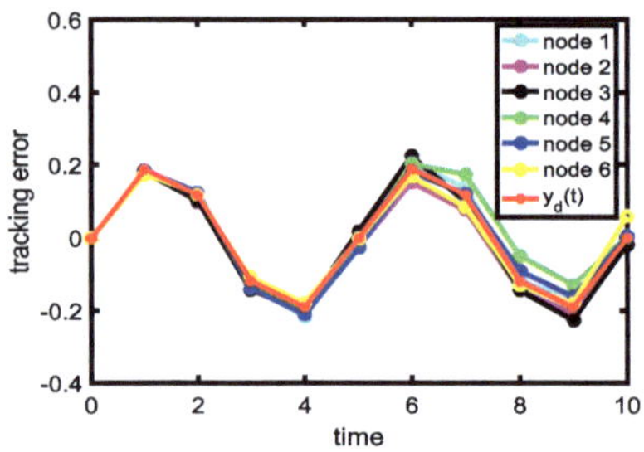

(a) Tracking trajectories at the 200th iteration

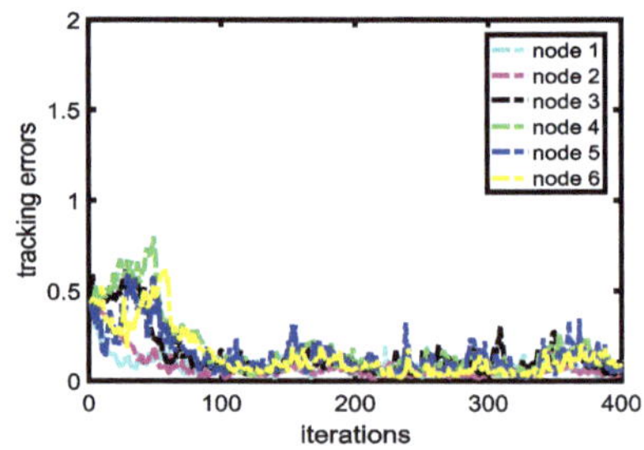

(b) Tracking error curve for each node

Fig. 12.3 Tracking trajectories and error curves of each node in a network system

And the interference term $\xi_i(t, k)$ in deception attack satisfies $\|\xi_i(t, k)\| \le \xi_{i,b} = 0.1$. The simulation results are shown in Fig. 12.3.

Figure 12.3a depicts the tracking trajectory of each node at the 200th iteration. Figure 12.3b describes the node tracking error curve, where the tracking error is defined as $e_i(k) = \left(\sum_{t=0}^{T} \|e_i(t, k)\|^2\right)^{1/2}$. As shown in Fig. 12.3, the tracking error of each node fluctuates around a smaller value as the number of iterations increases. This indicates that the tracking error eventually stabilizes within a certain range, resulting in network system security by employing our ILC scheme.

To demonstrate the impact of interference on network system security, we set different interference thresholds and then compare and verify them by using numerical simulation. The results are shown in Fig. 12.4.

Figure 12.4a, b and c describes the trend of tracking error of each node under different threshold interferences. It can be seen from Fig. 12.4a and b that the fluctuation of the tracking error also increases as the interference in the deception attack increases. It implies that the interference factor in the spoofing attack has a positive relationship with the error convergence. Figure 12.4c depicts the convergence of tracking errors for each node when there is no deception attack in the network system. As a result, it is clear that the designed iterative learning strategy can achieve consensus when there is only DoS attacks causing data loss. This also means that DoS attacks have a relatively small impact on network system security.

In addition, Fig. 12.4d shows the trend of the network system's composite tracking error curve under two different ILC schemes. And the composite tracking error is defined as $e(k) = \left(\sum_{i=1}^{N} |e_i(k)|^2\right)^{1/2}$. Figure 12.4d demonstrates that the ILC scheme proposed in this chapter can keep the error curve in a relatively small range. It also implies the effectiveness and advantages of our iterative learning strategy.

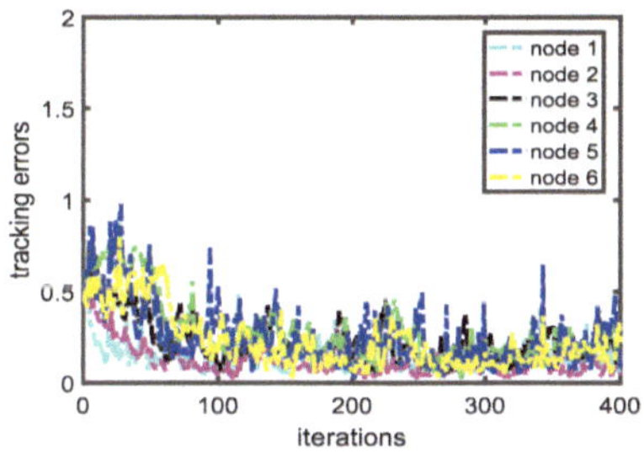

(a) Tracking error curve when $\xi_{i.b} = 0.2$

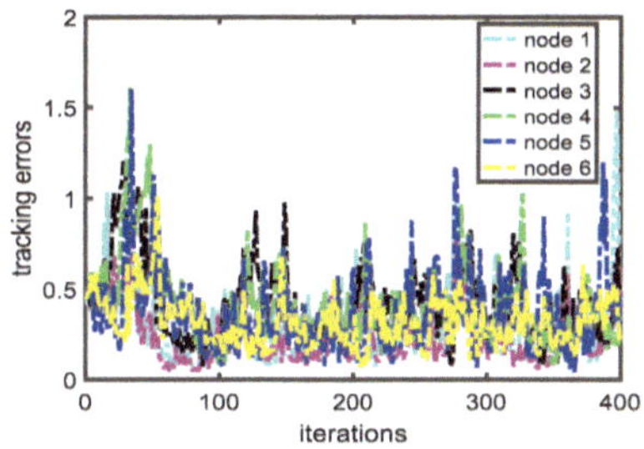

(b) Tracking error curve when $\xi_{i.b} = 0.4$

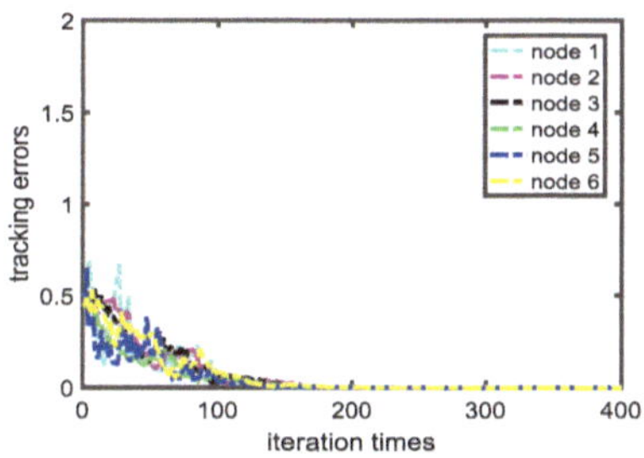

(c) Tracking error curve when $\xi_{i.b} = 0$

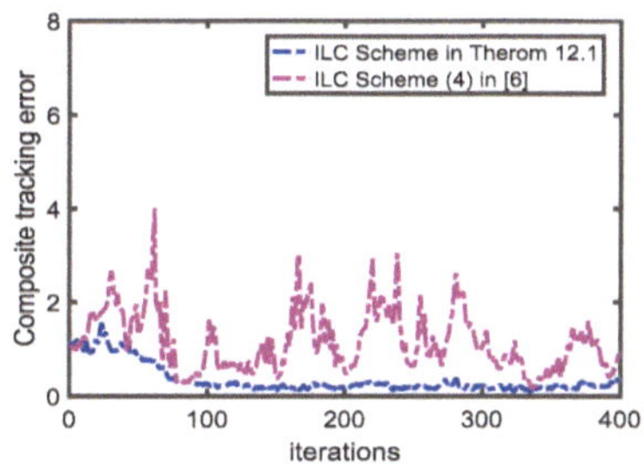

(d) Error curve under two different ILC schemes

Fig. 12.4 Tracking error curve under different interference

12.6 Summary

In this chapter, we have investigated the security problem of discrete-time systems under deception and DoS attacks. From the perspective of trajectory tracking, the security problem has been studied by designing iterative learning algorithms. The main results have been obtained by applying the norm estimation and matrix analysis theory. The results have shown that system security can be effectively achieved with our ILC approaches. Also, the conditions of our results are simple and can be easily checked in real-life applications. Finally, an example has been given to illustrate the effectiveness of the main results.

References

Bu XH, Yu FS, Hou ZS, Wang FZ (2013) Iterative learning control for a class of nonlinear systems with random packet losses. Nonlinear Anal 14(1):567–580

Deng C, Che WW, Shi P (2020) Cooperative fault-tolerant output regulation for multiagent systems by distributed learning control approach. IEEE Trans Neur Net Lear Syst 31(11):4831–4841

Fu Q, Gu PP, Li XD, Wu JR (2017) Iterative learning control approach for consensus of multi-agent systems with regular linear dynamics. Sci China Inf Sci 60(7):079202

Guo L, Cui TT, Yu H, Hao F (2021) Stability of networked control system subject to denial-of-service. Sci China Inf Sci 64(2):129203

He WL, Gao XY, Zhong WM, Qian F (2018) Secure impulsive synchronization control of multi-agent systems under deception attacks. Inform Sci 459:354–368

Li JN, Ren W (2021) Finite-horizon H_∞ fault-tolerant constrained consensus for multiagent systems with communication delays. IEEE Trans Cybern 51(1):416–426

Li AS, Li XC, Pan YC, Zhang W (2015) Strategies for network security. Sci China Inf Sci 58(1):012107

Liu J, Ruan XE (2017) Networked iterative learning control for discrete-time systems with stochastic packet dropouts in input and output channels. Adv Diff Equ 2017:1–21

Liu J, Ruan XE (2018) Networked iterative learning control design for nonlinear systems with stochastic output packet dropouts. Asian J Control 20(3):1077–1087

Meng DY (2019) Convergence conditions for solving robust iterative learning control problems under nonrepetitive model uncertainties. IEEE Trans Neur Netw Lear Syst 30(6):1908–1919

Meng DY, Moore KL (2017) Robust iterative learning control for nonrepetitive uncertain systems. IEEE Trans Auto Control 62(2):907–913

Meng DY, Zhang JY (2021) Convergence analysis of robust iterative learning control against non-repetitive uncertainties: system equivalence transformation. IEEE Trans Neur Net Lear Syst 32(9):3867–3879

Ren Y, Hou ZS, Sirmatel II, Geroliminis N (2020) Data driven model free adaptive iterative learning perimeter control for large-scale urban road networks. Transport Res Part C Emer Tech 115:102618

Ren W, Jungers RM, Dimarogonas DV (2022) Razumikhin and Krasovskii approaches for safe stabilization. Automaitca 146:110563

Tahoun AH, Arafa M (2021) Cooperative control for cyber-physical multi-agent networked control systems with unknown false data-injection and replay cyber-attacks. ISA Trans 110:1–14

Wan Y, Cao JD, Chen GR, Huang W (2017) Distributed observer-based cyber-security control of complex dynamical networks. IEEE Trans Circuits Syst 64(11):2966–2975

Wang M, Zhang YL, Dong HF, Yu JZ (2020) Trajectory tracking control of a bionic robotic fish based on iterative learning. Sci China Inf Sci 63(7):170202

Xiao F, Shi Y, Ren W (2018) Robustness analysis of asynchronous sampled-data multiagent networks with time-varying delays. IEEE Trans Auto Control 63(7):2145–2152

Xiong WJ, Xu L, Huang TW, Yu XH, Liu YH (2020) Finite-iteration tracking of singular coupled systems based on learning control with packet losses. IEEE Trans Syst Man Cyber Syst 50:245–255

Xiong WJ, Ho DWC, Wen SF (2021) A periodic iterative learning scheme for finite-iteration tracking of discrete networks based on FlexRay communication protocol. Inform Sci 548:344–356

Xiong WJ, Liu C, Wen GH, Qiu JL, Xu YJ (2022) On designing learning control scheme for multilayer supply chain networks with constraints. IEEE Trans Syst Man Cybern Syst 52(12):7422–7430

Xiong WJ, Gong K, Wen GH, Lü JH (2022) Security analysis of discrete nonlinear systems with injection attacks under iterative learning schemes. IEEE Trans Syst Man Cybern Syst 52(2):927–935

Xue MQ, Tang Y, Ren W, Qian F (2022) Stability of multi-dimensional switched systems with an application to open multi-agent systems. Automatica 146:110644

Yang Y, Li YF, Yue D (2020) Event-trigger-based consensus secure control of linear multi-agent systems under dos attacks over multiple transmission channels. Sci China Inf Sci 63(5):150208

Yu W, Bu XH, Hou ZS (2022) Security data-driven control for nonlinear systems subject to deception and false data injection attacks. IEEE Trans Net Sci Eng 9(4):2910–2921

Zhang JY, Meng DY (2021) Cooperative iterative learning for uncertain nonlinear agents in leaderless switching networks. Automatica 129:109692
Zhao D, Wang ZD, Ho DWC, Wei GL (2021) Observer-based PID security control for discrete time-delay systems under cyber-attacks. IEEE Trans Syst Man Cybern Syst 51(6):3926–3938
Zhou LM, Jia L, Wang YL (2019) A robust integrated model predictive iterative learning control strategy for batch processes. Sci. China Inf Sci 62(11):219202

MIX
Papier aus verantwortungsvollen Quellen
Paper from responsible sources
FSC® C105338

If you have any concerns about our products,
you can contact us on
ProductSafety@springernature.com

In case Publisher is established outside the EU,
the EU authorized representative is:
Springer Nature Customer Service Center GmbH
Europaplatz 3, 69115 Heidelberg, Germany

Printed by Libri Plureos GmbH
in Hamburg, Germany